Herbs: Challenges in Chemistry and Biology

ACS SYMPOSIUM SERIES **925**

Herbs: Challenges in Chemistry and Biology

Mingfu Wang, EDITOR
Rutgers The State University of New Jersey

Shengmin Sang, EDITOR
The University of Hong Kong

Lucy Sun Hwang, EDITOR
National Taiwan University

Chi-Tang Ho, EDITOR
Rutgers The State University of New Jersey

**Sponsored by the
ACS Division of Agricultural and Food
Chemistry, Inc.**

American Chemical Society, Washington, DC

Library of Congress Cataloging-in-Publication Data

Herbs: challenges in chemistry and biology / Chi-Tang Ho, editor ... [et al.] ; sponsored by the ACS Division of Agricultural and Food Chemistry.

p. cm. — (ACS symposium series ; 925)

"Developed from a symposium sponsored by the Division of Agricultural and Food Chemistry at the 227th National Meeting of the American Chemical Society, Anaheim, California, March 28–April 1, 2004"—Pref.

Includes bibliographical references and index.

ISBN-13: 978-0-8412-3930-2 (alk. paper)

1. Functional foods—Congresses. 2. Phytochemicals—Physiological effect—Congresses. 3. Herbs—Therapeutic use—Congresses. 4. Medicine, Chinese—Congresses.

I. Ho, Chi-Tang, 1944- II. American Chemical Society. Division of Agricultural and Food Chemistry. III. American Chemical Society. Meeting (227th : 2004 : Anaheim, Calif.). IV. Series.

QP144.F85H47   2005
615'.321—dc22
                                                2005048237

The paper used in this publication meets the minimum requirements of American National Standard for Information Sciences—Permanence of Paper for Printed Library Materials, ANSI Z39.48–1984.

Copyright © 2006 American Chemical Society

Distributed by Oxford University Press

ISBN 10: 0-8412-3930-4

All Rights Reserved. Reprographic copying beyond that permitted by Sections 107 or 108 of the U.S. Copyright Act is allowed for internal use only, provided that a per-chapter fee of $30.00 plus $0.75 per page is paid to the Copyright Clearance Center, Inc., 222 Rosewood Drive, Danvers, MA 01923, USA. Republication or reproduction for sale of pages in this book is permitted only under license from ACS. Direct these and other permission requests to ACS Copyright Office, Publications Division, 1155 16th Street, N.W., Washington, DC 20036.

The citation of trade names and/or names of manufacturers in this publication is not to be construed as an endorsement or as approval by ACS of the commercial products or services referenced herein; nor should the mere reference herein to any drawing, specification, chemical process, or other data be regarded as a license or as a conveyance of any right or permission to the holder, reader, or any other person or corporation, to manufacture, reproduce, use, or sell any patented invention or copyrighted work that may in any way be related thereto. Registered names, trademarks, etc., used in this publication, even without specific indication thereof, are not to be considered unprotected by law.

PRINTED IN THE UNITED STATES OF AMERICA

Foreword

The ACS Symposium Series was first published in 1974 to provide a mechanism for publishing symposia quickly in book form. The purpose of the series is to publish timely, comprehensive books developed from ACS sponsored symposia based on current scientific research. Occasionally, books are developed from symposia sponsored by other organizations when the topic is of keen interest to the chemistry audience.

Before agreeing to publish a book, the proposed table of contents is reviewed for appropriate and comprehensive coverage and for interest to the audience. Some papers may be excluded to better focus the book; others may be added to provide comprehensiveness. When appropriate, overview or introductory chapters are added. Drafts of chapters are peer-reviewed prior to final acceptance or rejection, and manuscripts are prepared in camera-ready format.

As a rule, only original research papers and original review papers are included in the volumes. Verbatim reproductions of previously published papers are not accepted.

ACS Books Department

Contents

Preface xi

Overview and Perspectives

1. **Botanical Quality Initiatives at the Office of Dietary Supplements, National Institutes of Health** 2
   Joseph M. Betz

2. **Current Trends and Future Prospects of Traditional Chinese Medicines in the 21st Century** 14
   Xinsheng Yao, Ke Hu, Jianghua Liu, Nali Wang, Chengbin Cui, and Xiangjiu He

3. **Bioactive Polyphenols from Foods and Dietary Supplements: Challenges and Opportunities** 25
   Navindra P. Seeram

4. **Instrumental Analysis of Popular Botanical Products in the U.S. Market** 39
   Mingfu Wang, Chia-Pei Liang, Qing-Li Wu, James E. Simon, and Chi-Tang Ho

5. **Challenges in Assessing Bioactive Botanical Ingredients in Functional Beverages** 55
   Catharina Y. W. Ang, Jin-Woo Jhoo, Yanyan Cui, Lihong Hu, Thomas M. Heinze, Qingyong Lang, Chien M. Wai, Jeremy J. Mihalov, Michael DiNovi, and Antonia Mattia

6. **Bioactive Natural Products from Chinese Tropical Marine Plants and Invertebrates** 73
   Y.-W. Guo, X.-C. Huang, W. Zhang, and Y.-Q. Sun

7. Curcumin: Potential Health Benefits, Molecular Mechanism
   of Action, and Its Anticancer Properties In Vitro and In Vivo 92
   Resmi Ann Matchanickal and Mohamed M. Rafi

Natural Product Chemistry and Analysis

8. Need for Analytical Methods and Fingerprinting: Total
   Quality Control of Phytomedicine *Echinacea* 110
   Jatinder Rana and Amitabh Chandra

9. Bioassay-Guided Isolation, Identification, and Quantification
   of the Estrogen-Like Constituent from PC SPES 117
   Shengmin Sang, Zhihua Liu, Robert T. Rosen, and Chi-Tang Ho

10. Intraspecific Variation in Quality Control Parameters,
    Polyphenol Profile, and Antioxidant Activity in Wild
    Populations of *Lippia multiflora* from Ghana 126
    H. Rodolfo Juliani. Mingfu Wang, Hisham Moharram,
    Julie Asante-Dartey, Dan Acquaye, Adolfina R. Koroch,
    and James E. Simon

11. Protein Tyrosine Phosphatases 1B Inhibitors from Traditional
    Chinese Medicine 143
    Tianying An, Di Hong, Lihong Hu, and Jia Li

12. Thioglucosidase-Catalyzed Hydrolysis of the Major
    Glucosinolate of Maca (*Lepidium meyenii*) to Benzyl
    Isothiocyanate: Mini-Review and Simple Quantitative
    HPLC Method 157
    Matthew W. Bernart

13. Studies on Chemical Constituents of Jiaogulan
    (*Gynostemma pentaphyllum*) 170
    Feng Yin and Lihong Hu

14. Chemical Components of Noni (*Morinda citrifolia* L.) Root 185
    Shengmin Sang and Chi-Tang Ho

15. Characterization of Chemical Components of *Ixeris denticulata* 195
    Hang Chen, Shiming Li, Zhu Zhou, Naisheng Bai,
    and Chi-Tang Ho

16. Bioavailability, Metabolism, and Pharmacokinetics
    of Glycosides in Chinese Herbs 212
    Pei-Dawn Lee Chao, Su-Lan Hsiu, and Yu-Chi Hou

17. Moringa, a Novel Plant Rich in Antioxidants, Bioavailable
    Iron, and Nutrients 224
    Ray-Yu Yang, Samson C. S. Tsou, Tung-Ching Lee,
    Leing-Chung Chang, George Kuo, and Po-Yong Lai

18. Stability and Transformation of Bioactive Polyphenolic
    Components of Herbs in Physiological pH 240
    Shiming Li and Chi-Tang Ho

19. Bioavailability and Synergistic Effects of Tea Catechins
    as Antioxidants in the Human Diet 254
    John Shi and Yukio Kakuda

Biological Activities

20. Targeting Inflammation Using Asian Herbs 266
    Haiqing Yu, Mohamed M. Rafi, and Chi-Tang Ho

21. Induction of Apoptosis by *Ligusticum chuanxiong* in HSC-T6
    Stellate Cells 281
    Yun-Lian Lin, Ting-Fang Lee, Young-Ji Shiao,
    and Yi-Tsau Huang

22. Effect of Combined Use of Isothiocyanate and Black Tea
    Extract on Dental Caries 290
    Hideki Masuda, Saori Hirooka, and Toshio Inoue

23. Cytotoxic Properties of Leaf Essential Oil and Components
    from Indigenous Cinnamon (*Cinnamomum
    osmophloeum* Kaneh) 299
    Tzou-Chi Huang, Chi-Tang Ho, Hui-Yin Fu, and Min-Hsiung Pan

24. Effect of Black Tea Theaflavins on 12-*O*-Tetradecanoylphorbol-13-acetate-Induced Inflammation: Expression of Proinflammatory Cytokines and Arachidonic Acid Metabolism in Mouse Ear and Colon Carcinogenesis in Min (Apc+/−) Mice 314
Mou-Tuan Huang, Yue Liu, Divya Ramji, Shengmin Sang, Robert T. Rosen, Geetha Ghai, Chung S. Yang, and Chi-Tang Ho

25. Shea Butter: Chemistry, Quality, and New Market Potentials 326
Hisham Moharram, Julie Ray, Sibel Ozbas, Hector Juliani, and James Simon

Author Index 341

Subject Index 343

Preface

There is an increasing use of botanicals, botanical extracts, and purified natural compounds for the treatment and prevention of disease, especially chronic disease, in North America and Europe—not to mention Asian countries, which already have a long history of applying herbal medicines. For example, in China, herbal medicine (Traditional Chinese Medicine) is viewed as a national treasure, and has equal status with Western medicine. Traditional Chinese medicine is well-known to be effective for all kinds of disease, especially for age-related diseases.

In the United States, research and product development in nutraceuticals and functional foods are producing many novel botanicals, vitamins, and/or mineral products that are appearing in the market. The reevaluation of botanical products is, however, accompanied by higher expectations from nutritionalists, physicians, pharmacists, and consumers who demand quality, safety, and efficacy from these products. With the continuing reports of the side effects and toxicity of popular botanical products such as Ephedra and Kava, consumers are starting to back away from botanical products and are beginning to lose faith in dietary supplements. Thus, it is time to present updated research relating to the quality control, chemical, and pharmacological respects of herbal products in order to regain the confidence of the public.

The symposium upon which this book is based was organized to bring together leading researchers from the United States, China, Taiwan, Japan, and Canada to discuss future research directions in herbal products and to share works in botanical research. The symposium was held in Anaheim, California from March 28 to April 1, 2004 as part of the 227th American Chemical Society National meeting.

This book is divided into four sections. In the overview (first) section, Dr. Betz first describes the National Institutes of Health's vision of clearing confusion in the marketplace via validated analytical methods and referenced botanical materials. This is followed by Chapter 2 on the development of patentable new drugs from Traditional Chinese Medicine, Chapter 3 about bioactive polyphenols from foods and dietary supplements, Chapter 4 about the application of instrumental methods to the analysis and the prevention of adulteration in popular dietary supplements in the market, and Chapter 5 about assessing bioactive botanical ingredients in functional beverages. Finally, Chapter 6 covers bioactive natural products from Chinese tropical marine plants and invertebrate, and Chapter 7 reviews the potential health benefits of curcumin. The other three sections cover natural product chemistry and analysis, natural product bioavailability and stability, and finally the biological activities of various herbal products and single natural compounds.

We thank all the authors for their contributions and efforts in the preparation of this book. May their insight and research lead us all to longer and healthier lives.

Mingfu Wang
Department of Botany
The University of Hong Kong
Pokfulam Road, Hong Kong

Shengmin Sang
Department of Chemical Biology
Ernest Mario School of Pharmacy
Rutgers, The State University of New Jersey
164 Frelinghyusen Road
Piscataway, NJ 08854

Lucy Sun Hwang
Institute of Food Science and Technology
National Taiwan University
1 Roosevelt Road, Sec. 4
Taipei, Taiwan 10672
Republic of China

Chi-Tang Ho
Department of Food Science
Rutgers, The State University of New Jersey
New Brunswick, NJ 08901

Herbs: Challenges in Chemistry and Biology

Overview and Perspectives

Chapter 1

Botanical Quality Initiatives at the Office of Dietary Supplements, National Institutes of Health

Joseph M. Betz

Office of Dietary Supplements, National Institutes of Health, Bethesda, MD 20892

Quality of botanical products is one of the biggest uncertainties that consumers, clinicians, regulators, and researchers face. Definitions of quality abound, and include specifications for sanitation, adventitious agents (pesticides, metals, weeds), and content of natural chemicals. Because dietary supplements (DS) are often complex mixtures, they pose analytical challenges and methods validation may be difficult. In response to product quality concerns and the need for validated and publicly available methods for DS analysis, the U.S. Congress directed the Office of Dietary Supplements (ODS) at the National Institutes of Health (NIH) to accelerate an ongoing methods validation process. The Dietary Supplements Methods and Reference Materials Program was created. The program was constructed from stakeholder input and incorporates several federal procurement and granting mechanisms in a coordinated and interlocking framework. The framework facilitates validation of analytical methods, analytical standards, and reference materials.

© 2006 American Chemical Society

The Dietary Supplement Health and Education Act (DSHEA) of 1994 (United States Public Law 103-417) amended the Federal Food, Drug and Cosmetic Act by defining as a dietary supplement any product (other than tobacco) that contains a vitamin, mineral, herb or other botanical, or amino acid and is intended as a supplement to the diet (1). The new law also established the Office of Dietary Supplements (ODS) within the Office of the Director at the U.S. National Institutes of Health (NIH). The mission of the ODS is to "strengthen knowledge and understanding of dietary supplements by evaluating scientific information, stimulating and supporting research, disseminating research results, and educating the public to foster an enhanced quality of life and health for the U.S. population."

The regulatory category into which a product falls is determined in the U.S. by the intended use of the product. The Act was significant because it established dietary supplements as a separate legal category and defined a framework for Food and Drug Administration (FDA) regulation of this category (2). It also established the regulatory framework for supplements as foods, not drugs, set rules for what information labels must contain, and gave FDA the authority to write supplement specific Good Manufacturing Practices based on a food model.

Briefly, companies that sell products intended for use as dietary supplements are prohibited from selling a product that is toxic or unsanitary, makes false or unsubstantiated claims, or claims to cure, treat, mitigate, or prevent a disease. Companies are also not permitted to introduce a new dietary ingredient into the marketplace without notifying FDA in advance, and are prohibited from selling a product that has not been produced according to current good manufacturing practices. Finally, companies must make sure that products are labeled properly (3).

When the DSHEA became law in 1994, there were an estimated 600 U.S. dietary supplement manufacturers producing about 4,000 products (4). According to Food and Drug Administration estimates, there were more than 29,000 different dietary supplement products on the market by the year 2000 with an average of 1,000 new products being added annually (5). A full overview of the regulatory history of botanical and other products is beyond the scope of this chapter. Readers interested in a brief but thorough overview of the subject should consult Israelsen and Barrett (6). Growth of the marketplace was fueled by increased consumer demand, often following publicity about the utility and efficacy of a particular herb. For example, in 1996 Linde *et al.* (7) published a meta-analysis of randomized clinical trials of the herb St. John's wort (*Hypericum perforatum* L.) in the prestigious *British Medical Journal* (BMJ) that concluded that the herb was more effective than placebo in treating certain types of mild to moderate depression. A June 27, 1997 broadcast of the popular television news magazine *20/20* (**Using Herb St. John's Wort To Treat Depression**) highlighted the BMJ article, and sales of the herb boomed. Other mainstream media outlets quickly followed with articles on herbal medicine (*8,9*) and the industry went through several years of rapid growth (*10*).

Only a short time later, the mainstream media began to take a closer look at the subject of herbs. In 1998, the Los Angeles Times commissioned a survey of St. John's wort products purchased from retail stores. The contract laboratory that did the work was asked to measure the amount of the phytochemical marker compound hypericin. The newspaper reported that 3 of 10 tested products contained no more than 50% of the hypericin content declared on the label and that another 4 of 10 contained less than 90% of claim (*11*). A later survey conducted by the Boston Globe in 2000 reported similar results (*12*). Publications in the peer-reviewed scientific literature have reported complementary findings. Draves and Walker observed that only two products (of 54) had a total naphthodianthrone concentration within 10% of label claim (*13*). Edwards and Draper examined levels of berberine and hydrastine in 20 goldenseal *Hydrastis canadensis*) products and found that only 10 of 17 root products met alkaloid content standards proposed by the United States Pharmacopeia (USP) and that 5 products contained little or no hydrastine, unusual berberine:hydrastine ratios, and additional peaks not observed with other products (*14*). Harkey *et al.* tested ginsenoside content of 24 commercial ginseng products and found that concentrations of marker compounds differed significantly from the amounts listed on the labels (*15*).

Unexpectedly, and in contrast to the picture for pharmaceutical products, there are frequently several analytical methods available for determination of phytochemicals in botanical products. An unpublished 2002 literature search performed for an AOAC International review panel found twenty-two different methods for constituents of St, John's wort. Many of these methods are based on different physical principles that would be expected to lead to different numerical results. Internal validation of methods is often lacking, as is cross-validation between methods. Lack of standard methods also leads individual investigators new to the field to invent a method when they become interested in a particular plant. This has led to the the publication in peer reviewed journals of results that can only be described as improbable. The report of the presence of colchicine in ginkgo and Echinacea products following discovery of this compound in the placental blood of women who were consuming dietary supplements is an example of this phenomenon (*16*), and repeated attempts to reproduce the results have failed (*17,18*).

Product Quality

Product quality is one of the biggest question marks facing consumers, clinicians, regulators, and researchers (*19*). As expected, there are some differences between definitions of quality for herbal preparations and for chemically synthesized products. Despite this, fundamental quality parameters are the same for both: identity, purity, content determination (i.e. strength) (*20*).

The DSHEA does not set a detailed framework for quality except to state that manufacturers are prohibited from introducing products posing "significant or unreasonable risk" into interstate commerce; must follow labeling regulations (accuracy, disclaimers, notification of claims); and must have substantiation that claims are truthful and not misleading (*1*). Compliance with quality standards set out in official compendia is voluntary. Working definitions of quality within the supplement industry vary. Quality parameters used by individual companies in the U.S. range from the simple to the complex. For example, for some companies, quality is ascertained from determination that the material was grown or wildcrafted organically, that the correct plant species and plant part (or extracts thereof) are in fact present in the product and that the product has been manufactured in a sanitary fashion. Other companies may set (or follow) explicit specifications for microbial load, adventitious agents (poisonous and otherwise), and content of desirable and undesirable natural chemicals (see the United States Pharmacopeia, American Herbal Pharmacopoeia (AHP), or European Pharmacopoeia (EP)). Swanson (*21*) provided a useful overview of the sorts of parameters that researchers and others should keep in mind when performing research or establishing quality assurance procedures for botanicals.

Analytical Challenges

Analytical challenges in quality assurance range from establishing the identity of the botanical source from which an extract was derived to measuring the amount of one or more desirable or undesirable natural constituents, such as pesticides, toxic elements, natural toxins, or marker compounds. Analytical methods are intended to generate reliable, accurate data for use by manufacturers or regulators for quality control or enforcement actions, respectively. Reliability, accuracy, precision and specificity are the keys to the utility of a method, but analysts must take steps to prove that any method they use has these features, especially if the method is to be used in a critical setting such as a quality control lab, a regulatory enforcement action, or a clinical laboratory. Fortunately, there are systematic approaches to validating that a particular method yields accurate and precise data. Manufacturers generally need methods applicable throughout the manufacturing process, while regulators require versatile methods that can be used for the same analyte(s) in a number of dissimilar finished products. The ability of a particular method to fit the specified purpose is one element of validation that is important but often overlooked.

In addition to the difficulties noted above, botanicals are complex mixtures that originate from biological sources. Such products and their ingredients pose particular analytical challenges for a number of reasons, and methods validation has proven particularly difficult. Raw materials are invariably "irregular" because their chemical composition depends on factors such as geographical origin, weather, harvesting practices, etc., while finished products frequently contain multiple botanical ingredients (20).

Role of the Office of Dietary Supplements

As noted above, the DSHEA empowered the FDA to establish current good manufacturing practices (CGMPs) for dietary supplements, and a proposed rule has been published (22). The law requires that any enforcement action taken against dietary supplement products use "publicly available" methods. In response to general concerns about the lack of properly validated publicly available methods, and general concerns about product quality, the U.S. Congress directed the Office of Dietary Supplements (ODS) at NIH to accelerate an ongoing methods validation process (23).

The ongoing process was a collaboration between AOAC International (the Scientific Society devoted to quality analytical measurements) and various representatives of the dietary supplements industry, regulatory and other governmental bodies, consumer groups, non-governmental organizations, and research scientists. The effort was embodied in a Dietary Supplements Task Group (DSTG) established by AOAC International. The group was originally proposed by several dietary supplement trade associations and was in existence before the congressional language. The DSTG met to select analytes and ingredients for study and to facilitate selection, solicitation, and validation of methods by AOAC International.

Parallel to the development of the DSTG by AOAC International, the ODS and FDA's Center for Food Safety and Applied Nutrition (CFSAN) had entered into an Interagency Agreement (IAG) for the purpose of funding validation of analytical methods for Ephedrine-type alkaloids and an FDA method for aristolochic acids in dietary supplements through the AOAC *Official Methods*^{SM} program. The task group was asked to decide which of several available ephedra methods were to be validated. Eventually, the group selected a LC/MS/MS method and a LC-UV method to be validated for botanical raw material, a commercial raw material extract, a finished ephedra product, a complex mixture of multiple supplement products, and an ephedra containing high protein drink. The methods were also to be validated on human specimens (serum and urine). Collaborative studies have been completed and study reports published (24,25,26). The LC/MS/MS method for six ephedrine alkaloids in biological fluids and the LC/UV method for ephedrine and pseudoephedrine were approved as "First Action" methods by AOAC's Official Methods Board.

The aristolochic acid method has not fared as well. The contract with AOAC called for validation of an LC/MS method for determination of aristolochic acids in *Aristolochia* spp. and *Asarum* spp, as well as in *Akebia* spp., *Clematis* spp., and *Stephania* spp. spiked with aristolochic acid. A single laboratory validation study failed, and AOAC is in the process of optimizing the method for another study, and a full collaborative study of the optimized method is scheduled for the Spring of 2005.

Once the Congressional language went into effect and the budget appropriation for ODS was signed, ODS was able to begin the new program. The first activities of the newly formed Dietary Supplement Methods and Reference Materials Program were two public meetings. The meeings were intended to gather stakeholders and get input on establishing goals and a direction for the new program and to solicit advice on mechanisms by which various groups select and validate methods. The meetings also began the process of establishing priorities (botanicals and others) and identifying potential research partners. The purposes of the program are fairly straightforward: to develop, validate, and share analytical methods and reference materials. The first public meeting was a Stakeholders meeting. It was convened in early 2002 at the Natcher Center on the NIH campus. Representatives of the supplements industry (manufacturers, suppliers, trade associations), the analytical laboratories industry, regulatory and other governmental entities, non-governmental organizations, and consumer groups met to identify needs that might be met by the new program. These stakeholders advised ODS that the new program should emphasize basic quality issues such as identity and contamination; accept the role of existing frameworks for methods vaidation (i.e. AOAC International; www.aoac.org); and accept the recommendations of the AOAC DS task group for prioritization of ingredients (*27*).

The second public meeting was an Analytical Methods Workshop held in the Spring of the same year. The workshop was intended to solicit ideas from stakeholder groups on the process that these groups use to select and prioritize methods for for their own uses. The groups were also invited to make formal presentations on the technical evaluation criteria they use for selecting analytical methods. Following these presentations, ODS hosted a series of breakout sessions to discuss needs, approaches, and capabilities. Details of the discussions and recommendations can be found in Saldahna et al. (*27*), but the key points were continuation of the DSTG to set broad priorities, the establishment of a small expert steering group for prioritorization of method selection, focus on qualitative medods for confirming botanical identity as well as quantitative methods for measuring individual constituents, and promoting basic research for identifying compounds of interest.

The operational program four major areas: methods development, methods validation, reference materials, and "other."

The bulk of the methods development work is being done by the Food Composition Laboratory at the United States Department of Agriculture (USDA) through an interagency agreement between the ODS and the National Heart, Lung, and Blood Institute (NHLBI) at NIH. The NHLBI had an existing contract with USDA to develop analytical methods for food constituents, including nutrients. ODS provided supplemental funds to allow USDA to develop validated methods for phenolic glycosides in foods and supplements. In addition, ODS has recently entered into agreements with experts at the FDA's CFSAN to develop or extend validated methods for determination of mycotoxins, pesticides, and toxic elements in Dietary Supplement raw materials and finished products.

The second and largest part of the program is analytical methods validation. As noted above, the problem facing manufacturers, regulators, and researchs for the top selling botanicals was not a lack of methods, but an overabundance of methods. At the start of the the ODS process, a literature review turned up no less than 22 different methods just for the constituents of St. John's wort *(Hypericum perforatum)*. The methods make different analytical assumptions, use different chemical or physical principles, or measue different chemical entities. In addition, methods had not undergone validation to assure accuracy or precision. As a result, no two analyses of the same product could be expected to be in agreement. In order to facilitate availability of valid methods to the user community, ODS expanded the original FDA/AOAC contract mentioned above to support infrastructure development and maintenance at AOAC. The stated deliverable for the contract is 20 AOAC Official Methods of Analysis after 5 years. However, the main purpose of the contract is to allow AOAC International to build its programs so that it has the capacity to process many more than the bare minimum of 20 methods.

Specific areas implemented by the contract include increases in AOAC staff for the purposes of coordinating and planning the DSTG and the the small steering group for methods prioritization recommended by the stakeholders (see above) (the Ingredient Ranking Subcommittee, or IRS), and the establishment and maintenance of a series of ad hoc Expert Review Panels (ERP). The latter is modeled after NIH technical peer review panels and serve to evaluate the technical merits of methods prior to AOAC selection for collaborative study. Ingredients are ranked by the IRS and weighted scores are assigned based on elements such as ingredient market share, safety concerns, positive public health implications of better measurements, and availability of methods that are sufficiently well developed to undergo the collaborative study process. The contract also supports AOAC volunteer committee functions, development and implementation of an AOAC short course on Single Laboratory Validation and on designing and conducting Collaborative Studies. Development, implementation, and maintenance of an electronic peer-review process for AOAC protocols and completed collaborative studies was also provided in the contract.

At present, AOAC International is in the process of convening Expert Review panels, conducting Single Laboratory Validation studies or collaborative studies for one or more constituents of SAMe, β-carotene, chondroitin sulfate, glucosamine, St. John's wort, ginkgo, and saw palmetto. The Ingredient Ranking Subcommittee has begun to solicit methods for L-carnitine, B vitamins, black cohosh, Ω-3 fatty acids, soy isoflavones, green tea catechins, lutein, turmeric, ginger, milk thistle, African plum, and flax seed. Unprioritized ingredients include hawthorne, biotin, feverfew, and pyrrolizidine alkaloids.

As suggested by its title, the third part of the Dietary Supplements Methods and Reference Materials Program provides for the production of reference materials. The stakeholder's meetings pointed out that there are several different types of reference material. These range from the pure chemical entity (analytical standard) used for determination of chromatographic retention time and quantitative instrument calibration, to matrix materials for evaluating method performance, to reference plant material used for comparison with unknown biomass for the purpose of plant identification.

One of the most important features of any analytical testing protocol is the ability of the analyst to verify whether or not the analytical instrument is operating properly and whether the assay has been performed correctly. One of the most useful ways of making this determination is to perform the method on a material that has had values for the analyte of interest assigned to it through a formal certification process. If the analyst succeeds in reproducing the certified values, then he or she can have confidence that the analysis was properly performed. The ODS has funded a five year Interagency Agreement with the NIST (National Institute of Standards and Technology, U.S. Department of Commerce) for the production of "suites" of Standard Reference Materials (SRM^{TM}). Each suite consists of properly identified dried, powdered botanical raw material, a commercial raw material extract, and one or more representative commercial products. The NIST process involves obtaining authentic botanical raw material, developing and validating analytical methods (if none exist) to determine compounds to be certified, using two or more methods and laboratories to analyze for the compounds of interest, and assigning values to be written into oa certificate of Analysis. Materials are then appropriately packaged and made available for purchase. The original goal was to develop SRM suites for 6-8 botanicals (beginning in 2002), but the program is progressing more rapidly than anticipated. By 2006, suites of ephedra, Saw palmetto, St. John's wort, ginkgo, green tea, β-carotene, tocopherols, bitter orange, and black cohosh should be available. In addition to these materials, NIST is in the process of certifying Ω-3 fatty acid values in an existing cod liver oil SRM and developing a multivitamin/multimineral SRM.

Other projects within the ODS program include production by the National Research Council (Canada) of Certified Reference Materials for *Panax* spp. (ginseng) and production of pure analytical reference standards. In the past, inhibitions to methods development and validation have been that with some exceptions (caffeine) highly purified plant secondary metabolites are very rare and therefore very expensive. The usual research practice was for the individual investigator to isolate pure compounds from the plant of interest, but this process is time consuming, expensive, and has low yields. In addition, the compounds may also be unstable in pure form. In order to expedite development and validation of methods, ODS has sponsored research into small- to medium-scale isolation methods for production of pure compounds as well as acquisition of these compounds for use in collaborative studies. It has also sponsored research into methods for stabilizing labile compounds. While quantities of the materials produced for collaborative are large by historical standards, they remain quite small. In practice, at the moment most part these compounds are valuable only to analytical chemists, and the marketing incentive for production of kilogram quantities of pure natural products (as for drugs such as paclitaxel) does not exist. ODS has therefore funded production of larger quantities of very high purity standards for national standard setting bodies such as the United States Pharmacopeia.

Additional projects include funding of Single Laboratory Validation studies for ingredients that are deemed important by NIH or FDA but are not highly ranked by the IRS (e.g. constituents of bitter orange, anthocyanins in berries) as well as a study for validation of thin layer chromatographic fingerprinting methods for determining botanical identity. This latter project is in keeping with stakeholder recommendations about pursuing methods for verification of plant identity and is complemented by funding for an electronic herbarium pilot project and for production of a handbook of botanical microscopy to replace the botanical drug microscopy texts of the late 19th and early 20th Century. An additional project related to plant identification is the development of a system for the identification, isolation, and characterization of compounds in plants that may indicate the presence of undesirable plant species (adulterants) in the botanical raw material or finished product. Identification of these "negative marker" compounds by bioassay-directed fractionation will be followed by development and validation of analytical methods for these compounds.

In conclusion, one of the explicit goals in the original and the current ODS strategic plan (*28*) was that the Office would "promote and support the development and improvement of methodologies appropriate to the scientific study of dietary supplement ingredients." In 2002, the stimulus provided by the language from the Congressional Appropriations Committee (23) as well as an increase in the ODS appropriation permitted a substantial expansion of efforts in this area, and the Dietary Supplements Methods and Reference Materials Program was created.

There remains an enormous challenge in developing, validating, and disseminating methods and reference materials for the projected 40,000 or so supplement products projected to be on the U.S. market by 2010 (5). But the goal of the U.S. Congress when it created the ODS was to provide an organization to coordinate and conduct basic and clinical research, develop education and communication programs directed to all segments of the public and private sectors with an interest in dietary supplements, and coordinate federal efforts related to issues associated with dietary supplements. These directions, especially the last, give the ODS a great deal of latitude in leveraging its resources across the NIH and other federal departments (28). The Dietary Supplement Methods and Reference Materials program has begun to lay the groundwork for addressing this challenge, but the challenge cannot be met by ODS alone. Congressional appropriations language (29) continues to support the program, and the program will grow, but only for as long as stakeholders from government agencies, the dietary supplements industry, and the academic world step forward, embrace the need for rigorous pursuit of excellence in analytical measurements, and actively participate in the process of achieving that excellence.

References

1. United States Public Law 103-417. 103rd Cong., 25 October 1994. *Dietary Supplement Health and Education Act of 1994.*
2. Hoffman, F.A. Regulation of dietary supplements in the United States: Understanding the Dietary Supplement Health and Education Act. *Clin. Obstet. Gynecol.* **2001,** *44*, 780-788.
3. Soller, R.W. Regulation in the herb market: The myth of the "unregulated industry". *HerbalGram* **2000,** *49*, 64-67.
4. Commission on Dietary Supplement Labels. *Report of the Commission on Dietary Supplement Labels*; U.S. Government Printing Office: Washington, DC, 1997; p 17.
5. Sarubin, A. *The Health Professional's Guide to Popular Dietary Supplements*;. The American Diatetic Association: Chicago, IL, 2000; p 3.
6. Israelsen, L.D.; Barrett, M. "History and Regulation of Botanicals in the United States" In: *The Handbook of Clinically Tested Herbal Remedies*; M. Barrett, Editor; Haworth Press: Binghamton, NY; 2004; pp 3-12.

7. Linde, K.; Ramirez, G;, Mulrow, C.D.; Pauls, A.; Weidenhammer, W.; Melchart, D. St John's wort for depression--an overview and meta-analysis of randomised clinical trials. *BMJ* **1996**, *313*, 253-258.
8. Fenyvesi, C. Herbal Tonic. *U.S. News and World Report* **1998**, January 19.
9. Greenwald, J. Herbal Healing. *Time* **1998**, *152*.
10. Blumenthal, M. *The ABC Clinical Guide to Herbs*; Thieme: New York, NY; 2003; p xviii.
11. Monmaney, T. Remedy's U.S. Sales Zoom, but Quality Control Lags St. John's wort: Regulatory vacuum leaves doubt about potency, effects of herb used for depression. *Los Angeles Times* **1998**, *August 31*.
12. Foreman, J. St. John's wort: Less than meets the eye. *Boston Globe* January 10, 2000.
13. Draves, A.H.; Walker, S.E. Analysis of the hypericin and pseudohypericin content of commercially available St John's Wort preparations. *Can J Clin Pharmacol* **2003**, *10*, 114-118.
14. Edwards, D.J.; Draper, E.J. Variations in alkaloid content of herbal products containing goldenseal. *J Am Pharm Assoc* **2003**, *43*, 419-423.
15. Harkey, M.R.; Henderson, G.L..; Gershwin, M.E.; Stern, J.S.; Hackman, R.M. Variability in commercial ginseng products: an analysis of 25 preparations. *Am J Clin Nutr* **2001**, *73*, 1101-1106.
16. Petty, H.R.; Fernando, M.; Kindzelskii, A.L.; Zarewych, B.N.; Ksebati, M.B.; Hryhorczuk, L.M.; Mobashery S. Identification of colchicine in placental blood from patients using herbal medicines. *Chem Res Toxicol* **2001**, *14*, 1254-1258.
17. Li, W.; Sun, Y.; Fitzloff, J.F.; van Breeman, R.B. Evaluation of commercial ginkgo and echinacea dietary supplements for colchicine using liquid chromatography-tandem mass spectrometry. *Chem Res Toxicol* **2002**, *15*, 1174-1178.
18. Li, W.; Fitzloff, J.F.; Farnsworth, N.R.; Fong, H.H. Evaluation of commercial Ginkgo biloba dietary supplements for the presence of colchicine by high-performance liquid chromatography. *Phytomedicine* **2002**, *9*, 442-446.
19. Ernst, E. Risks of herbal medicinal products. *Pharmacoepidemiol Drug Saf* **2004**, *13*, 767-71.
20. Busse, W. The significance of quality for efficacy and safety of herbal medicinal products. *Drug Information J* **2000**, *34*, 15-23.
21. Swanson, C.A. Suggested guidelines for articles about botanical dietary supplements. *Am J Clin Nutr* **2002**, *75*, 8-10.
22. Department of Health and Human Services. Food and Drug Administration. Proposed rule for the establishment of Good Manufacturing Practices for Dietary Supplements. *Federal Register* **2003**, *68*, 12157-12263.

23. Harkin, T. Committee Report 5 of 100 - Senate Rpt.107-084- DEPARTMENTS OF LABOR, HEALTH AND HUMAN SERVICES, AND EDUCATION, AND RELATED AGENCIES APPROPRIATION BILL, U.S. Government Printing Office, Washington, DC; 2002; pp183-184.
24. Roman, M.C. Determination of ephedra alkaloids in urine and plasma by HPLC-UV: collaborative study. *J AOAC Int* **2004,** *87,* 15-24.
25. Roman, M.C. Determination of ephedrine alkaloids in botanicals and dietary supplements by HPLC-UV: collaborative study. *J AOAC Int* **2004,** *87,* 1-14.
26. Trujillo, W.A., Sorenson, W.R. Determination of ephedrine alkaloids in dietary supplements and botanicals by liquid chromatography/tandem mass spectrometry: collaborative study." *J AOAC Int* **2004,** *86,* 657-668.
27. Saldahna, L.G.; Betz, J.M.; Coates, P.M. Development of the Analytical Methods and Reference Materials Program for Dietary Supplements at the National Institutes of Health. *J AOAC Int* **2004,** *87,* 162-165.
28. Office of Dietary Supplements. *Promoting Quality Science in Dietary Supplement Research, Education, and Communication: A Strategic Plan for the Office of Dietary Supplements.* NIH Publication Number 04-5533,Department of Health and Human Services, National Institutes of Health, Bethesda, MD; 2004; 20 pp.
29. Specter, A. Committee Report 2 of 13 - Senate Rpt.108-081 - DEPARTMENTS OF LABOR, HEALTH AND HUMAN SERVICES, AND EDUCATION, AND RELATED AGENCIES APPROPRIATION BILL, U.S. Government Printing Office, Washington, DC; 2004; pp170-171.

Chapter 2

Current Trends and Future Prospects of Traditional Chinese Medicines in the 21st Century

Xinsheng Yao[1,2], Ke Hu[3], Jianghua Liu[4], Nali Wang[1], Chengbin Cui[5], and Xiangjiu He[2]

[1]Traditional Chinese Medicines and Natural Products Research Center at Shenzhen, Shenyang Pharmaceutical University, Shenzhen 518057, China
[2]Institute of Traditional Chinese Medicine and Natural Products, Jinan University, Guangzhou 510632, China
[3]Department of Pharmaceutical Sciences, School of Pharmacy and Pharmaceutical Sciences, State University of New York at Buffalo, Buffalo, NY 14214
[4]21st Century Herbs and Health Company, Diamond Bar, CA 91765
[5]Institute of Pharmacology Toxicology, Academy of Military Medical Sciences, Beijing, China

Current well-marketed and non-patented traditional Chinese medicine (TCM) products are good candidates for new drug development. Clarification of bioactive components and functional mechanisms of TCM is a basis to reach this goal. To face the challenge of foreign companies, and to modernize and globalize the traditional medicines, high-tech R&D has to be applied in the Chinese pharmaceutical industry.

Since the People's Republic of China enrolled in WTO in November 2001, the Chinese government has made efforts to commit to the articles of WTO and adjust domestic laws to protect intellectual properties. To follow WTO rules, copies of foreign patented products such as softwares, videos and publications are forbidden. In response to this change, previous regulations of *'Brand-Name Drugs Protection'* and *'Traditional Chinese Medicines Products Protection'* that protected illegal copies of patented Western drugs and non-patented traditional Chinese medicine (TCM) products will be abolished soon. Meanwhile, the State Drug Administration also amended the "Drug Registration and Administration Law".

In the revised law (*1*), concept of New Drug is demarcated to two categories:

Drugs that have never been marketed in China.

Drugs that have been marketed in China, but their formulations, manufacturing procedures, or administration routes are newly improved or modified.

Pharmaceutical companies in China are now required to provide the patents for the registered drugs and manufacturing procedures and to be responsible for any violations. In this new situation, pharmaceutical companies can no longer survive by manufacturing Western patented drug. Therefore, many of them are currently challenged because the shortage of modern R&D staffs and facilities for new drug development make them hard to compete with these giant foreign pharmaceutical companies. The Chinese pharmaceutical industry is in dire need to find a way-out.

TCM is a Rich Resource for New Drug Development

As a distinctive national feature, TCM is a rich resource to explore and develop new drugs. According to reports, 10,000 TCM products in 26 preparation forms are currently available in the Chinese markets. Most of them have been used for treatment of diseases for thousands or hundreds of years in China, and their efficacies and toxicities have been well known. However, most of their principle bioactive components and therapeutic mechanism are still unknown. Most of the manufacturing procedures are not under the state-of-the-art quality control and have no protection of intellectual properties. These situations are potential obstructions to the development of traditional pharmaceutical industry in China and to the globalization of TCM products. Some suggestions are proposed to solve these problems. Here we mainly introduce approaches to modernization and globalization of TCM:

1. Increase government's policy and financial support to TCM research.
2. Increase collaboration between pharmaceutical companies and universities or research institutes.
3. Establish or upgrade high-tech R&D.
4. Integrate sources and form industrial chains.
5. Merge companies to form strategic partners.

Diversity of Chemical Components in TCMs

Each TCM has many different kinds of components. The components can be classified as general chemical compounds, bioactive compounds, pharmacologically active compounds, and indicative compounds for quality control. The following is an example to search for pharmacologically active compounds.

Procedures of Screening Bioactive Components from TCMs

Study on bioactive components in TCMs has experienced two stages: a classical stage and a bioactivity-oriented isolation stage. The classical stage is systematic isolation. The crude materials were systematically extracted, fractionated and purified. The bioassay was performed only to the isolated pure compounds for their bioactivity. This mode has been out-dated because the chance to find bioactive compounds is almost random.

The current mode is bioactivity-oriented isolation. Different from the classic mode, bioassays are applied to isolation in every step till pure compounds are isolated. In this approach, reliability and sensitivity of the assay is critical for no loss of activity and bioactive compounds in each isolation step. The disadvantages of this method are:

1. The method is labor intensive because assay was performed for all fractions and compounds at each isolation step.
2. Single rather than multiple assays are generally used.
3. High liability to obtain pseudo-positive and pseudo-negative result due to interference of impurities in the crude extracts or fractions.
4. *In vitro* assay results are not always representative of *in vivo* test results.

As an example for this approach, an assay of cell cycle and apoptosis was performed to screen crude extracts of 137 TCMs. The result showed that 37 of them exhibited bioactivities. Some of them inhibited G0/G1 or G2/M phase. The others induced apoptosis. One of the TCMs was Hei Guo Huang Pi. (stem bark of *Clausena dunniana*). By bioactivity-guided isolation, 29 alkaloids with anticancer activities were isolated and identified. The result of the induction of apoptosis and inhibition of cell cycle by compounds 4 is shown in Figure 1. In addition, active compounds were also tested in xenograft mice. Based on those findings, patents for these bioactive compounds were applied (2).

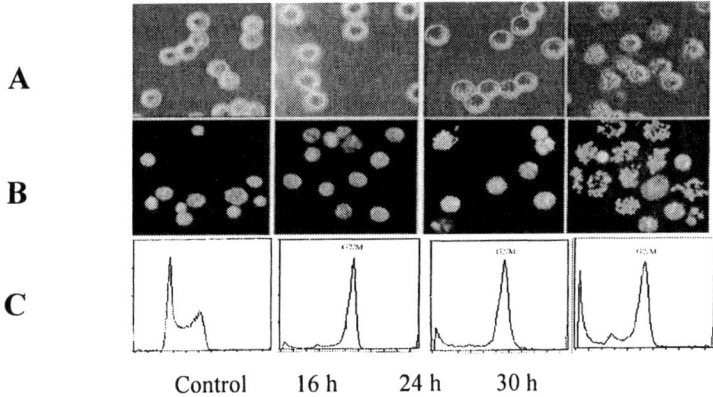

Figure 1. Induction of apoptosis and inhibition of cell cycle by compound 4 (12.5 µg/mL) isolated from Hei Guo Huang Pi. A. Under optical microscope; B. Under fluorescence microscope; C. DNA distribution analyzed by cytometer.

Major difficulties for Research on Bioactive Components from TCMs

Each TCM has diverse therapeutic functions. So far, bioassay model is not available for all of these functions. Therefore, the biological research on bioactive compounds from TCMs is confined in some respects. Moreover, thousands of formulas from hundreds of TCMs are used for treatment of all kinds of diseases, highlighted by the philosophy *"Different treatments for one disease'* and *'One treatment for different diseases'*. Thus, in many cases, the bioactive components screened out by *in vitro* bioassay could not ideally explain the functional diversity of TCMs. Even a single animal model can hardly indicate the total efficacies of a TCM or TCM formula. It is also hard to develop appropriate models to specifically evaluate the effects of trace bioactive compounds, or its acting mechanisms in human body.

Diverse Components of TCMs Work on Multiple Targets

It should be recognized that TCMs exhibit efficacy *via* different actions of bioactive compounds on different targets. It is impossible for a single assay to systemically evaluate the specificity and diversity of TCMs and TCM fomulus used in clinic. The diversity of pharmacological efficacy suggests that it is necessary to apply multiple-target assays to screen various biologically active compounds that act on different targets *via* different mechanisms. Because the limited knowledge about the working mechanism of our body, and the complicated composition of TCMs, such investigation is very complicated and challenging, yet it brings us chances.

The relationship between the diverse components and multiple targets are illustrated in Figure 2, 4A represent the mixture of a TCM extract. Red blocks stand for those compounds exhibiting pharmacological effects. Blue blocks stand for those compounds responsible for toxicity. While, 4B represents the human systems, including CNS, endocrinal, immune, digestive, and respiratory systems *etc*. Each compound may act on one or several systems. In each system, tissues, organs, cells, proteins, and genes are their targets. Most of the mechanisms are unknown, like a black box. Our task is to reveal the black box to explore the mechanism.

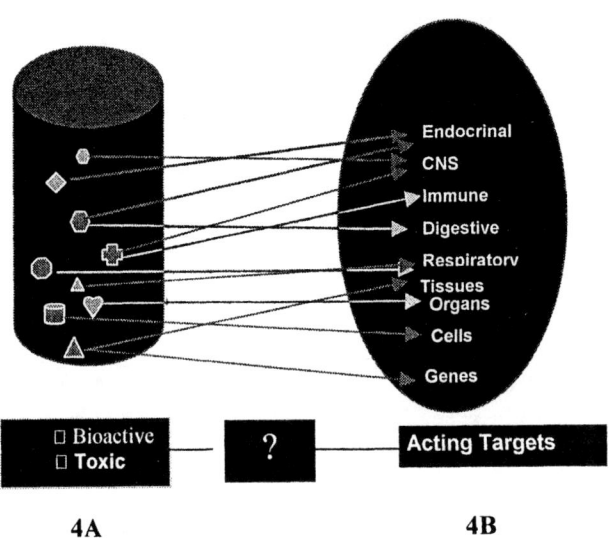

Figure 2. Diverse Components of TCMs work on multiple targets in human body.

Table I reveals some samples of different bioactive components acting on different targets. *Actonium* is a TCM having cardiotonic, diuretic, euphoria and analgesic functions. Screened by a cardiotonic assay, higenamine was obtained from its extract. By an analgesic assay, aconitine was isolated. Another example is *Arnebia euchroma* extract, in which, shikonin derivatives were screened to be cytotoxic against cancer cells, and arnebinol, arnibinone, and arnebifuranone were isolated as inhibitors of PGE2 biosynthesis.

Table I. Different Bioactive Components Acting on Different Targets

TCM	Bioactivities	Assays	Compounds
Wu Tou (*Aconitum spp.*)	*Cardiotonic, Diuretic, Euphoria, Analgesic*	Yagi-Hartung Assay (*de nova* frog' heart□ Analgesic	Higenamine Aconitine
Da Huang (*Rheum coreanum* and *R. palmatum*)	Stomachtonic, Antidiarrheal	Diarrhea Induction Assay (mice)	Sennoside
Ying Chen Hao(*Artemisia capillaris*)	Protection of gall bladder, Anti-inflammation	Bile Induction Assay (Rat)	Capillarisin
Bei (*Babylonia japonica*)	Thirsty, Sight weakness, Mydriasis, Dysphonia, Astriction	Atropine Assay (Mice)	Surugatoxin
Ruan Zi Cao (*Arnebia euchroma*)	Hemostasia, Anti-inflammation, Anti-microbe, Anti-virus, Anti-cancer	Biosynthesis of PGE2 Cytotoxicity	Arnebinol Arnibinone Arnebifuranone Shikonins

Investigation on TCM Formulas is More Complicated

To the more complex TCM formulas, the diverse bioactive compounds may interfere each other to cause drug-drug interaction, including synergic, additive and antagonistic effects. Some researchers take a stand on studying the crude extracts of intact prescriptions rather than the individual TCMs and bioactive compounds. Others agree that individual compounds can represent the total efficacies of TCMs and TCM formulas. In our opinion, though it is difficult to do a thorough investigation of formulas, it is worthy of systemic research to know the whole picture.

Multiple-Target and High-Throughput Screening Assays

As mentioned above, TCMs and TCM formulas should be studied by multiple-target assays in animal, tissue, organ, cell, protein, enzyme, receptor, and gene levels. If available, high-throughput screening should be the first choice to efficiently screen the bioactive compounds.

High-throughput screening can be used for both crude extracts and compounds. But it should be noted that crude extracts sometimes are not easy to screen because some universal impurities therein may lead to pseudo-positive and pseudo-negative results. Isolated single bioactive compound may not represent the whole efficacy of a TCM, in which interaction may happen between different compounds.

Libraries for Extracts, Fractions and Compounds from TCMs for High-throughput Screening

The concept of libraries for extracts, fractions and compounds from TCMs for high-throughput screening is given in Figure 3. We target those TCM products that have multiple effects, good market but without patents. We carry on 4 steps isolation in a row. In each step, four fractions are obtained. We collect all fractions without any bias. At the 4th step, 256 fractions are available and applied to multiple-target assays. All fractions are fingerprint labeled and all the isolation procedures are recorded. If 2000 TCMs are collected and extracted, 500,000 fractions by this method will be stored in the library.

Significances of TCM Library

Unlike the general bioassay-guided isolation, all the isolated fractions in the library are non-arbitrary selected. Thus, all constituents in the stored extracts and fractions are available to meet the demands of high-throughput screening.

Comparing to the crude extracts, composition of the fractions by four-steps isolation become much more simplified, and the bioactive compounds in it are more concentrated. So they are easier for high-throughput screening. The fractions are applied to different kinds of bioassays based on the clinical efficacy of TCM. Once the assay result is positive, further isolation is performed till a single active compound is obtained. This will benefit mechanism elucidation of diverse bioactive components more efficiently. It is very critical for the development of patented TCM products and their market globalization.

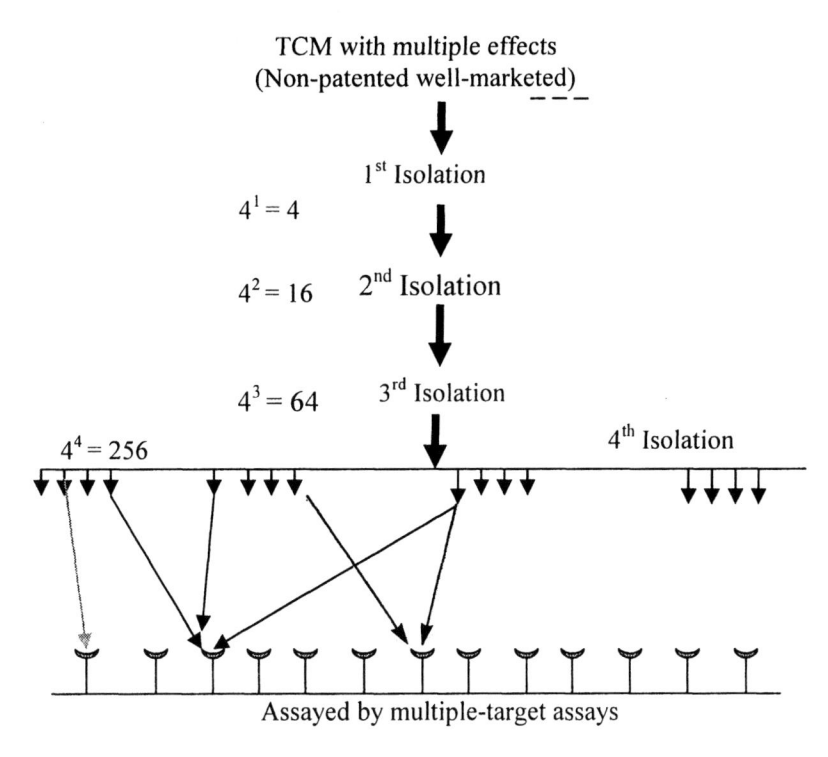

Figure 3. Four-steps isolation procedures for non-patented, well-marketed TCM with multiple effects.

Proposal of TCM R&D

As mentioned before, TCM is a rich source for new drug development. Here are some suggestions for R&D to pharmaceutical companies.

1. Select the non-patented, well marketed TCM or TCM formulas.
2. Clarify the functional mechanisms of these pharmacologically active compounds by assaying the samples in the library.
3. Apply for patents for these bioactive fractions or components.
4. Develop new TCM products or upgrade quality levels of marketed ones.

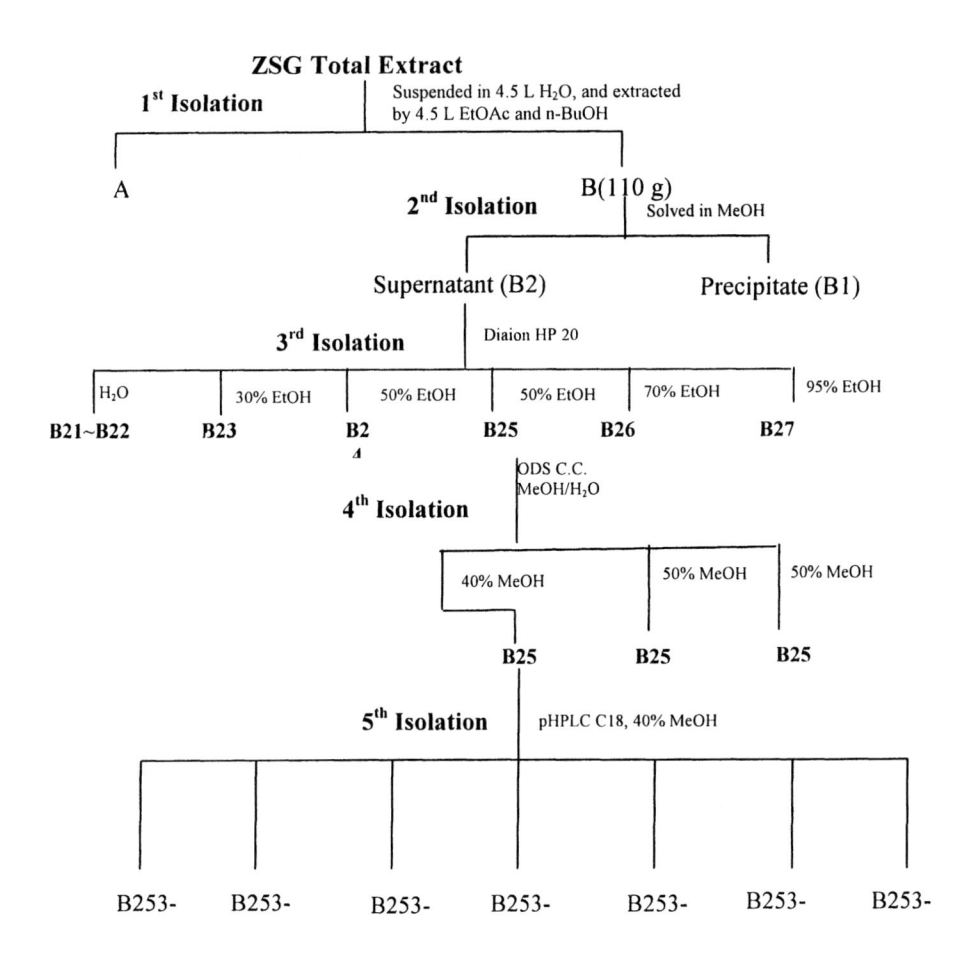

```
                        ZSG Total Extract
   1st Isolation    Suspended in 4.5 L H₂O, and extracted
                    by 4.5 L EtOAc and n-BuOH
   ┌────────────────────────┴────────────────────────┐
   A                                              B(110 g)
                     2nd Isolation                Solved in MeOH
                  ┌──────────────┴──────────────┐
                  Supernatant (B2)          Precipitate (B1)
         3rd Isolation       Diaion HP 20
   ┌──────┬──────────┬──────────┬──────────┬──────────┬──────────┐
   H₂O   30% EtOH   50% EtOH   50% EtOH   70% EtOH           95% EtOH
   B21~B22  B23       B24        B25        B26                B27

                                    ODS C.C.
         4th Isolation              MeOH/H₂O
                        ┌──────────────┼──────────────┐
                        40% MeOH    50% MeOH      50% MeOH
                          B25          B25            B25

         5th Isolation      pHPLC C18, 40% MeOH
   ┌────┬────┬────┬────┬────┬────┬────┐
   B253- B253- B253- B253- B253- B253- B253-
```

Figure 4. Isolation procedures of ZSG.

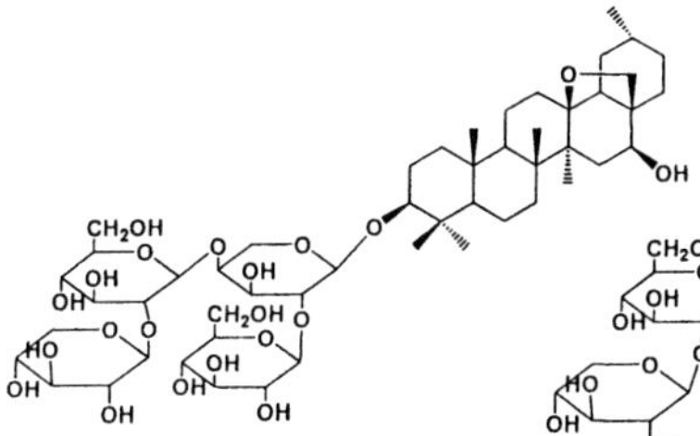

Figure 5. Bioactive compounds isolated from ZSG extract.

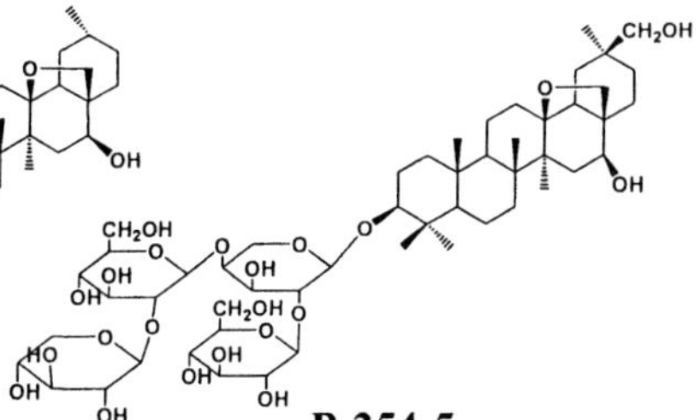

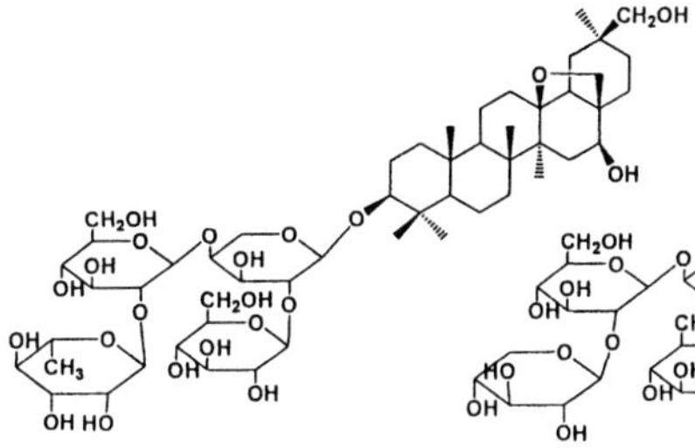

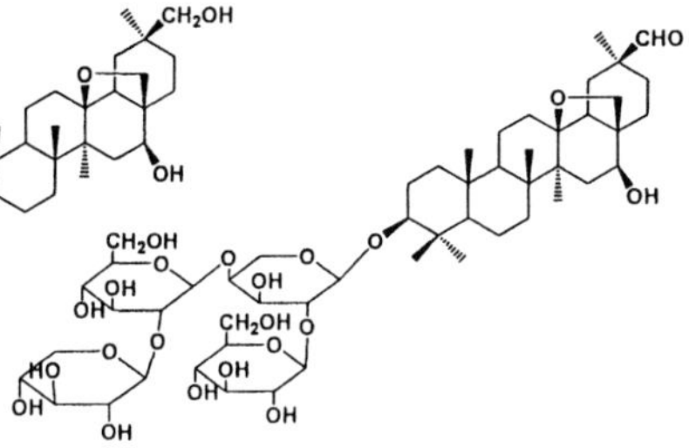

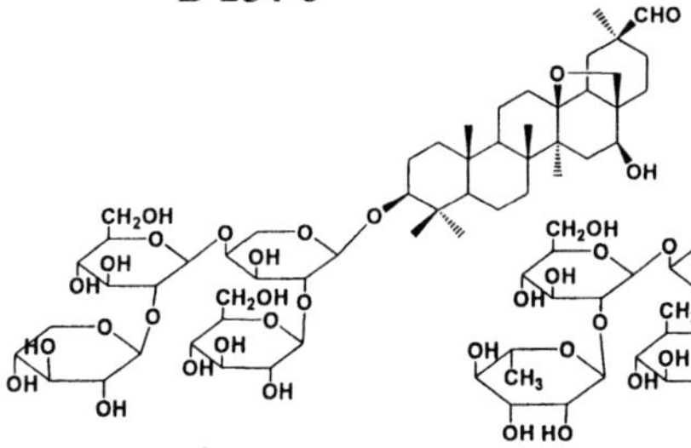

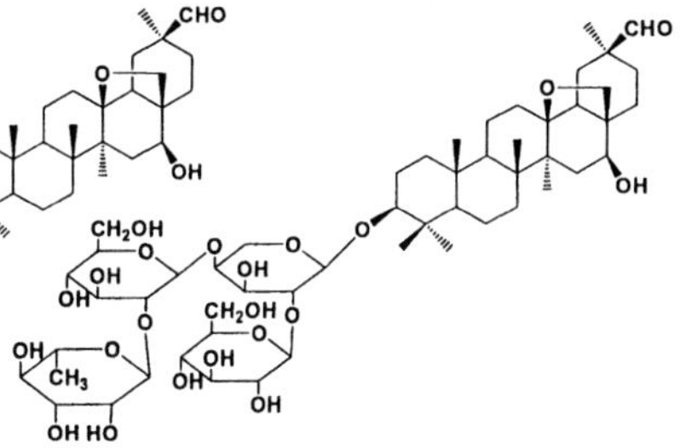

An anticancer TCM is taken as example. ZSG is a well-marketed anticancer TCM. Its efficacy has been proved in clinic, but its bioactive components and functional mechanism are unknown. The product has not been patented. TSG was extracted and fractionated (Figure 4). Tested by assays of cytotoxicity, apoptosis, anti-angiogenesis, and tubulin models, fractions B25, B26 and B27 showed bioactivities, among which, B25 showed the highest. Several compounds were isolated from this fraction. Their chemical structures were identified by NMR analysis as shown in Figure 5.

Summary

The efficacies and toxicities of TCM have been verified through hundreds of years. Therefore, they are rich source for new drug development. We suggest that clarification of the biological active components in TCMs and their effective mechanism are the way-out to develop new drug to Chinese pharmaceutical industry and the basis to modernize and globalize TCM products.

References

1. Chinese National Bureau of Medicine Supervision and Administration, Order No. 35: Medicine Registration and Administration (Trial, 20021030).
2. Cui, C.B.; Yan, S.Y.; Cai, B.; Yao, X.S. *J. Asian Natural Prod. Res.* **2002**, *4*, 233-241.
3. Directions of Patent Invention and Application from PR China National Bureau of Intelletual Properties (Application No. 01123988_3; Public No. CN1357327A)

Chapter 3

Bioactive Polyphenols from Foods and Dietary Supplements: Challenges and Opportunities

Navindra P. Seeram

Center for Human Nutrition, David Geffen School of Medicine, University of California, Los Angeles, CA 90095

Polyphenols are the most abundant phytochemicals in our diet and include flavonoids, tannins, lignans and stilbenes. Phenolic acids which are found in their free forms and as polyphenol metabolites, are also of significant importance in human nutrition. Tannins, grouped into condensed tannins (CTs) [also known as proanthocyanidins (PAs)], and hydrolyzable tannins (HTs) [ellagitannins (ETs) and gallotannins (GTs)], are complex polyphenol polymers with distinct chemical and biological properties, extensive structural diversity and high molecular weights. Examples of dietary sources of tannins are fruits (apples, grapes, strawberries, blueberries, cranberries, raspberries, plums, pomegranates), nuts (almonds, pistachios, walnuts, pecans, hazelnuts) and food-derived products such as beverages and snacks (fruit-juices, wines, tea, cocoa, cider, beer, chocolate). Because of their potent antioxidant, anticancer, anti-neurodegenerative and anti-atherosclerotic properties, tannins are used as botanical ingredients in dietary supplements for their potential human health benefits. However, due to their polymeric nature, ill-defined structures and unavailability of commercial standards, their analyses and estimation is difficult. Hence there is limited information on the dietary intake of tannins in their native forms. In addition, knowledge of their absorption, distribution, metabolism, bioavailability and pharmacokinetics in humans is limited. This overview examines the challenges, opportunities and recent progress made in herbal research of bioactive dietary polyphenols, focusing on the ubiquitous polyphenols, PAs and ETs.

Introduction

Phytochemicals (*phyto* meaning plant) having a polyphenolic structure (i.e. aromatic rings bearing several hydroxyl, -OH, groups) are ubiquitous in edible plants and are classified as secondary metabolites. Natural products or secondary metabolites are widely accepted to be involved in plant defense mechanisms against harmful ultraviolet (UV) radiation, pathogens, herbivorous predators etc. Dietary polyphenols are the most abundant antioxidants in human diet and are associated with health benefits against diseases mediated by oxidative stress including cancer, inflammation, cardiovascular and neurodegenerative diseases [reviewed in (*1-6*)]. Polyphenols are therefore the most abundant phytochemicals in our diet and because of their potent bioactivities, are of significant importance in human nutrition and health.

Types of Dietary Polyphenols

The considerable structural diversity of polyphenols imparts unique biological properties to each class which affects their absorption, distribution, metabolism, bioavailability and excretion in humans. The structural diversity of dietary polyphenols can be observed by the varying types and oxidation levels of their heterocycle ring, their substitution patterns of hydroxylation, the existence of steroisomers, their glycosylation by sugars and/or acylation by organic and phenolic acids, and by conjugation with themselves to form polymers etc. The main classes of dietary polyphenols are based on their skeletal structures and include the most abundant group, the flavonoids. Flavonoids have a basic skeletal structure of C6-C3-C6 and based on their degree of oxidation and substitution in the 3-position can be further sub-divided into six main classes: flavonols (e.g. quercetin), flavones (e.g. luteolin), isoflavones (e.g. genistein), flavanones (e.g. naringenin), flavanols or catechins (e.g. epigallocatechingallate, EGCG), and anthocyanidins (e.g. cyanidin). Figure 1 shows the structures of examples of polyphenols from the main sub-classes of flavonoids. Of these sub-classes, the anthocyanins (glycosides of anthocyanidins), which are pigments responsible for the attractive red, blue and purple colors of flowers, fruits and and vegetables, deserve special structural mention since as flavylium ions, their chemistry specifically influences their biological properties (*5*).

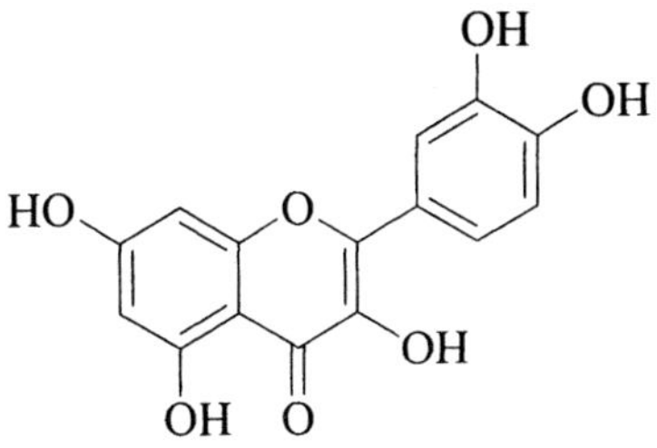

Quercetin (a flavonol)

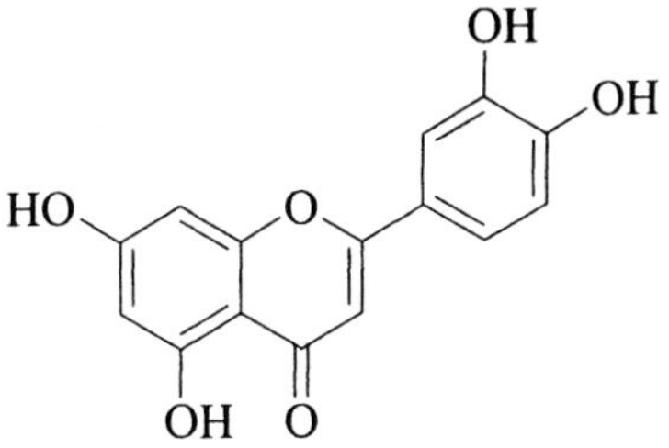

Luteolin (a flavone)

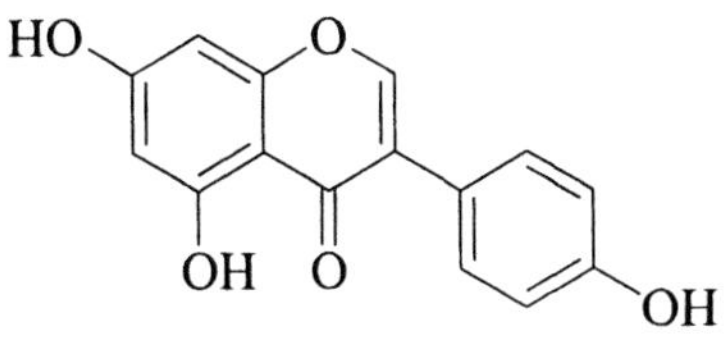

Genistein (an isoflavone)

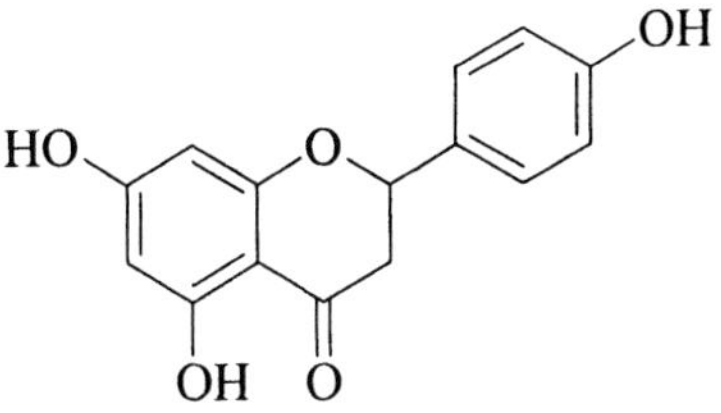

Naringenin (a flavanone)

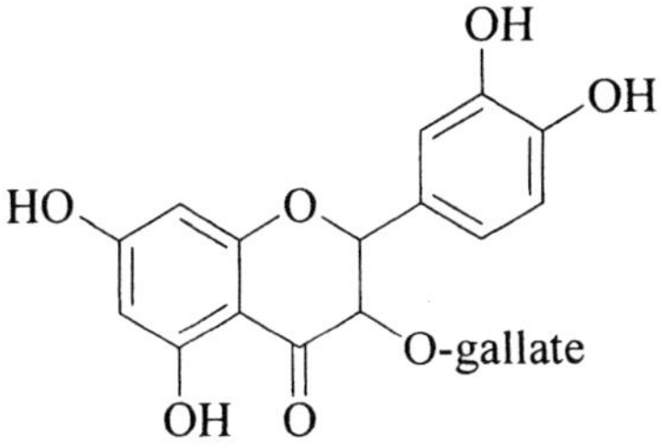

Epigallocatechingallate (a flavanol)

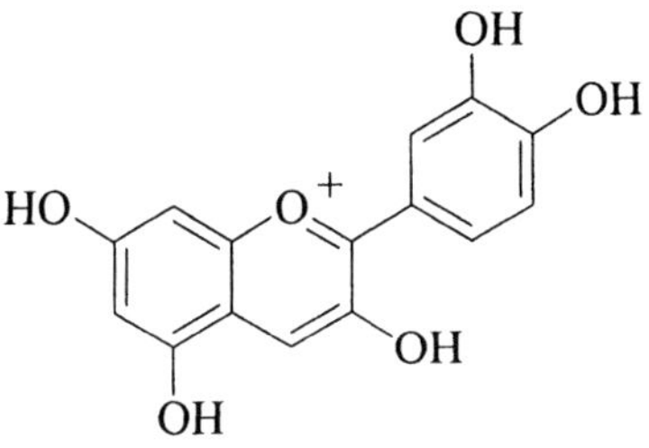

Cyanidin (an anthocyanidin)

Figure 1. Structures of the most abundant group of dietary polyphenols, flavonoids: a) flavonol (e.g. quercetin); b) flavone (e.g. luteolin); c) isoflavone (e.g. genistein); d) flavanone (e.g. naringenin); e) flavanol (e.g. epigallocatechingallate); d) anthocyanidin (e.g. cyanidin).

Other classes of dietary polyphenols include lignans found in flaxseed and other cereals and grains, e.g. secoisolariciresinol, which is abundant in linseed (6), and stilbenes, e.g. resveratrol and its analogs, found in a variety of dietary sources including grapes (and hence wine), and a number of berry fruits (7). Of considerable dietary importance and closely associated with polyphenols are phenolic acids, which include derivatives of hydroxybenzoic acid (e.g. gallic acid) and cinnamic acid (e.g. caffeic acid). Phenolic acids occur commonly as ferulic and caffeic acids and their derivatives for e.g. chlorogenic acid or caffeoyl ester, widely consumed from coffee (6). Although plant phenolics occur in their free forms, they can be considered as moieties on polyphenolic rings. Hence they are widely believed to be released from polyphenols *in vivo* and then further conjugated in the liver before excretion. Figure 2 shows chemical structures of common compounds of the lesser abundant groups of dietary polyphenols, lignans, stilbenes and phenolic acids.

Whereas much progress has been made into the research of flavonoids, lignans and stilbenes, the tannins constitute a group of structurally ill-defined polyphenols, which are lesser understood. These compounds are formed from the polymerization of polyphenols, either biogenetically by the plants or as a result of processing. The former group is referred to as tannins proper and are categorized into condensed and hydrolyzable tannins (CTs and HTs, respectively), which are the focus of this discussion. The latter group of dietary polyphenols are formed from the oxidative and enzymatic manipulations and processing of plants into foods such as oolong and black teas for e.g. theaflavins and thearubigins. These complex compounds are commonly referred to as derived tannins or 'tannin-like' compounds and are reported to have significant benefits to human health (6).

Dietary poyphenols, albeit ubiquitous in the human diet, are too wide a group to be examined in detail in a single chapter. This discussion focuses on the challenges and recent progress made in studies of the chemistry and biology of the dietary polyphenols, tannins i.e. CTs and HTs.

Tannins: An Overview

Tannins are polymers of polyphenols reaching high molecular weights (sometimes >3000 amu) and extensive structural diversity that make up a large proportion of phytochemicals consumed by humans. Dietary sources of tannins include fruits, legumes, seeds, grains, and food-derived products such as beverages and snacks (fruit-juices, wine, tea, cocoa, cider, beer, chocolate).

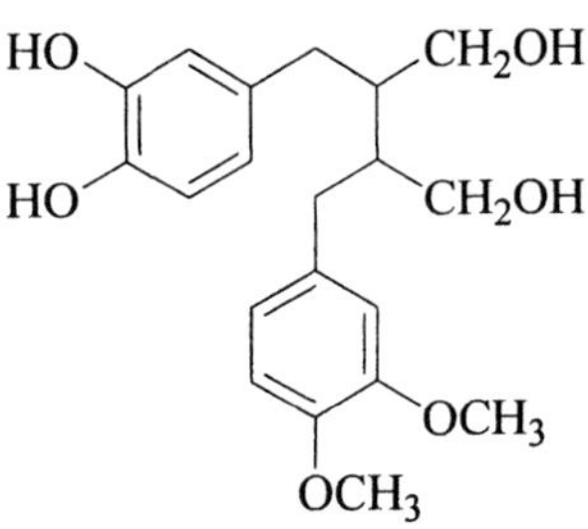

Secoisolariciresinol (a lignan)

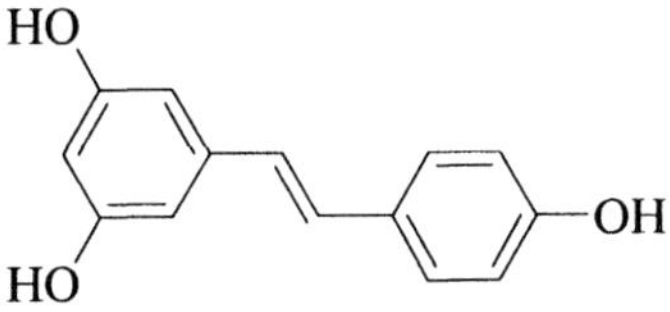

Resveratrol (a stilbene)

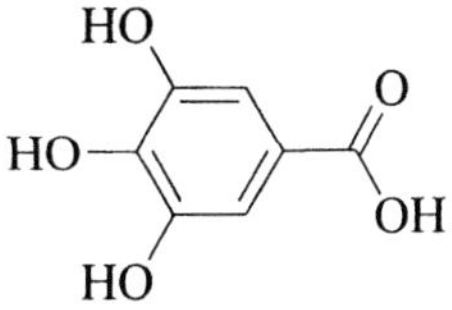

Gallic acid (a phenolic acid)

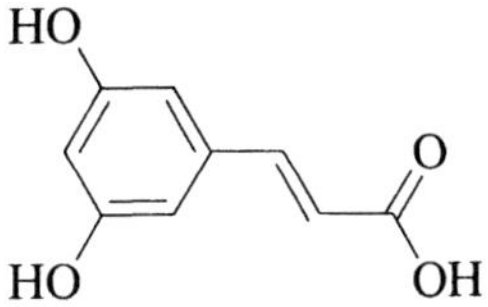

Caffeic acid (a phenolic acid)

Figure 2. Structures of lesser abundant groups of polyphenols: a) lignan (e.g. secoisolariciresinol); b) stilbene (e.g. resveratrol); c) phenolic acid [e.g. gallic acid (a hydroxybenzoic acid) and caffeic acid (a hydroxycinnamic acid)].

Tannins are divided into two classes which show distinct chemical and biological properties: condensed tannins (CTs) also known as proanthocyanidins (PAs), and hydrolyzable tannins (HTs) [consisting of ellagitannins (ETs) and gallotannins (GTs)] (*8,9*).

Tannins or 'vegetable tannin extracts' gained their original popularity in the brewery and leather 'tanning' industries which capitalized on the ability of these plant derived natural products to precipitate alkaloids, gelatin and other proteins (*10,11*). It is this property, and in particular precipitation of the parotid saliva and proline-rich proteins in the oral cavity, which is responsible for the astringent taste associated with the consumption of tannin-rich foods and beverages such as wines (*12,13*). The astringency of tannins is essential to explain the protective deterrence to herbivores from feeding on tannin-rich plants correlating to the protective roles of plant secondary metabolites. Other roles of tannins in plants include their reducing ability and strong affinity for complexation with metal ions (*8*). The latter abilities confer potent *in vitro* antioxidant capacities to these compounds which is thought to be borne out *in vivo*. Epidemiological studies have suggested that a phytochemical rich diet, that includes fruits and vegetables, may reduce the incidence of diseases mediated by oxidative stress such as cancer, inflammation, neurodegenerative and heart disease (*1-3*).

The chemical and biological properties of tannins have been intensively studied for more than a century. However, their complex polymeric nature, ill-defined structures, wide structural diversity and unavailability of commercial standards have made their chemical and biological studies extremely challenging. Structurally, tannins possess 12-16 phenolic groups and 5-7 aromatic rings per 1000 units of relative molecular mass (*14*). Because of their high molecular weights and structural diversity, their estimation in foods, food-derived products and dietary supplements is extremely difficult. Hence information on their dietary intake as well as data on their bioavailability and pharmacokinetics in humans is limited. However, in the past few decades significant progress has been made in tannin research with the advent of modern chromatographic and analytical techniques including high performance liquid chromatography (HPLC), coupled with sensitive detectors for e.g. UV (PDA, photodiodearray or DAD, diodearray detector), electrochemical detector (ECD) and mass spectrometry (LCMS) and its tandem applications (LC-MSn), gas chromatography (GC), gas chromatography mass spectrometry (GCMS) and nuclear magnetic resonance (NMR) spectroscopy. Only recently have appropriate analytical methods using HPLC and LC-MSn been reported for the measurement of CTs i.e. PAs (*15-18*). These studies showed that PAs account for a major proportion of polyphenols ingested by humans. However much still needs to be done with regard to HTs (as ETs) which, albeit not as common as PAs, are of equivalent significance in human nutrition. Ellagic acid (EA) can be considered as a chemical marker of ETs since it is thought to be a hydrolytic product of ETs (*19*) formed from the spontaneous rearrangement of hexahydroxydiphenic acid (HHDP). Traditionally, spectrophotometric methods were used to detect EA from foods such as strawberries (*20*). However, since the last decade, there have been reports of HPLC methods (*21,22*) and very recently, LC-MSn methodologies (*23*) to detect EA from food sources. Due to the lack of knowledge of the actual identities and structures of ETs from many foods, as well as the unavailability of commercial standards, estimation of their levels and studies on their *in vivo* fate after absorption is difficult.

Tannin research focusing on PAs and ETs have been reviewed (*8,9,14,24-26*). This current overview updates prior research on CTs (henceforth referred to as PAs) and HTs (focusing on ETs), in the context of human nutrition and disease prevention, and outlines recent progress made in the author's laboratory on research of human pharmacokinetic studies of ETs.

Condensed Tannins: Proanthocyanidins (PAs)

PAs are dimers, oligomers and polymers of flavan-3-ols linked by C-C and C-O-C bonds whose key feature is that they release anthocyanidin units on hydrolysis (Figure 3), resulting in the origin of their name. The flavanol units can be linked through C4-C6 or C4-C8 bonds established from 'upper' and 'lower' units referred to as B-type linkages. However in addition to C-C linkages, PAs can also have ether linkages between C2-O5 or C2-O7, referred to as A-type linkages. PAs with A-type linkages are common to tea, cocoa, cranberries, plums, peanuts, avocado, cinammon and curry powder (curcumin or turmeric) *(15,24)*. PAs can also be classified based on their constituent units which are produced on acid hydrolysis. In this case they are referred to by the nomenclature system established for anthocyanidins (shown in Figure 3). Hence, PAs consisting of individual (epi)catechin units are referred to as procyanidins. Similarly, PAs consisting of (epi)afzelechin units are known as propelargonidins and those with (epi)gallocatechin units are known as prodelphinidins. Propelargonidins and prodelphinidins are heterogenous in their constituent units and coexist with the most commonly occuring PAs, procyanidins.

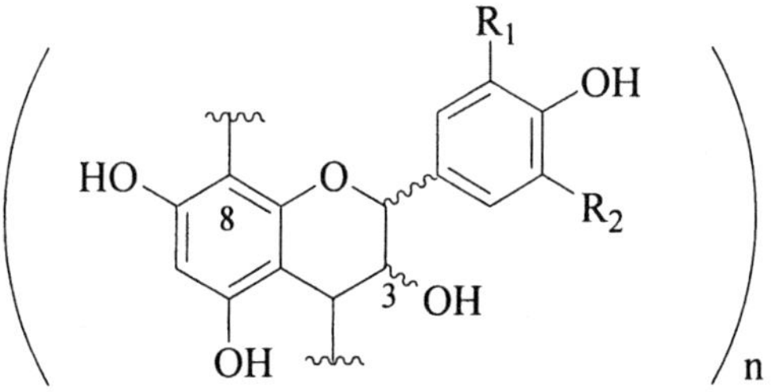

$R_1, R_2 = H$, propelargonidins
$R_1 = H, R_2 = OH$, procyanidins
$R_1, R_2 = OH$, prodelphinidins

Figure 3. Basic structure of condensed tannins (proanthocyanidins).

The size of PA molecules can be described by their degree of polymerization (DP) *(24)*. However, the mean DP of PAs in foods has rarely been determined although they are known to vary widely and can reach up to DP17 *(8)*. A major challenge into PA research arises because of the lack of data on their levels in foods, food-derived products and dietary supplements, due to their ill-defined structures and unavailability of commercial standards. The lack of reliable values for PA content in food can also be explained by the differences in assays used for their estimation or in the nature of the samples analysed *(8)*.

Their DP can vary widely due to oxidative and/or enzymatic action, especially when foods rich in PAs are processed. For example, green tea, one of the most popularly consumed beverages second only to water, on steaming, causes deactivation of oxidative enzymes which prevents the conversion of its flavanols into theaflavins and thearubigens, as found in black and oolong teas.

Recent progress has been made in the measurement of PAs resulting from the development of more sensitive analytical methods and their levels in common foods have been estimated (*15-18*). Foods which contain PAs include fruits (apples, grapes, cranberries, blueberries, raspberries, strawberries, chokeberries, plums, cherries, blackcurrants, peaches, blackberries, avocados), cereals (sorghum), beans (pinto, red kidney), nuts (hazelnuts, pecans, pistachios, almonds, walnuts), spices (cinnamon, curry-powder), beverages and snacks (fruit-juices, wines, beer, chocolate and chocolate milk) (*8,15*).

Estimates of consumed levels of PA vary in populations depending on cultural and dietary habits which may or may not include tannin-rich foods such as tea, wine, berry fruits, grapes, apples, pears and plums. Tea which is consumed widely by English and Asian cultures, provides high concentrations of polyphenols including PAs. Epidemiological studies have shown an inverse relationship between tea consumption and diseases such as cancer. The French population consume substantial amounts of red, white and sherry wines and hence intake large amounts of PAs. This is key to account for the now widely accepted phenomenon known as the "French Paradox" whereby epidemiological data have shown that these populations have a lower incidence of cardiovascular diseases despite a high fat diet similar to Americans (*7,27*). It is believed that the consumption of significant levels of polyphenols present in wines contributes towards the observed health benefical effects. The major source of PAs in American diet are apples, chocolate and grapes (*15*). Recently, the mean dietary intake of PAs for infants 6-10 mos old was estimated to be higher than that of adults >20 y old (3.1 and 0.77 mg/Kg body weight, respectively). Increasing evidence suggests that exposure to these nutrients in early infancy may have long term beneficial effects on later life (*28*). As research progress continues to be made, agencies such as the United States Department of Agriculture (USDA) periodically updates their database on flavonoid contents of selected foods (*29*).

In conclusion, PAs are responsible for a considerable proportion of polyphenols present in human diet due to their ubiquitous existence. PAs are of significant interest in human nutrition due to their potent bioactivities and possible protective effects on human diseases (*8*). For example, studies on cranberry PAs have shown that they have benefical effects against urinary tract infection (UTI) and antiproliferative effects against human tumor cell lines (*30,31*). Tannins from grape seeds and berry fruits, such as blueberries and strawberries, are widely accepted to have potent antioxidant properties with implications against diseases mediated by oxidative stress such as cancer and neurodegenerative diseases (*32,33*). Tea polyphenols including its PAs have been shown to have protective effects against atherosclerosis and cancer (*34,35*). However, the lack of knowledge of levels of PAs present in foods results in significant disadvantages to current research especially due to the unavailability of comprehensive food-composition tables. These data are necessary given the potential health benefits of these phytochemicals and their ubiquitous presence in our diet. Knowledge of PA intake is necessary for epidemiological studies for correlation with the incidence of early markers of oxidative-stress mediated diseases such as cancer, neurodegenerative and heart disease. Nevertheless, it is noteworthy that with the development of modern research methodologies, and the focus of more research attention from laboratories towards solving these problems, recent progress are being made. Hence, with the advent of more sophisticated instrumentation with increased sensitivity in detectors, improved analytical methodologies, and increased scientific focus on this important aspect of human nutrition and disease prevention, advances in research will continue to be achieved.

Hydrolyzable Tannins: Ellagitannins (ETs)

Hydrolyzable tannins (HTs) differ from PAs both in their biology and in their chemistry and are categorized into gallotannins (GTs) and ellagitannins (ETs). ETs are esters of hexahydroxydiphenic acid (HHDP: 6,6'-dicarbonyl-2,2',3,3',4,4'-hexahydroxybiphenyl moiety) and a polyol, ususally glucose or quinic acid (*9,25,36*). The defining structural characteristic unit of ETs is the HHDP moiety which originates biogenetically from oxidative C-C coupling of phenolic galloyl groups *in vivo*. A key feature of ETs is their ability to release the bislactone, ellagic acid (EA), which is formed from the hydrolytic release of HHDP esters groups which undergo rapid, facile and unavoidable conversion (Figure 4). ETs show immense structural diversity depending on the variation, position, frequency and stereochemistry of the HHDP units and also by the galloylation extent and anomeric stereochemistry of the glucose core (*25*). In addition, the myriad of structures can be further complicated by oligomerization of ET monomers which arises from oxidative C-O couplings between galloyl and HHDP moieties of precursor molecules.

A main challenge to ET research lies in the lack of knowledge of the actual identities of these polyphenols as well as to the unavailability of commercial standards which make their dietary estimation difficult. The presence of ETs in dietary sources has been estimated by their hydrolysis into EA and then detection of the generated EA by spectrophotometric methods (*19*). However, more identification and quantification of EA. Using the 'EA-chemical marker method', commonly identified dietary sources of ETs include fruits (raspberry,

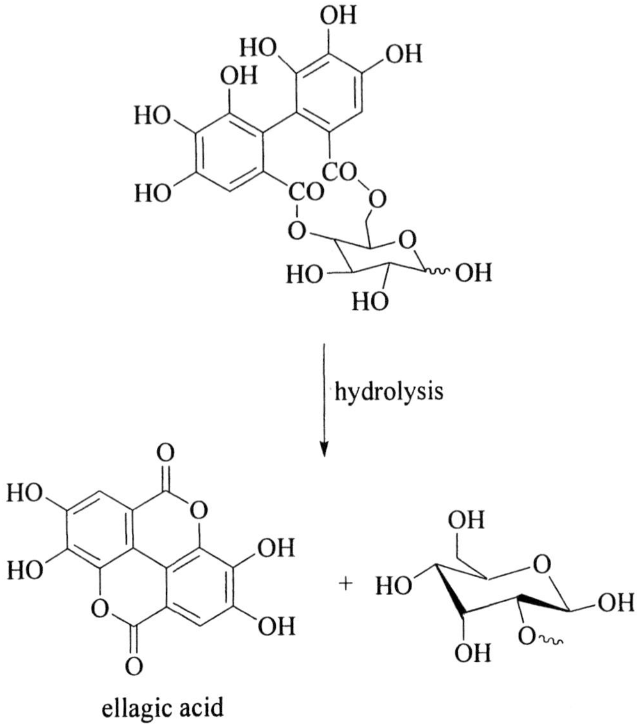

Figure 4. Formation of ellagic acid from hydrolyzable tannins (ellagitannins).

strawberry, blackberry, black raspberry, pomegranate, plum, apricot, peach, pineapple, blackcurrant, muscadine grapes), nuts (hazelnut, walnut, pistachio, chestnut, cashew nuts, pecans) and beverages (wines which are aged in oak barrels, fruit-juices e.g. pomegranate juice) (*9,19,22*). However, data on the daily dietary intake of EA by humans is not available and there is no information on the proportion ingested in the form of ETs (*9*).

ETs and EA have been implicated with potent biological activities including antioxidant, anticancer, antiatherosclerotic and anti-neurodegenerative diseases (*37-41*). ETs present in pomegranates were shown to contribute significantly towards the potent antioxidant and antiatherosclerotic properties of the juice (*41-43*). Strawberries, which are known to be a rich source of EA, have shown antioxidant, anti-neurodegenerative and anti-inflammatory properties (*32,41,44*). However, the absorption, bioavailability and pharmacokinetics of ETs and EA administered orally have not been adequately investigated (*9*) and until recently there were no reports of definitive studies on the absorption and metabolism of

ETs in humans (*42,43*). Previous to these reports of human studies, knowledge on the bioavailability of EA and ETs was confined to animal studies. The low levels of free EA detected in plasma were attributed to its low solubility in water and its ability to bind irreversibly to cellular DNA and proteins (*42*, and references cited therein). However with the use of modern chromatographic techniques such as HPLC and LC-MSn, free EA (*42*) as well as its metabolites (*43*), have been recently detected in human plasma and urine. Because ETs are relatively easily hydrolyzed, the *in vivo* action of physiological pH and/or enzymatic action by gut microflora could cause them to break down to release EA units. These recent advances in tannin research have showed that EA and its metabolites may be considered as biomarkers for human studies involving the consumption of ETs from food sources (*42,43*).

Conclusions

Dietary polyphenols are widely consumed due to their abundance in foods, food-products and beverages, and are implicated with human health benefits including antioxidant, anti-inflammatory, anticancer, anti-neurodegenerative and anti-atherosclerotic properties. Polyphenols are also increasingly being consumed as botanical ingredients from herbal medicines and dietary supplements for a variety of nontoxic therapeutic effects. However, little is known concerning the dietary levels of particular classes of polyphenols for example, tannins, due to their ill-defined structures, lack of availability of commercial standards, and lack of reliable and sensitive assays and analytical methodologies. Hence, continued research is necessary to establish the dietary intake of polyphenols by humans. These data are of high importance in the context of human nutrition for correlation in epidemiological studies with the incidence and early markers of oxidative stress-mediated diseases. In addition, there is limited knowledge concerning the bioactive forms of these polyphenols *in vivo* and the mechanisms by which they contribute towards disease prevention. Hence, future studies should focus on the bioavailability of dietary polyphenols, not only through their uptake and urinary excretion of conjugated forms, but via their metabolites formed *in vivo* by physiological changes, enzymatic action of gut microflora and subsequent absorption. Also, since these biologically active polyphenol molecules have been shown to have important effects on human health, further studies on the detection and bioactivities of their metabolites, methods to increase their bioavailability, and strategies to increase their concentration in blood and target tissues are warranted. It is noteworthy that recent progress has been made in the *in vitro* and *in vivo* analyses of tannins resulting from the development of more sensitive analytical methods and use of state of the art instrumentation such as HPLC, LCMS and its tandem methods

(LC-MSn), and NMR. It is imperative that research efforts continue to focus on studies of the chemistry and biology of dietary polyphenols, for understanding of their impact on human health and diseases

Acknowledgments

Funding was provided by the Center for Dietary Supplement Research in Botanicals (CDSRB) at the UCLA Center for Human Nutrition from NIH/NCCAM grant P50AT00151.

References

1. *Flavonoids in health and disease*; Rice-Evans, C.A.; Packer, L., Eds.; New York: Marcel Dekker Inc., 1998.
2. *Natural antioxidants and food quality in atherosclerosis and cancer prevention.* Kumpulainen, J.T.; Salonen, J.T., Eds.; The Royal Society of Chemistry: Cambridge, 1995.
3. *Phenolic compounds in food and their effects in health I + II.* Ho, C.-T.; Lee, C.V.; Huang, M.T.; Eds.; American Chemical Society: Washington, DC, 1992.
4. Scalbert, A.; Williamson, G. *J. Nutr.* **2000**, *130*, 2073S-2085S.
5. Clifford, M.N. *J. Sci. Food. Agric.* **2000**, *80*, 1063-1072.
6. Manach, C.; Scalbert, A.; Morand, C.; Remesey C.; Jimenez L. *Am. J. Clin. Nutr.* **2004**, *79*, 727-747.
7. Aggarwal, B.B.; Bhardwaj, A.; Aggarwal, R.S.; Seeram, N.P.; Shishodia, S.; Takada, Y. *Anticancer Res.* **2004**, *24*, 2783-840.
8. Buelga-Santos, C.; Scalbert, A. *J. Sci. Food. Agric.* **2000**, *80*, 1094-1117.
9. Clifford, M.N.; Scalbert, A. *J. Sci. Food. Agric.* **2000**, *80*, 1118- 1125.
10. *Chemistry of vegetable tannins*; Haslam E., Ed.; London: Academic Press, 1966.
11. *Plant polyphenols, vegetable tannins revisited*; Haslam E., Ed.; Cambridge: Cambridge University Press, 1989.
12. *Practical polyphenolics*; Haslam E., Ed.; Cambridge: Cambridge University Press, 1998
13. Bacon, J.R.; Rhodes, M.J.C. *J. Agric. Food Chem.* **2000**, *48*, 838-843.
14. Haslam, E. *Practical Polyphenolics-from Structure to Molecular Recognition and Physiological Action.* Cambridge University Press: Cambridge, 1998.
15. Gu, L.; Kelm, M.A.; Hammerstone, J.F.; Beecher, G.; Holden, J.; Haytowitz, D.; Gebhardt, S.; Prior, R.L. *J. Nutr.* **2004**, *134*, 613-617.

16. Hammerstone, J.F.; Lazarus, S.A.; Schmitz, H.H. *J. Nutr.* **2000**, *130*, 2086S-2092S.
17. Gu, L.; Kelm, M.A.; Hammerstone, J.F.; Beecher, G.; Cunningham, D.; Vannozzi, S.; Prior, R.L. *J. Agric. Food Chem.* **2002**, *50*, 4852-4860.
18. Gu, L.; Kelm, M.A.; Hammerstone, J.F.; Beecher, G.; Holden, J.; Haytowitz, D.; Prior, R.L. *J. Agric. Food Chem.* **2003**, *51*, 7513-7521.
19. Lin T.Y.; Vine R.P. *J. Food Sci.* **1990**, *55*, 1607-1609.
20. Williner M.R.; Pirovani M.E.; Guemes D.R. *J. Sci. Food Agric.* **2003**, *83*, 842-845.
21. Hakkinen S.H.; Karenlampi, S.O.; Heinonen, I.M.; Mykkanen, H.M.; Torronen, A.R. *J. Sci. Food Agric.* **1998**, *77*, 543-551.
22. Amakura Y.; Okada M.; Tsuji A.; Tonogai Y. *J. Chromatog. B* **2000**, *896*, 87-93.
23. Mullen W.; Yokota T.; Lean M.E.; Crozier A. *Phytochemistry* **2003**, *64*,617-624.
24. Porter, L.J. In *The Flavonoids-Advances in Research since 1986*, Harborne, J.B.; Ed.; Chapman and Hall: London, 1994; p. 23-55.
25. Quideau, S.; Feldman, K.S. *Chem. Rev.* **1996**, *96*, 475-503.
26. Okuda, T.; Yoshida, T.; Hatano, T. *Phytochemistry* **1993**, *32*, 507-521.
27. Carando, S.; Teissedre, P.L. In: *Plant Polyphenols 2. Chemistry, Biology, Phrmacology, Ecology*. Gross, G.G., Hemingway, R.W.; Yoshida, T.; Eds.; Kluwer Academic/Plenum Publishers, New York, p. 725-737.
28. Lucas, A. *J. Nutr.* **1998**, *128*, 401S-406S.
29. US Department of Agriculture.USDA database for the flavonoid content of selected foods can be accessed on the website. Internet:http://www.nal.usda.gov/fnic/foodcomp (Accessed June 26 2004).
30. Cunningham, D.G.; Vannozzi, S. A.; Turk, R.; Roderick, R.; O'Shea, E.; Brilliant, K. In ACS Symposium Series (Nutraceutical Beverages) **2004**, *871*, 35-51.
31. Seeram, N.P.; Adams, L.S.; Hardy, M.L.; Heber, David. *J. Agric. Food Chem.* **2004**, *52*, 2512-2517.
32. Bickford, P.C.; Gould, T.; Briederick, L.; Chadman, K.; Pollock, A.; Young, D.; Shukitt-Hale, B.; Joseph, J. *Brain Res.* **2000**, *866*, 211-217.
33. Aldini, G.; Carini, M.; Piccoli, A.; Rossoni, G.; Facino, R.M. *Life Sci.* **2003**, *73*, 2883-2898.
34. Kaul, D.; Sikand, K.; Shukla, A. R. *Phytother. Res.* **2004**, *18*, 177-179.
35. Saleem, M.; Adhami, V.M.; Siddiqui, I.A.; Mukhtar, H. *Nutr. Cancer* **2003**, *47*,13-23.
36. Mueller-Harvey, I. *Animal Feed Sci. Tech.* **2001**, *91*, 3-20.
37. Fukuda, T.; Ito, H.; Yoshida, T. *Phytochemistry* **2003**, *63*, 795-801.

38. Castonguay, A. Gali, H.U.; Perchellet, E.M.; Gao, X.M.; Boukharta, M.; Jalbert, G.; Okuda, T.; Yoshida, T.; Hatano, T.; Perchellet, J-P. *Intl. J. Oncol.* **1997**, *10*, 367-373.
39. Okuda, T.; Yoshida, T.; Hatano, T. *Planta Med.* **1989**, *55*, 117-22.
40. Joseph, J.A.; Denisova, N.A.; Bielinski, D.; Fisher, D.R.; Shukitt-Hale, B. *Mech. Aging Dev.* **2000**, *116*, 141-153.
41. Aviram, M.; Aviram, R.; Fuhrman, B. In: *Natural Antioxidants and Anticarcinogenesis in Nutrition, Health and Disease* Kumpulainen J.T.; Salonen J.T.; Eds.; Royal Society of Chemistry: Cambridge, U.K., 1999, pp. 106-113.
42. Seeram, N.P.; Lee, R.; Heber, D. *Clin. Chim. Acta* **2004**, *348*, 63-68.
43. Cerda, B.; Espin, J.C.; Parra, S.; Martinez, P.; Tomas-Barberan, F.A. *Eur. J. Nutr.* **2004**, *43*, 205-220.
44. Seeram, N.P.; Momin, R.A.; Bourquin, L.D.; Nair, M.G. *Phytomedicine*, **2001**, *8*, 362-369.

Chapter 4

Instrumental Analysis of Popular Botanical Products in the U.S. Market

Mingfu Wang[1], Chia-Pei Liang[2], Qing-Li Wu[1], James E. Simon[1], and Chi-Tang Ho[2]

Departments of [1]Plant Biology and Pathology and [2]Food Science, New Use Agriculture and Natural Plant Products Program, Rutgers, The State University of New Jersey, New Brunswick, NJ 08901

A large number of botanicals are used in various products including beverages, dietary supplements, foods, healthcare and personal care products. Dietary supplements and functional foods are among the most rapidly growing sectors in the food and personal care products industry. The demand for safe and effective dietary supplement in the US market has grown exponentially and consumers, the government, and the media, which has been a major force in the market, have each begun to seriously question the quality of dietary supplements. Given the consumer-driven nature of the dietary supplements industry and the regulatory framework in the US, the quality control standards for botanical products are still found lacking and/or unevenly implemented, despite major gains achieved by the industry as the private sector seeks to improve quality and meet governmental expectations under the current policy of largely self-regulation. While major accomplishments continue to be achieved by industry to ensure improved quality of the product including packaging, GMP issues, improved labeling clarity, the addition of product expiration dates and claims, and more, for those standardized botanical extracts, many problems still remain. In this paper, we evaluated the quality of popular botanical extracts including artichoke, bilberry, soy any tribulus using HPLC fingerprinting, LC-MS analyses, and other analytical techniques and focus purposefully not on the

© 2006 American Chemical Society

notable gains in overall quality that have been achieved in the botanical and herbal industry over the last decade but on specific issues relating to the natural product content in botanicals, particularly relative to the products own label claim and standardization. The improvement in quality control and detection of adulteration is also discussed.

With the increasing of demand for the prevention of diseases and improvement of human and animal health, herbal products have become increasingly popular in the international market, and continue to be even more important in those countries with a long history of using herbal medicines such as China, India and Japan. Currently, there are about 75,0000 plant species, 30,0000 registered species, 30,000-75,000 medicinal plants, 20,000 medicinal plants listed by WHO and 400 medicinal plants widely traded in the world.

In the US, herbal products are largely marketed and thus, regulated as dietary supplements and sold in a variety of forms including but not limited to fresh plant products, dried botanical products, partially or fully processed dried powders, crude extracts, liquid botanical extracts, soft extracts, dry extracts, tinctures and partially-purified and purified natural compounds, as single plant-based products and as multi-ingredient plant-based products. As an open market for the entire world, one can find a plethero of botanical products available in the US with ginseng, ginkgo, saw palmetto, echinacea, soy, bilberry, grape seed and green tea extracts as top-selling products. Botanical products are also allowed to be sold in other commercial channels such as a botanical drug following IND (Investigative Drug Application) or NDA (New Drug Application), OTC (Over-the-Counter) drugs, as functional food ingredients, or simply as spices, teas and herbal teas, and flavor and fragrance ingredients. In general, all botanicals must be authenticated as to their origin and species and as needed chemotype. Botanicals need to be safe to use, and fall within specific limits related to cleanliness (freedom from foreign materials non-product debris), and freedom from heavy metals, aflatoxins and/or other microbiological contamination, and pesticides. For many botanicals, the actual pH value, ash contents, moisture contents, particle size and other relevant physicochemical parameters are known and provide support for the cleanliness and an adherence to a specific botanical description. Sales and product entry of botanicals then which meet those quality parameters provide an assessment of quality for that entry level. Quality however extends beyond basic issues of safety, product purity and consistency. Quality of botanicals can also include, and often need to include a base understanding of the natural products profile upon which standardized botanical products can be developed and manufactured. While this appears to be a reasonable approach, the standardization of botanicals is actually a far more complex issue and thus a challenge to those involved in developing standards of quality. We have found that the specific standards for a wide range herbal products is often lacking, and in other cases is questionable as to what needs to be screened.

Instrumental methods, especially HPLC (high pressure liquid chromatography) are well known to be reliable, stable and provide this needed foundation for quality control and the chemical fingerprinting of botanical products directed into the phytopharmaceutical and/or pharmaceutical products. HPLC coupled with photodiode array detector, MS/MS and ELSD (Evaporative light scattering detector) are suitable for analysis of various phytochemicals in botanical products and GC (Gas Chromatography) and GC/MS are excellent instruments for analysis of volatile compounds, fatty acids and sterols from herbs. In this paper, the applications of instrumental methods for quality control of several popular herbal products in the US market are discussed.

Materials and Instruments

Analytical HPLC analyses were performed on a Hewlett-Pachard 1100 modular system equipped with an auto-sampler, a quaternary pump system, a photodiode array detector, and a HP Chemstation data system. Negative and positive ESI-mass spectra were measured on Agilent 1100 LC-MSD system (Agilent Technologies, Germany) equipped with an electrospray source, Bruker Daltonics 4.0, and Data analysis 4.0 software. Colorimetric methods were preformed on a Hewlett-Pachard 8453 spectrometer. HPLC columns were purchased from Phenomenex (Torrance, CA) and all the organic solvents are HPLC grade and were obtained from Fisher Scientific Inc.

Artichoke

Artichoke, an ancient herbaceous perennial plant, has been long appreciated as a tasty vegetable with known health promoting properties for a long time. Currently, artichoke is widely cultivated around the world with Italy as the world's leading producer (*1*). Artichoke extracts, especially the leaf extracts have been shown to produce various pharmacological effects, including the inhibition of cholesterol biosynthesis and of LDL oxidation (*2*). The chemical components of artichoke leaves have been well-studied and found to be a rich source of phenolic compounds, with mono- and dicaffeoylquinic acids and flavonoids as the major chemical components in both heads and leaves (*3*). The main caffeoylquinic acid derivatives in artichoke are chlorogenic acid and 1.5-dicaffeoylquinic acid (Figure 1), which can be converted into 1,3-dicaffeoyquinic acid during processing and preparation of artichoke extracts. Sesquiterpenoid lactones and flavonoids (luteolin-7-glucoside, luteolin-7-rutinoside etc.) have also been identified in artichoke and luteolin-7-glucoside is also found in appreciable amounts in artichoke.

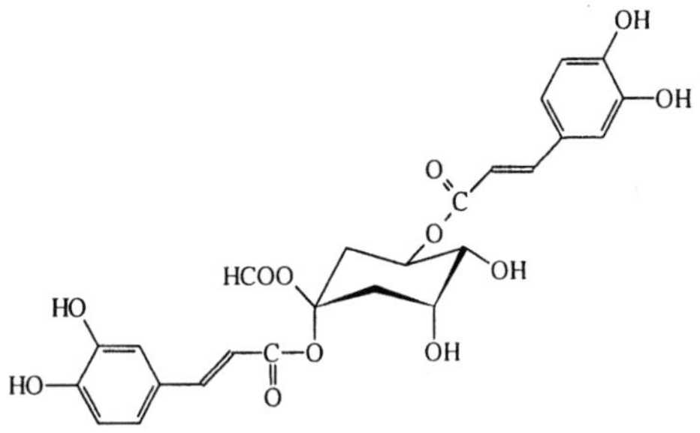

Figure 1. Structure of 1,5-dicaffeoylquinic acid.

In the US nutraceutical market, the caffeoylquinic derivatives are used as marker compounds for quality control of artichoke leaf extracts. The artichoke leaf extract generally is assumed to contain 13-18% caffeoylquinic acid derivatives (some companies claim that the total caffeoylquinic acid is calculated against chlorogenic acid). Reverse-phase HPLC method has been the widely used method of choice for analysis of phenolic compounds in artichoke, and recently several strong HPLC and LC-MS methods have been published (*1,3,4*). We recently described an HPLC method to analyze the phenolic compounds in artichoke leaves and head, the edible unopened flower which we consume as a vegetable (*3*). A Phenomenex Prodigy (ODS3, 5 μm, 100Å, 3.2 x 150 mm) column was selected and the column temperature was ambient. The mobile phase included water (containing 0.2% phosphoric acid, solvent A) and acetonitrile (solvent B) in the following gradient system: initial 94% A and 6% B, linear gradient to 70% A and 30% B in 20 min, held at 30%B for 5 min. The total running time was 25 min and the post-running time was 10 min. The flow rate was 1.2 mL/min, the injection volume was 10 μL, and the detection wavelength was set at 330 nm and sample was extracted using 60% aqueous methanol.

To exam the quality of commercial artichoke extracts, three bottles of artichoke herbal supplements were purchased from local health food stores and supermarkets. The capsules were extracted by 60% methanol and then analyzed using the HPLC method described above. The HPLC profiles of two samples (B and C) matched with the HPLC profile of artichoke leaf with chlorogenic acid and 1,5-dicaffeoylquinic acid as major phenols (Figure 2). The HPLC profile of sample A is different from B and C as shown in Figure 3. Except chlorogenic acid and 1,5-dicaffeoylquinic acid, other caffeoylquinic acid derivatives with molecular weight 354 and 514 were found to exist in significant amounts. However, none of these compounds are 1,3-dicaffeoylquinic acid, which is well-known to be produced from 1,5-dicaffeoylquinc acid during extraction of artichoke (1 and 4). Thus, the quality of this sample is questionable or minimally we may conclude that we have a highly isomerized extract which will generate difficulty for quality control.

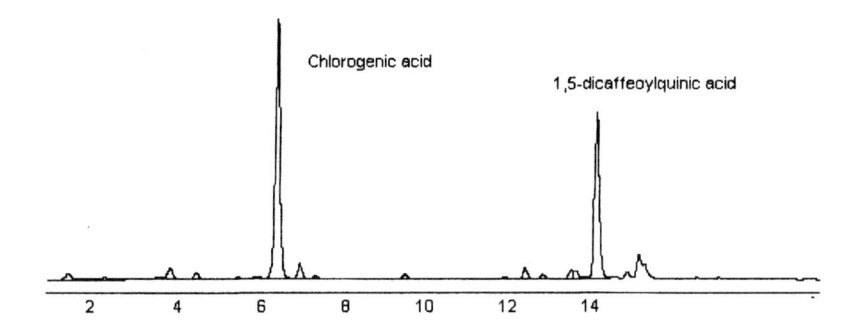

Figure 2. HPLC chromatogram of Sample B.

The caffeoylquinic acid derivatives in these samples were tested by quantification HPLC using chlorogenic acid and 1,5-dicaffeoylquinic acid as standards. The results show that each matched or almost matched their own label claims just based on chlorogenic acid and 1,5-dicaffeoylquinic acid contents (Figure 4).

Bilberry

Anthocyanins are a group of naturally occurring water-soluble flavonoids responsible for the pink and scarlet pigmentation of various of plant flowers, fruits, and leaves such as bilberry, blueberry, elderberry, red grape, red cabbage, cornflower and purple basil. Anthocynins usually exist in plant as glycosylated polyhydroxy and polymethoxy derivatives of flavylium salts. Some plants only contain one or two major anthocyanins while others contain over 20 different anthocyanins such as blueberry fruits. The pharmacological activities of anthocyanins have been long recognized. Anthocyanins are found to be antioxidant, anti-inflammatory, would-healing and vasoprotective agents and used to treat diseases characterized by capillary bleeding associated with increased capillary fragility (*5,6,7*). Anthocyanins also showed cancer chemopreventive activity related to cancer suppression (*8*). Anthocyanin extracts may also be used natural food coloring agents.

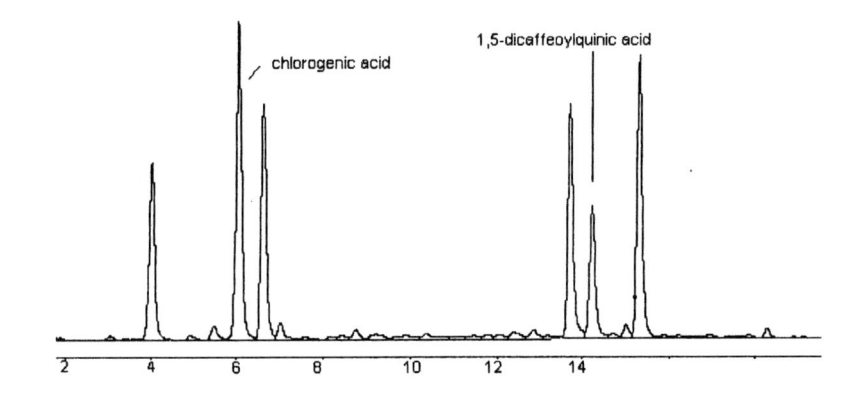

Figure 3. HPLC chromatogram of sample A.

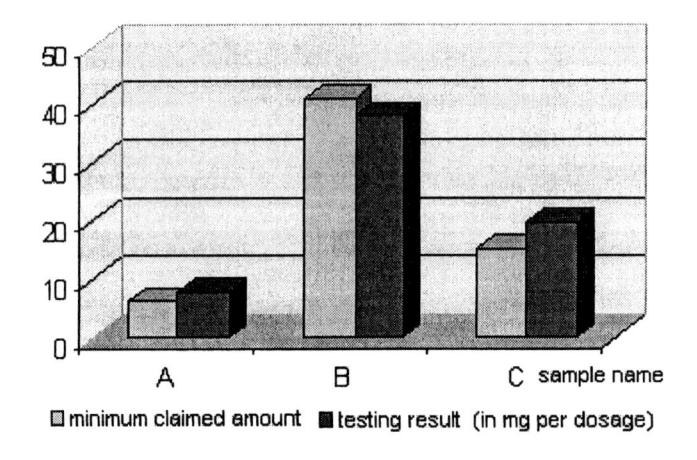

Figure 4. Testing and claimed caffeoylquinic acid in three commercial artichoke sample.

Among the anthocyanin-contained plants, the extract of bilberry has been widely used as dietary supplements in the US nutraceutical industry. About 20 clinical trials with bilberry have demonstrated positive effects for visual problems and circulatory disorders with the anthocyanins as the bioactive compounds. Several groups of compounds such as flavonoids including anthocyanins, iridoids, organic acids, sugars, and terpenes have been found in bilberry and recently resveratrol, a well-known cancer chemopreventive agent was also detected in bilberry (*9*). The ripe fruit of bilberry contains anthocyanidins (delphinidin, cyanidin, petunidin, peonidin, malvidin type), each of which can be linked with three different sugar units, glucose, galactose, and arabinose. Thus, there are at least 15 anthocyanins in bilberry with delphinidin type as the major ones.

The HPLC has been widely used for analysis of anthocyanins and this method provides anthocyanin profiles of different plants which are very useful for authentication, quality control and examination of consistency in raw materials. Recently, several good papers detailing HPLC methods for analysis of anthocyanins in bilberry have been published (*10,11,12*). In the US industry, a colorimetric method is still widely used for quality control of bilberry extracts (www.nsfina.org) which measures the total anthocyanins. However this method can not be used to differentiate bilberry extract from other less expensive anthocyanin-contained extracts such as grape skin and elderberry extract. An HPLC method is still necessary for actual authentication purposes.

We recently tested the quality of bilberry extracts purchased from the local health food stores and supermarkets. We purchased seven different brands of products and then subjected each one to a series of chemical tests including the colorimetric method for total anthocyanins (www.nsfina.org) and an HPLC method for identification purposes. The HPLC method was modified from that of Chandra and Rana (*10*). The HPLC analysis was run on a Phenomenex phenyl-hexyl 150 x 4.6 mm, 3 μm column and the mobile phase was water(0.5% phosphoric acid, mobile phase A) and water/acetonitrile/acetic acid/phosphoric acid (50: 48.5: 1.0: 0.5, v/v/v/v, mobile phase B) gradient, 20% B gradient to 60% B in 26 minutes, then to 20% in 4 minutes and keep at 20% for 5 minutes. The flow rate was 0.8 mL/min and injection volume was 10 μL. The column temperature was maintained at 30°C, with the detection wavelength at 520 nm.

Among the seven samples, only five claimed to contain anthocyanins and the range is 6.26-57.6 mg per dosage. The total anthocyanins were tested by the colorimetric method and the results illustrated in Figure 5. Commercial product Samples 1 to 4 met their own label claims but Sample 5 did not meet. Samples 6 and 7 did not claim to contain anthocyanins, and at least in this way these products did not 'fail' to meet their own label description, as we found that both of those commercial products contained only trace amounts of anthocyanins.

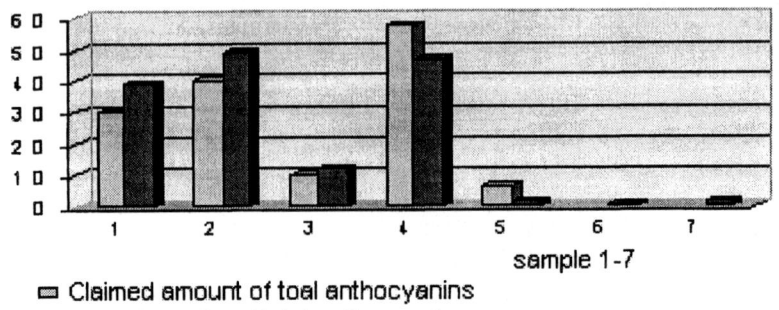

Figure 5. Total anthocyanin contents from seven commercial samples

When these same samples from the seven commercial products were subjected to HPLC analysis, only five of the products contained the HPLC profiles of bilberry (Figure 6), while Samples 6 and 7 showed instead the expected HPLC profile of elderberry with cyanidin-3-sambuboiside and cyanidin-7-glucoside as its major anthocyanins (Figure 7).

In summary with our brief screening in bilberry, we found two of the seven bilberry commercial samples to be of low quality and actually were not even the correct botanical, rather they were elderberry extracts.

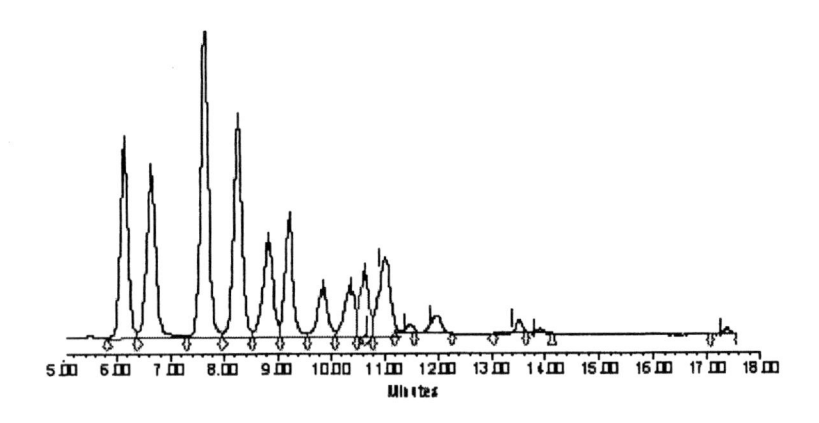

Figure 6. HPLC chromatogram of bilberry extract.

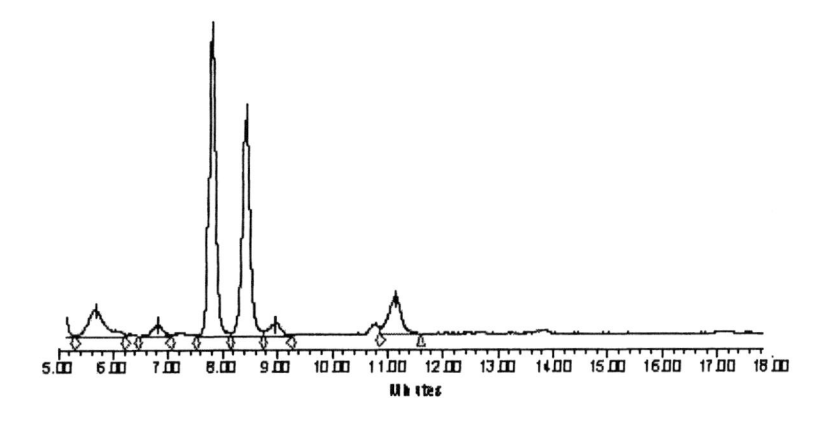

Figure 7. HPLC chromatogram of samples 6 and 7, matching with the HPLC profile of elderberry.

Soy Isoflavones

Soy is one of the most valuable crops in the world and is an excellent source of proteins, lipids, sterols, isoflavones and saponins. Soy products have shown many health benefits and currently the Food and Drug Administration permits food manufacturers to put labels on products high in soy protein indicating that these foods may help lower heart disease risk. Soy isoflavones, a group of unique compounds in soy, has become popular as dietary supplements in the US market. Isoflavones are known to be phytoestrogens from natural resources on the basis of the structural resemblance to the endogenous estrogen, 17-β-estrodiol. Isoflavones have potential health benefits related to age-related and hormone dependent diseases, including cancer, menopausal symptoms, cardiovascular disease, and osteoporosis (*13*). There are 12 major isoflavones in different soy products, which are divided into three groups basing on the aglycone moieties, including daidzein, genistein and glycitein. In each group, there are four different types, isoflavone aglycone, glycoside, 6"-acetylglycoside, and 6"-malonyl glycoside. In nature 6"-malonyl glycoside is the most popular form, but it is unstable and susceptible to rapid degradation by heat. Isoflavone 6"-acetylglycoside does not exist naturally, but appears to be generated by heat-induced degradation of 6"-malonylglycoside during drying and processing. In addition to these 12 compounds, other isoflavones have also been identified in soy products (*14,15*). In a soy concentrated extract, genistin and daidzin are the two major isoflavonoids.

Soybeans also contain 5-6% soysaponins (*16*). The soysaponins can be divided into two groups (group A and B) based on their aglycone structures. Soysaponins (group A) are bidesmosidic saponins with two glycosylation sites on their aglycone (soyasapogenol A). Soyasaponins (Group B) are monodesmosidic saponins, with only one glycosylation site on the aglycone (soysapogenol B and E). The genuine soysaponins of group B in the legume are conjugated with 2,3-dihydro-2,5-dihydroxy-6-methyl-4-pyrone (DDMP) which will degrade to no-DDMP counterparts during heat treatment and extraction (*17,18*). Soysaponins also shown various health benefits for example soyasaponins are potential cancer chemopreventive agents (*19,20*),

Various methods have been used to analyze the isoflavone contents in soy products and HPLC/UV method is the most widely used. Isoflavones and soyasaponins are usually analyzed separately. However, at present newer methods can provide simultaneously analyze both groups by using a combination of liquid chromatography with an UV and evaporative light-scattering detector (ELSD). Six isoflavones (detected at 254 nm) and four triterpene saponins (monitored with the ELSD) could be baseline separated using a reversed-phase C18 column and a mobile phase consisting of water and acetonitrile, both containing 0.025% trifluoroacetic acid (*21*).

We also recently published an HPLC method to identify the isoflavones in Edamame and Tofu soybeans (*13*). The method uses a Phemomenex Prodigy ODS3 column (5μm, 3.2 x 150 mm), with the mobile phase containing solvent A and B in gradient, where A was 0.1% formic acid (v/v) in water and B was 0.1% formic acid (v/v) in acetonitrile. The linear gradient profile was from 10 to 35% B in 40 min. The wavelength of UV detection was 254 nm. Column compartment was set at 25 °C. We have been using this method to assess the quality of commercial soy isoflavone extracts. HPLC profile of a typical soy isoflavone extract is illustrated in Figure 8, and genistin and daidzin are found to the major isoflavones. In contrast, Figure 9 illustrates the HPLC profile of a questionable soy extract, possibly spiked with synthetic genistein and daidzein.

Tribulus

Tribulus terrestris L., an annual herb found growing world-wide has long been used in folk medicine, and used in traditional health care in China, India and Bulgaria and other countries against sexual impotency, edema, abdominal distention, eye trouble, skin itching, and cardiovascular diseases (*22,23*). Tribulus extracts standardized to total saponins are available and marketed in the US as dietary supplements and claimed to treat impotence and as a restorative tonic for vigor to improve performance in sports. The best quality of tribulus extract appears to come from Bulgaria, where it has been well-studied.

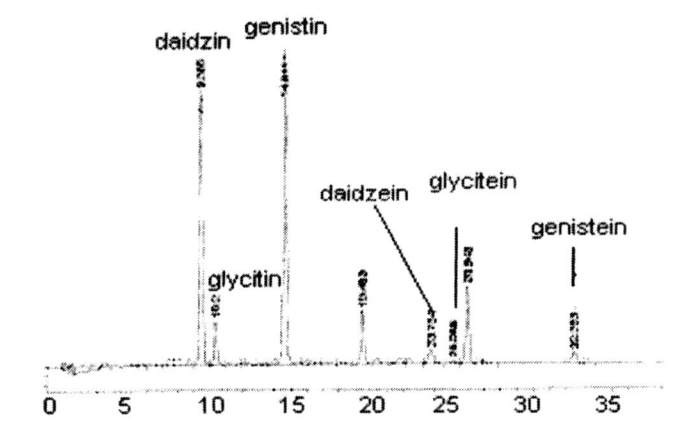

Figure 8. HPLC profile of a typical soy isoflavone extract.

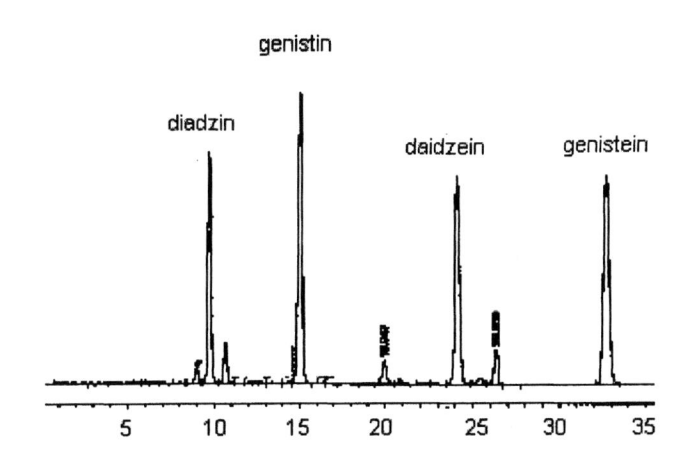

Figure 9. HPLC Profile of a questionable soy extract.

Based on the literature, a wide range of compounds have been found in tribulus including alkaloids, amides, flavonoids and saponins. Saponins are the major chemical components of tribulus and various saponins have been identified (*22-30*). However, the group of saponins is difficult to analyze and to quantify, and it appears that the Tribulus coming from different origins contain different saponins. As the specificity of the saponins can be related to their bioactivity, then it becomes of particular interest to better understand the specific saponins present for specific applications. Currently, in the US market, protodioscin is used as a marker compound for the identification of Tribulus extract. At present, the use of protodioscin appears to be the only good saponin marker compound for Bularian Tribulus extract. Chinese and Indian Tribulus extracts tested did not contain protodioscin. It now appears that even for saponins from the Bulgarian Tribulus, greater research is needed to ensure the proper identification of the saponins is achieved. For example protogracillin was reported in Bulgarian Tribulus and used as a marker compound in Bulgarian Tribulus (www.tribestan.net and *31*). However, it appears that protogracillin is not actually in Bulgarian Tribulus. The structures of saponins from Bulgarian Tribulus were only solved recently. Protodioscin, prototribestin, 5,6-dihydroprotodioscin (neoprotodioscin) and 5,6-dihydroprototribestin (neoprototribestin) were found to be the major saponins (*22*) in Bulgarian Tribulus. The actual structures of protodioscin and prototribestinare presented in Figure 10. Three additional saponins were reported in Tribulus (*23*). Although methyl protodioscin and methylprototribestin were also reported in Bulgarian Tribulus (*27*), both of these compounds are considered to as artifacts produced from protodioscin and prototribestin, respectively.

Upon closer examination, we find very few methods published on the best analytical protocols to identify and also quantitate the saponins in Tribulus. Ganzera et al. described the first analytical method suitable for the determination of steroidal saponins in Tribulus using a reversed-phase (RP-18) column, evaporative light scattering (ELS) detection, and a water/acetonitrile gradient as the mobile phase (*32*). However this method is designed to only quantitate protodioscin, not other saponins. In addition, this method may not be able to separate 5,6-dihydroprotodioscin from protodioscin. Mulinacci et al. reported an HPLC-ESI-MS method for an estimation of Tribulus saponins (*33*). Yet, the saponin profile they present is totally different from the expected reported profile of Bulgarian Tribulus. De Combarieu *et. al.* reported one HPLC/ELSD method for analysis of furostanol saponins in Bulgarian Tribulus and this method is suitable for fingerprinting Bulgarian Tribulus (*22*).

The HPLC is run on a X Terra RP 18 (5 µm, 250 x 4.6 mm) column with water (0.1% formic acid) and acetonitrile gradient. 0-10 min, 80% A/20% B to 75% A/25% B, 10-20 min, isocratic at 75% A/25% B, 20-40 min. 75%A/25% B to 50% A/50% B, 40-41 min, 50% A/50% B to 0% A/100% B and 41-45 min. isocratic at 0% A/100% B. The flow rate is 1.0 mL/min and the column temperature is at 30°C and the detector is ELSD.

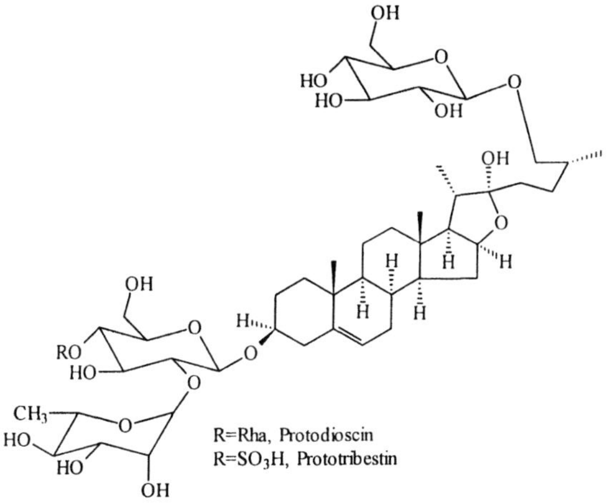

Figure 10. Structures of Protodioscin and prototribestin.

Due to the confusion in the proper identification and quantitation of these compounds, we also recently developed an HPLC/MS method for identification of Tribulus saponins. The HPLC was run on a Phenomenex Luna C18 (2) column (4.6 x 250 mm, 5 μm). The mobile phase is water (0.1% formic acid) and acetonitrile gradient. The mobile phase starts with 24% B and gradient to 35% B in 20 minutes, and then to 60% B in 20 minutes and finial linear gradient to 70% B in 20 minutes. The electrospray mass spectrometer (ESI-MS) was operated under negative ion mode and optimized collision energy level of 60%, scanned from m/z 200 to 1600. ESI was conducted using a needle voltage of 3.5 kV. High-purity nitrogen (99.999%) was used as dry gas and flow rate at 10 mL/min, capillary temperature at 350 °C. Helium was used as Nebulizer at 50 psi. The Total ion chromatogram of an authenticated Bulgarian herb is presented in Figure 11. We found an excellent match with the HPLC profile reported by De Combarieu *et al.* (*22*).

Two commercial Tribulus extracts were then purchased from a local health store and evaluated for saponins using this LC/MS method we developed and described above. Neither of the commercial products contained protodioscin, 5,6-hydroprotodiscin, prototribestin or 5,6-hydroprototribestin. Apparently, both commercial products are not Bulgarian Tribulus extracts.

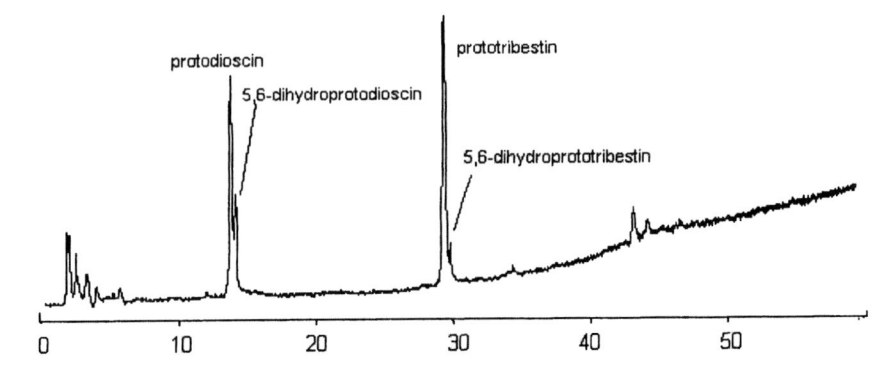

Figure 11. Total ion chromatogram of one Bulgarian tribulus extract under negative mode.

Discussion and Conclusion

Several popular herbal dietary supplements were comparatively screened for quality related to their natural products content by instrumental methods (colorimetric, HPLC, LC/MS). The HPLC and LC/MS are suitable methods for quality control including the chemical fingerprinting of plant powders and extracts. The inclusion of such instrumental methods of analysis, in addition to the typical review of published literature, should be a standard and critical component in the assessment of quality and in botanical standardization for all plant-based dietary supplements. Too often, the literature is used in place of, rather than as a compliment to, actual analytical testing. Companies and even research groups may lack specific (or accurate) standards needed for compound identification and quality assessment in specific botanicals due to their unavailability and/or expense and instead rely on previously reports which may illustrate elution of compounds pictured in labeled peaks from GC and/or LC analysis. False positive compound identifications can result. Many botanicals are "quality challenged" in that: (i) natural products may vary significantly from their own stated label claim (if there is one); (ii) another plant part or plant species is used either on purpose or inadvertently; (iii) adulteration is more common than acknowledged; and (iv) some botanical products can be spiked with synthetic chemicals or compounds from other botanicals.

To overcome these problems now facing the botanical industry, we recommend that reference materials and reference standards for these botanicals should be available through governmental and research institutes or non-profit organizations. Universal analytical methods should continue to be developed and validated for the most popular botanicals, and in particular for those botanicals that are potentially problematic from a safety perspective. Detailed genetic and chemical fingerprinting for every popular herb is needed and software programs to integrate the various chemical and genetic fingerprinting systems are long overdue. Using instrumental analysis as a core and standard part of the quality control process in botanicals though well recognized needs further implementation and a continued focus.

References

1. Schuetz, K.; Kammerer, D.; Carle, R.; Schieber, A. *J. Agric. Food Chem.* **2004**, *52*, 4090-4096.
2. Lupattelli, G.; Marchesi, S.; Lombardini, R.; Roscini, A. R.; Trinca, F.; Gemelli, F.; Vaudo, G.; Mannarino, E. *Life Sci.* **2004**, *76*, 775-782.
3. Wang, M.; Simon, J.E.; Aviles, I. F.; He, K.; Zheng, Q. Y.; Tadmor, Y. *J. Agric. Food Chem.* **2003**, *51*, 601-608.
4. Mulinacci, N.; Prucher, D.; Peruzzi, M.; Romani, A.; Pinelli, P.; Giaccherini, C.; Vincieri, F. F. *J. Pharm. Biomed. Anal.* **2004**, *34*, 349-357.
5. Morazzoni, P.; Bombardelli E. *Fitoterapia* **1996**, *67*, 3-29.
6. Dugo, P.; Mondello, L.; Errante, G.; Zappia, G.; Dugo, G. *J. Agric. Food Chem.* **2001**, *49*, 3987-3992.
7. Skrede, G.; Larsen, V. Bryhn; Aaby, K.; Jorgensen, A. Skivik; Birkeland, S.-E. *J. Food Sci.* **2004**, *69*, S351-S356.
8. Du, Q.; Jerz, G.; Winterhalter, P. *J. Chromatog. A* **2004**, *1045*, 59-63.
9. Rimando, A.M.; Kalt, W.; Magee, J.B.; Dewey, J.; Ballington, J.R. *J. Agric.Food Chem.* **2004**, *52*, 4713-4719.
10. Chandra A; Rana J; Li, Y. *J. Agric.Food Chem.* **2001**, *49*, 3515-3521.
11. Ichiyanagi, T.; Hatano, Y.; Matsugo, S.; Konishi, T. *Chem. Pharm. Bull.* **2004**, *52*, 628-630.
12. Zhang, Z.; Kou, X.; Fugal, K.; McLaughlin, J. *J. Agric. Food Chem.* **2004**, *52*, 688-691.
13. Wu, Q.L.; Wang, M.; Simon, J.E. *J. Agric. Food Chem.* **2004**, *52*, 2763-2769.
14. Hosny, M.; Rosazza, J.P.N. *J. Nat. Prod.* **2002**, *65*, 805-813.
15. Gu, L.; Gu, W. *Phytochem. Anal.* **2001**, *12*, 377-382.
16. Dalluge, J.J.; Eliason, E.; Frazer, S. *J. Agric. Food Chem.* **2003**, *51*, 3520-3524.
17. Hu, J.; Lee, S.O.; Hendrich, S.; Murphy, P.A. *J. Agric.Food Chem.* **2002**, *50*, 2587-2594.

18. Gu, L.; Tao, G.; Gu, W.; Prior, R. L. *J. Agric. Food Chem.* **2002**, *50*, 6951-6959.
19. Hsu, C.C.; Lin, T.W.; Chang, W.W.; Wu, C.Y.; Lo, W.H.; Wang, P. H.; Tsai, Y.C. *Gynecologic Oncology* **2005**, *96*, 415-422.
20. Ellington, A. A.; Berhow, M.; Singletary, K. W. *Carcinogenesis* **2005**, *26*, 159-167.
21. Ganzera, M.; Stuppner, H.; Khan, I. A. *J. AOAC Internat.* **2004**, *875*, 1189-1194.
22. De Combarieu, E.; Fuzzati, N.; Lovati, M.; Mercalli, E. *Fitoterapia* **2003**, *74*, 583-591.
23. Conrad, J.; Dinchev, D.; Klaiber, I.; Mika, S.; Kostova, I.; Kraus, W. *Fitoterapia* **2004**, *75*, 117-122.
24. Huang, J.W.; Tan, C.H.; Jiang, S. H.; Zhu, D.Y. *J. Asian Nat. Prod. Res.* **2003**, *5*, 285-290.
25. Deepak, M.; Dipankar, G.; Prashanth, D.; Asha, M. K.; Amit, A.; Venkataraman, B. V. *Phytomed.* **2002**, *9*, 753-756.
26. Sun, W.; Gao, J.; Tu, G.; Guo, Z.; Zhang, Y. *Nat. Prod. Lett.* **2002**, *16*, 243-247.
27. Kostova, I.; Dinchev, D.; Rentsch, G. H.; Dimitrov, V.; Ivanova, A. *Zeitschrift für Naturforschung, C: J. Biosci.* **2002**, *57*, 33-38.
28. Xu, Y.; Xie, S.; Zhao, H.; Han, D.; Xu, T.; Xu, D. *Yaoxue Xuebao* **2001**, *36*, 750-753.
29. Bedir, E.; Khan, I. A. *J. Nat. Prod.* **2000**, *63*, 1699-1701.
30. Xu, Y.X.; Chen, H.S.; Liang, H.Q.; Gu, Z.B.; Liu, W.Y.; Leung, W.N.; Li, T.J. *Planta Med.* **2000**, *66*, 545-550.
31. Obreshkova, D.; Pangarova, T.; Mitkov, S.; Dinchev, D. *Farmatsiya (Sofia)* **1998**, *45*, 10-12.
32. Ganzera, M.; Bedir, E.; Khan, I.A. *J. Pharm. Sci.* **2001**, *90*, 1752-1758.
33. Mulinacci, N.; Vignolini, P.; la Marca, G.; Pieraccini, G.; Innocenti, M.; Vincieri, F. F. *Chromatographia* **2003**, *57*, 581-592.

Chapter 5

Challenges in Assessing Bioactive Botanical Ingredients in Functional Beverages

Catharina Y. W. Ang[1], Jin-Woo Jhoo[1], Yanyan Cui[1], Lihong Hu[1,4], Thomas M. Heinze[1], Qingyong Lang[3,5], Chien M. Wai[2], Jeremy J. Mihalov[3], Michael DiNovi[3], and Antonia Mattia[3]

[1]U.S. Food and Drug Administration, National Center for Toxicological Research, 3900 NCTR Road, Jefferson, AR 72079
[2]Department of Chemistry, University of Idaho, Moscow, ID 83844
[3]U.S. Food and Drug Administration, Center for Food Safety and Applied Nutrition, College Park, MD 20740
[4]Current address: National Center for Drug Screening, Shanghai Institute of Materia Medica, Shanghai 201203, China
[5]Current address: Nutritional Laboratories International, Missoula, MT 59801

Bioactive botanicals are used as functional ingredients in conventional foods to be marketed as functional foods. Among these functional foods, teas and non-alcoholic, non-carbonated bottled beverages are especially popular. The analytical assessment of these ingredients in functional foods and beverages presents a new challenge because of low levels of analytes and the matrix complexity. This article summarizes recent development of analytical methods for the determination of bioactive constituents from St. John's wort (SJW; *Hypericum perforatum* L.), ginkgo *(Ginkgo biloba* L.) and kava *(Piper methysticum,* Forst.) in functional beverages. The constituents selected as marker compounds were hyperforin, adhyperforin, hypericin and pseudohypericin for SJW, five terpene trilactones (ginkgolides A, B, C and J, and

Disclaimer: The views presented in this article do not necessarily reflect those of the U.S. Food and Drug Administration.

© 2006 American Chemical Society

bilobalide) for ginkgo products, and six kavalactones (methysticin, dihydromethysticin, kavain, dihydrokavain, yangonin and desmethoxyyangonin) for kava products. The methods developed include improved sample preparation, liquid chromatography with ultra violet/photodiode array detection (LC/UV/PDA), LC with evaporative light scattering detection (ELSD) and LC with mass spectrometry (MS). The stability of SJW constituents in non-alcoholic, non-carbonated, fruit-flavored beverages was also discussed.

The passage of the Dietary Supplement Health and Education Act (DSHEA) in 1994 resulted in an increase of herbal dietary supplements in the USA. St. John's wort, echinacea, ginseng and ginkgo are among the top selling herbal supplements in American health food stores. These supplements may be in the form of dried root or leaf, liquid extract, capsule, tablet, and may be as a mixture with other ingredients. Although consumers are interested in "natural" supplements, many of these supplements have pharmacological actions and are often regarded as drugs in many parts of the world.

Recently, interest has grown for food ingredients that provide health benefits. These food ingredients are commonly referred to as functional food ingredients, and have been used in conventional foods (e.g., juices, drinks, cereals, tea blend or bags and snacks). A number of these functional foods are manufactured by the incorporation of bioactive botanicals in their formulations. Although DSHEA has established regulations for dietary supplements, the use of bioactive botanicals as an ingredient in functional food is a different issue. Ingredients added to conventional foods are regulated as food additives. Many bioactive botanicals are not approved for use in food or considered as "generally recognized as safe" (GRAS). Information is generally not available on the amounts of botanicals present in functional foods. The analysis of functional foods for bioactive components derived from added botanicals presented a new challenge. In addition to the complex composition of specific herbs, functional food products generally contain other ingredients, such as food additives, sugars, fruit juices, artificial colors, natural or artificial flavors and other herbal constituents. Furthermore, the bioactive ingredients may change in the final product depending on food formulation, processing and storage conditions. The quality and biological implications of these functional food products are not yet fully understood.

Investigation of the bioactive botanical composition of functional foods is essential to evaluate their quality and safety as food additives. This article will highlight the use of some popular botanicals i.e., St. John's wort (SJW; *Hypericum perforatum* L.), ginkgo *(Ginkgo biloba* L.) and kava *(Piper methysticum,* Forst.), and the recent development in quantitative assessment of functional ingredients in herbal teas and fruit-flavored drinks. The effect of pH and light exposure on SJW constituents in aqueous systems is also explored.

St. John's Wort (SJW) in Functional Beverages

SJW *(Hypericum perforatum,* L.) is an herbaceous perennial plant. The leaf and flower are traditionally used for many purposes, including treatment of burns, ulcers, respiratory and urinary tract inflammations, gastroenteritis, and food preservation (*1*).

Biological Activities and Phytochemical Constituents

Commercially available SJW extracts in capsule or tablet forms are used in Europe as drugs for the treatment of depressive disorders. This use is supported by clinical evidence (*2*); SJW extracts demonstrated an increased effect compared to placebo for the treatment of mild to moderately severe depression (*3*). A possible mechanism of the antidepressant activity is by inhibition of the uptake of serotonin and norepinephrine (*4*).

In the USA, SJW extracts are used as a dietary supplement for treatment of antidepression. However, two large-scale clinical studies in the USA reported that SJW was not effective for the treatment of major or moderately severe depression (*5*). Some reported side effects of SJW supplementation were nausea, rash, fatigue, restlessness, photosensitivity and acute neuropathy. The most severe cases were the herbal-drug interactions when SJW supplements were administered simultaneously with prescription drugs such as indinavir, cyclosporine, warfarin, and digoxin (*6*).

The major SJW active constituents are naphthodianthrones (hypericin and pseudohypericin), phloroglucinols (hyperforin and adhyperforin) (Figure 1), flavonol glycosides and biflavones (*7*). Many pharmacological activities appear to be attributable to hypericin, hyperforin and flavonoid constituents. Studies suggest that hypericin is responsible for photosensitive reactions, and that hyperforin is the major constituent with antidepressant activity (*3*). However, hyperforin is also suggested as the constituent responsible for SJW-drug interactions. Hyperforin was found to cause induction of liver isoforms such as human CYP2C9 (*8*) and CYP3A4 (*9*), the major liver enzymes for drug metabolism.

Analysis of SJW Constituents in Teas

Among the SJW active constituents, hypericins and hyperforins are uniquely present in SJW. They are seldom found in other plants, and their potential pharmacological and adverse effects have been studied more often than other constituents. Thus, hypericin and hyperforin are commonly used as biomarker compounds for product standardization and quality control purposes. Several methods are reported for the determination of marker compounds from SJW plant material and dietary supplements (*10, 11*). However, for the analysis

Hypericin   R=CH$_3$
Pseudohypericin R=CH$_2$OH

Hyperforin   R=H
Adhyperforin R=CH$_3$

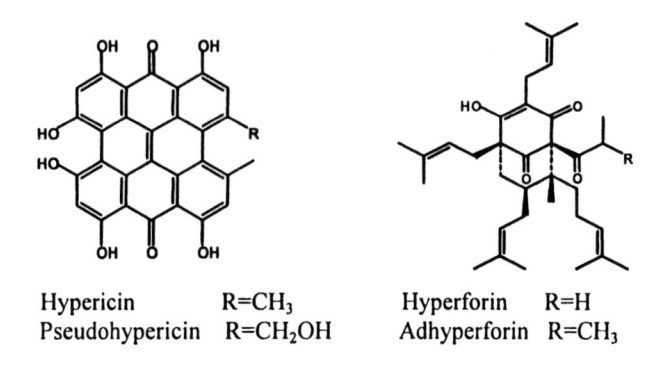

Figure 1. Structures of major active constituents of St. John's wort.

of SJW constituents in functional foods, our laboratory developed a rapid extraction technique followed by an isocratic high performance liquid chromatographic (HPLC) method for the determination of four marker compounds (hypericin, pseudohypericin, hyperforin and adhyperforin) in a variety of matrices, including tea bags, snack bar and fruit-flavored drinks (*12*).

Generally there are two types of herbal tea-bag products. One contains only the specific herb and the other contains other ingredients in addition to the specific herb. In our previous study (*12*), sample Tea-1 was a product that apparently contained only SJW while the sample Tea-2 contained many other plant materials as described on the labels. Thus, the concentrations of SJW constituents in Tea-2 were much lower than that of Tea-1. In that study, tea samples were extracted with methanol by sonication. The purpose was to determine the actual content of SJW constituents in the dry herbal tea samples.

In the present study, we also evaluated two types of tea product (Tea-A contained the SJW herb only and Tea-B contained various other plant ingredients). The brands used in this study differed from those in the previous study. Since tea beverages are usually prepared by steeping tea bags in hot water for consumption, it is of interest to assess the amount of herbal components in the hot water infusion. Hence in this study, we prepared hot water infusions of tea bags in addition to the methanolic extraction.

For the methanolic extraction method, tea bags were prepared and analyzed as described earlier (*12*). For preparation of hot water infusion, one tea bag was steeped in about 100 mL boiling water for 5 min and then the tea bag was removed. After cooling down to room temperature, the infusion volume was adjusted to 100 mL with water and analyzed by LC method directly. No internal standard [benzo(k)fluoroanthene] was used because it was not soluble in water, and no filtration was made before the LC analysis to avoid any losses of SJW components which could occur by filtering of the aqueous solutions (*12*).

Typical chromatograms of tea-bag infusion (Tea-B) are shown in Figure 2. A comparison of the analysis of SJW constituents between two tea-bag products

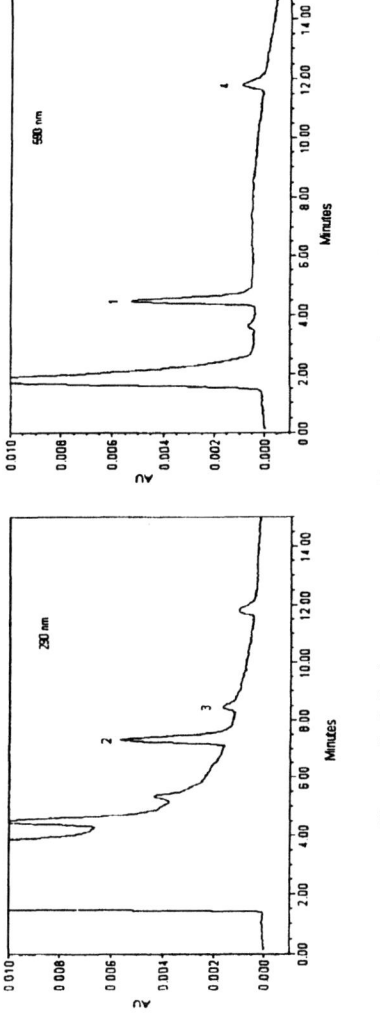

Figure 2. LC chromatograms of hot water infusion of tea bag.
1 = pseudohypericin; 2 = hyperforin; 3 = adhyperforin; 4 = hypericin

and between two extraction methods (methanol extraction and hot water infusion) is presented in Table I. The average net weight of Tea-A was 1.5 g/bag and Tea-B was 2.2 g/bag. On a dry weight basis (μg/g) for each tea product, the methanolic extraction resulted in much higher levels of all four marker components than those obtained by hot water infusion. Apparently the SJW components were more soluble in methanol than in hot water. When the two tea products were compared for the methanolic extraction, it was found that Tea-A had much higher levels of all marker compounds than Tea-B. However, when the data from hot water infusion were compared, it was interesting to note that more hyperforin and adhyperforin but less hypericin and pseudohypericin were found in Tea-B than Tea-A. It appeared that hyperforin and adhyperforin were not extracted by hot water from Tea-A to the same extent as from Tea-B.

A study by Jürgenliemk and Nahrstedt (13) showed that the solubility of pure hypericin in water increased upon addition of some phenolic constituents. Tea-B of the present study could contain more phenolic compounds than Tea-A, because Tea-B contained many other plant ingredients besides SJW. The dissolved phenolic compounds in the Tea-B infusion might have aided the solubility of hyperforin and adhyperforin in hot water. Since the net weight of Tea-B per bag was higher than Tea-A, the amounts of hyperforin and adhyperforin obtained by hot water infusion per serving of Tea-B were even more than Tea-A (Table I). Consumers are likely to ingest more hyperforins but less hypericins from Tea-B than Tea-A.

Table I. St. John's Wort Constituents in Tea Bags Extracted with Methanol (MeOH) or Hot-water Infusion.

Tea[a] & Method	Hyperforin	Adhyperform	Hypericin	Pseudo-hypericin
Tea-A				
MeOH	8,765 (3.9)	442.2 (17.2)	674.2 (8.4)	1,383 (4.5)
Water-1	13.8 (10.6)	4.0 (10.3)	31.5 (16.2)	194.6 (17.5)
Water-2[c]	20.8 (10.6)	6.0 (10.3)	47.2 (16.2)	291.9 (16.2)
Tea-B				
MeOH	672.8 (8.8)	ND[d]	80.3 (6.6)	176.9 (4.5)
Water-1	37.4 (18.4)	10.5 (10.9)	5.6 (3.1)	25.1 (15.1)
Water-2[c]	82.5 (18.4)	23.1 (10.9)	12.3 (3.1)	55.1 (15.1)

Mean, μg/g (RSD, %)[b], n=3

[a]Products from different manufactures. Tea-A=1.5 g/bag; Tea-B=2.2 g/bag.
[b]RSD, %: Relative standard deviation
[c]Unit expressed in μg/bag
[d]ND: not detectable, limit of detection (LOD) = 60 μg/g

Analysis of SJW Constituents in Fruit-flavored Beverages

A test portion of the beverage sample was centrifuged but not filtered. The supernatant was mixed with methanol (9:1, v/v) and analyzed by LC (*12*). The limit of detection (LOD) was 10 ng/mL for hyperforin, adhyperforin and hypericin and 5 ng/mL for pseudohypericin in drinks. One drink product was found to contain approximately 10 ng/mL of pseudohypericin (*12*). We hypothesized that SJW constituents were not stable under acidic and normal lighting conditions. The pH of drink samples was 2.7 - 3.0. The products were in clear glass bottles and displayed under fluorescent light in supermarkets and food stores. Results of our subsequent study using aqueous model systems have suggested that pH 2.65 was detrimental to SJW components especially for hyperforin and adhyperforin under light exposure (*14*).

The changes of SJW constituents in drink samples fortified with known amounts of the marker compounds are shown in Figure 3 (*14*). Initial increase of pseudohypericin was due to the photo-activated conversion of protopseudohypericin to pseudohypericin. Light exposure accelerated the change of all components. The reductions of hyperforins were > 97% after 4 hr of light exposure. From the control drink sample, a hyperforin degradation product was identified as furohyperforin isomer *a* by LC/MS/MS (*14*). The biological and toxicological effects of the changes of SJW in functional fruit-flavored beverages are yet to be evaluated.

Ginkgo in Functional Beverages

Ginkgo tree *(Ginkgo biloba* L.) is one of the world's oldest living trees and has been cultivated throughout the world. Ginkgo has been used in China as a traditional medicine for thousands of years. The seed is edible, and it is regarded as an effective medicine for the treatment of asthma and bronchitis (*15*). Large doses of cooked ginkgo seeds can be toxic due to the presence of ginkgotoxin (4'-*O*-methylpyridoxine) (*16*). This compound is also present in ginkgo leaves, and symptoms of this toxin are convulsions, loss of consciousness and even death (*17*). In Western countries, ginkgo leaves are used in the production of a concentrated, standardized *Ginkgo biloba* extract (GBE). The properties and usages of GBE are described below.

Biological Activities of Ginkgo Leaf Extracts

Over the last three decades pre-clinical and clinical studies extensively tested the biological activities of GBE (*18*). GBE products are widely used in Europe as drugs to alleviate a range of symptoms including cognitive disorders, dementia, vertigo, short term memory loss, dizziness, tinnitus and headache (*19*). Interestingly, two recent clinical studies in the USA found that GBE could

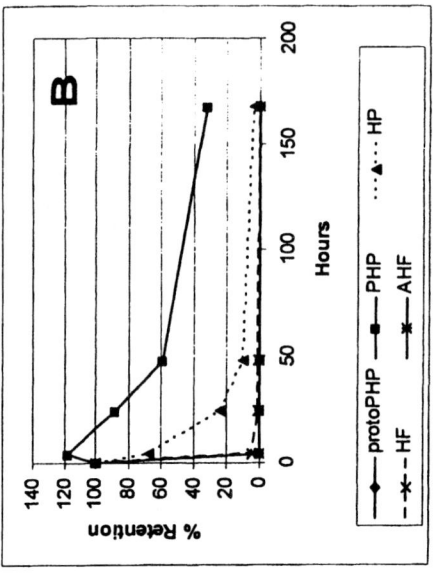

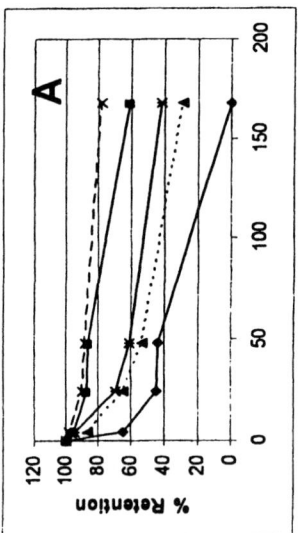

Figure 3. Changes of St. John's wort constituents in fruit-flavored drink product during storage: (A) in the dark, (B) under light.

delay the progression of symptoms in patients with mild to severe Alzheimer's disease and multi-infarct dementia by about 6 months (*20*).

A published study suggests that ginkgolic acids found in ginkgo fruit pods and leaves may be allergenic, although the finding is not conclusive (*21*). In Germany, ginkgolic acids and ginkgols are limited to a maximum of 0.0005% (5 µg/g) in standardized GBE (*22*). Ginkgo's antiplatelet effect may cause spontaneous bleeding and interact with drugs with a similar pharmacologic actions (*23*). Thus, it is advised that ginkgo should not be used in patients taking warfarin, triticlofidine or clopidogrel (*23*). Adverse effects by high dosages of ginkgo leaf extract include diarrhea, nausea, vomiting, allergic skin reactions, headache, gastrointestinal disturbances, rashes and other illness (*24*).

Phytochemical Constituents and Analysis

The isolation and determination of six terpene trilactones (i.e. ginkgolides A, B, C, J, and M, and bilobalide) and more than 40 flavonoids were discussed in a published report (*25*). The most important active flavonoids are the glycosides of kaempferol, quercetin, and isorhamentin. Other flavonoids include acacetin, amentoflavone, apigenin, D-catechin, luteolin and rutin (*25*). The major terpene trilactones (Figure 4) are the bilobalide and ginkgolides A, B, C, and J. The constituents of standardized GBE usually are 24% flavonol glycosides and 6% terpene trilactones (*21*). Since studies demonstrate that terpene trilactones are a potent platelet-activating factor receptor antagonist (*26*) and those flavonoids are an antioxidant (i.e., free radical-scavenger) (*27*), these compounds, especially the trilactones, are usually used as biomarkers for quality assurance purposes.

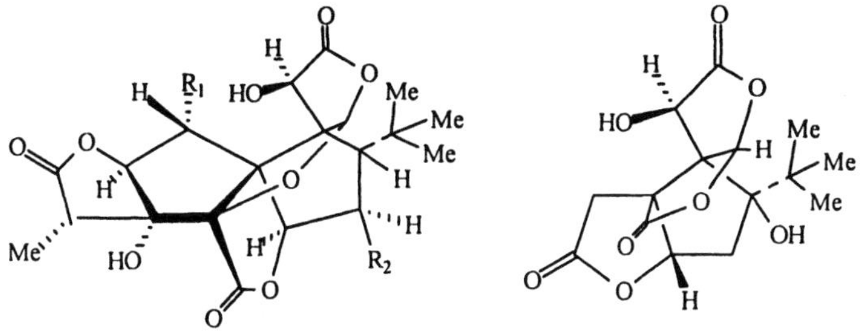

Ginkgolide A: $R_1 = R_2 = H$
Ginkgolide B: $R_1 = OH$, $R_2 = H$
Ginkgolide C: $R_1 = R_2 = OH$
Ginkgolide J: $R_1 = H$, $R_2 = OH$

Bilobalide

Figure 4: Chemical structures of ginkgo terpene trilactones.

A comprehensive review of the chemical analysis of ginkgo leaves or GBE was reported by van Beek (*21*). For the analysis of terpene trilactones, the key step is the sample clean-up procedure. Following the extraction with suitable solvent (e.g., methanol or methanol/water mixture), the sample extract is cleaned up by solid phase extraction or by partitioning with solvents and then determined by LC-evaporative light scattering detection (ELSD), LC-refractive index detection (RID) or gas chromatography-flame ionization detection (GC-FID) after silylation (*21*). For the analysis of flavonoids, an acidic hydrolysis of the flavone glycosides is required before the LC determination. Recently, two methods were also reported for simultaneous determination of terpene trilactones and flavonoids with GC-MS (*28*) and LC-ELSD (*29*).

Analysis of Ginkgo Constituents in Teas and Fruit-flavored Beverages

For quantitative determination of terpene trilactones in fruit-flavored drink products, an effective sample clean-up step must be performed. Recently, we developed a liquid-liquid extraction for clean-up and a LC-ELSD determination procedure for the quantification of terpene trilactones in fruit-flavored drinks and tea products (*30*). Figure 5 shows the chromatograms of two ginkgo-containing drink samples. One contained non-detectable amounts of the ginkgo constituents while the other contained significant levels of the active ginkgo constituents.

Identification of chromatograph peaks was made by comparison of retention times and reconfirmed by LC/electrospray ionization-tandem mass spectrometry (LC/ESI-MS/MS). Quantification of terpene trilactones was achieved using an external standard method and reconfirmed with a GC-FID procedure using an internal standard method. Ginkgo tea bags were extracted by steeping in hot water, while non-alcoholic, non-carbonated drinks were not treated prior to analysis. Both the tea infusion and drink samples were either filtered through filter paper or centrifuged to remove the suspended particles. After addition of NaCl to a concentration of 30% (w/v), the pH of each sample was adjusted to pH 3 – 5 with a small amount of KH_2PO_4, and a test aliquot was extracted with ethyl acetate/tetrahydrofuran (4:1, v/v). The organic solvent was removed, and the residue was re-dissolved in 0.5 or 1.0 mL of methanol for LC-ELSD analysis.

Using a standard addition method, the recoveries of bilobalide, ginkgolides A, B, and C were examined. Addition of individual terpene trilactones (1 – 2 μg/mL), resulted in recoveries between 93.8 – 108.2% with the relative standard deviations (RSD) in the range of 2.2 – 9.8% (*30*). Several brands of tea bags and drink products were analyzed and a wide variation was observed in terpene trilactone content. Two examples that represent the high and low concentrations of ginkgo constituents are shown in Table II. These data suggest that the quality of ginkgo-containing beverages varies significantly and some samples do not contain any detectable ginkgo terpene trilactones.

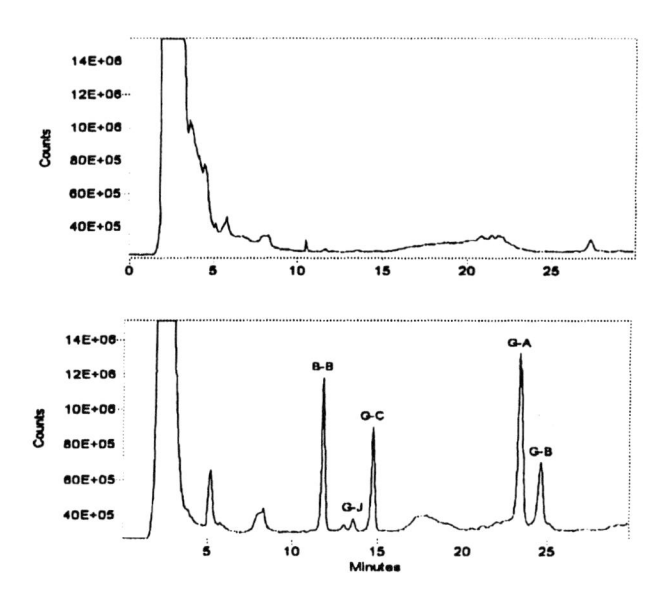

Figure 5. LC-ELSD analysis of ginkgo terpene lactones in two fruit -flavored drink products. Ginkgo components were not detectable in one product (top chromatogram) while they were present at high levels in another product (bottom). (Reproduced with permission from reference 30. Copyright 2004 AOAC International).

Table II. Analysis of Ginkgo Terpene Tilactones in Fruit-flavored Drinks and Hot Water Infusion of Tea Bags Containing *Ginkgo biloba*.

Product[a]	Mean, µg/serving (n=5) (RSD, %)[b]			
	B-B[c]	G-A[c]	G-B[c]	G-C[c]
Drink-1	835.2 (4.2)	1,140 (8.0)	458.4 (5.8)	900.0 (6.1)
Drink-2	39.5 (16.2)	ND[c]	ND[c]	35.6 (12.5)
Tea-1	1,728 (7.9)	1,185 (9.2)	867.1 (4.6)	1,392 (1.1)
Tea-2	213.9 (2.2)	ND[d]	171.9 (0.8)	109.8 (0.8)

[a] From different manufactures. Serving size: drink, 240 mL; tea bags, 1.5 - 2.2 g.
[b] RSD, %: Relative standard deviation
[c] B-B: Bilobalide; G-A: Ginkgolide-A; G-B: Ginkgolide-B; G-C: Ginkgolide-C
[d] ND: not detectable, LOD = 0.1 µg/mL for drink; LOD = 20 µg/g for tea.
(Adapted with permission from reference 30. Copyright 2004 AOAC International.)

Kava in Functional Beverages

Kava *(Piper methysticum,* Forst.) is a perennial shrub native to islands of the South Pacific. Kava root is used by the people of the South Pacific to prepare an important beverage for social ceremonies. Kava is also used as a traditional medicine for treatment of various medical symptoms *(31).* Since the introduction of kava into Western countries, commercial kava root extracts gained wide use for the treatment of nervous anxiety, tension, restlessness, and as an alternative treatment for mental disorders.

Biological and Toxicological Activities

The psychoactivity of kava, specifically the anti-anxiety and mild sedative action, is attributed to kavalactones via an effect on central nervous system receptors and sodium and calcium ion channels *(32).* A variety of biological activities are reported, such as local anesthesia, analgesia, ocular effects, and anticonvulsive effects. Kava is regarded as effective, non-addictive, and safe. However, heavy users of kava reported mild and reversible gastrointestinal disturbance and kava dermatopathy (scaly skin eruption) *(33).* A recent outbreak of severe hepatotoxicity associated with kava consumption raised a warning on the use of kava products. Several European countries began removal of kava containing products from the market. In 2002, the U.S. FDA issued a consumer advisory statement summarized the status of potential risk of liver injury associated with kava dietary supplements *(34).*

Several hypotheses are suggested for kava hepatotoxicity, such as genetic polymorphism in the consumers (e.g., Caucasian vs. Polynesian), different kava preparation methods (e.g., standardized solvent extraction vs. traditional aqueous extraction), toxic alkaloids in kava, and inhibition of cytochrome P450 (CYP) enzymes by kavalactones *(35-38).* However, it is still unclear whether or not kavalactones are responsible for the liver toxicity. Moreover, the hepatotoxic constituents of kava are not identified, and a mechanism for the hepatotoxicity of kava is not proven.

Phytochemical Constituents and Analysis

Kavalactones are regarded as pharmacologically active constituents of the kava plant and they are present as a mixture of more than 18 different forms. The six major kavalactones are (+)-methysticin (M), (+)-dihydromethysticin (DHM), (+)-kavain (K), (+)-dihydrokavain (DHK), yangonin (Y), and desmethoxyyangonin (DMY) (Figure 6) *(39).* These kavalactones are characterized by a variation in the presence of a double-bond at positions 5, 6

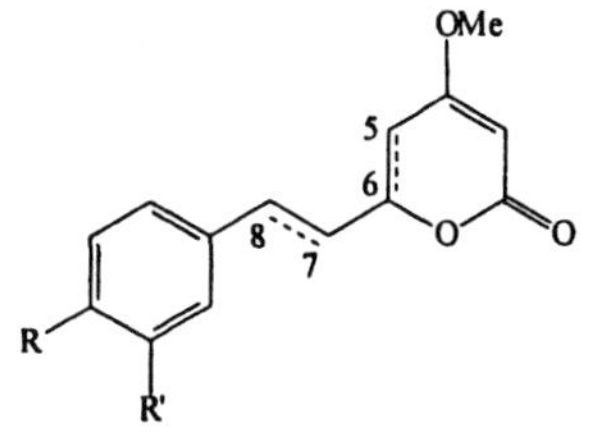

	R	R'	C_5-C_6	C_7-C_8
Kavain	H	H	—	=
Dihydrokavain	H	H	—	—
Methysticin	—O—CH$_2$—O—		—	=
Dihydromethysticin	—O—CH$_2$—O—		—	—
Yangonin	—OMe	H	=	=
Desmethoxyyangonin	H	H	=	=

Figure 6. Six major kavalactones in kava root.

and 7, 8. The total kavalactone content varies (3-20% dry weight) with dependence on plant age and cultivar (*31*). The concentration of kavalactones is generally highest in the roots and is lower in the aerial portions of the plant. The early TLC method (*40*) is replaced by GC and HPLC methods for quantitative analysis of kavalactones from kava plant. Precaution was needed since high temperature of the GC injection port may cause degradation of kavalactones, especially methysticin (*41*). Thus, HPLC methods seem to be a good alternative method for quantification of kavalactones.

Several reports described the analysis of kavalactones using normal phase HPLC (*42,43*), and more recently, a quantitative determination of six major kavalactones with use of a ChiraSpher NT (Merck, Germany; 44) and Nucleosil 50 column (*45*) and isocratic elution. However, a ChiraSpher NT column requires a long separation time for kavalactones. Other methods suggest the use of a YMC Basic (Waters, MA) reversed phase column (*46*) with isocratic elution, and a Luna C$_{18}$ column with linear gradient elution at 60°C (*47*). It is noteworthy that photoisomerization of yangonin in an aqueous methanol solution was reported (*43*). This can lead to inaccurate analytical results. Thus, studies recommend that light exposure be avoided during sample preparation and analysis.

Analysis of Kava Constituents in Teas and Fruit-flavored Beverages

In the course of the analytical method development for kavalactones in functional foods, we observed a complete separation of kavalactones with a Luna C$_{18}$-2 column and an isocratic mobile phase (*48*). This technique was then applied to the analysis of tea-bag beverages as described below.

(1) Kavalactone analysis of tea-bag products using HPLC

Two different tea-bag products that contain kava were tested. Tea-1 was a mixture of kava and other herbal ingredients, and was labeled to contain 22.9 mg of kavalactones per bag. Tea-2 contained ground kava root. However, Tea-2 was not specifically labeled for kavalactone concentrations. To determine the total kavalactones in each tea bag product, samples were pooled, and extracted with 100% methanol using an ultrasonic extraction technique. Then, the sample extract was analyzed by HPLC under the following conditions: isocratic mobile phase (2-propanol/acetonitrile/water/acetic acid=16:16:68:0.1; v/v/v/v) at a flow rate of 1 mL/min, a Luna C_{18}-2 column, column temperature: 60°C, UV absorbance detection at 240 nm. An internal standard, 5, 7-dihydroxyflavone, was used for construction of a standard calibration curve of kavalactone concentrations from 1 µg/mL to 100 µg/mL (48).

Tea-bags are usually prepared by hot water infusion for consumption. Therefore, we were interested in the quantity of kavalactones extracted into hot water from tea-bag products. Hot tea infusions were prepared according to the product label instructions. Specifically, tea-bags were steeped with 240 mL of boiling water for 8 min. The resulting solutions were analyzed (Figure 7) by the HPLC procedure described above. Methanol was a more efficient solvent for extraction of kavalactones than hot water. Tea-1 contained kavalactones at 92% of the label claim (Table III). However, the total amount of kavalactones in Tea1 extracted by hot water was only about 36% of those extracted by methanol. The total quantity of kavalactones extracted from Tea-2 by hot water was 20% of those extracted by methanol (48).

(2) Kavalactone analysis of fruit-flavored drinks using GC-FID.

Two non-alcoholic, non-carbonated fruit-flavored beverages were tested. Both beverage-1 and -2 contained water, fruit juice concentrates, citric and ascorbic acids, flavors, and kava. Beverage-1 was labeled to contain 50 mg of standardized kavalactones in 297 mL of beverage, and beverage-2 was labeled to contain 100 mg of "kava kava" in 591 mL, however, the label did not specify whether it contained a standardized kava extract or another form of kava. Some beverage products may contain high concentrations of sugars, various flavors, and color components. These additional ingredients may affect kavalactone analysis. Removal of these compounds from the sample solution was accomplished by solid phase extraction (SPE) with Sep-Pak C-18 and silica cartridges (Waters, MS). Samples were analyzed by GC-FID equipped with a DB-5 (30 m) capillary column. The GC analysis conditions differed between beverage-1 and -2. The chromatograms for beverages-1 and -2 are showed in Figure 8. An external standard calibration curve was constructed with a series of concentrations of kavalactones.

The total amount of six kavalactones in beverage-1 was determined to be 154.3 µg/mL (equivalent to 45.8 mg/297 mL) (Table III). This value was comparable to the amount stated on the product label (50 mg/297 mL of kavalactones). For the analysis of beverage-2, only two kavalactones (Y and DMY) were detected (Table III). Beverage-2 was labeled to contain 100 mg of "kava kava". It did not specify whether the kava was a standardized kava extract

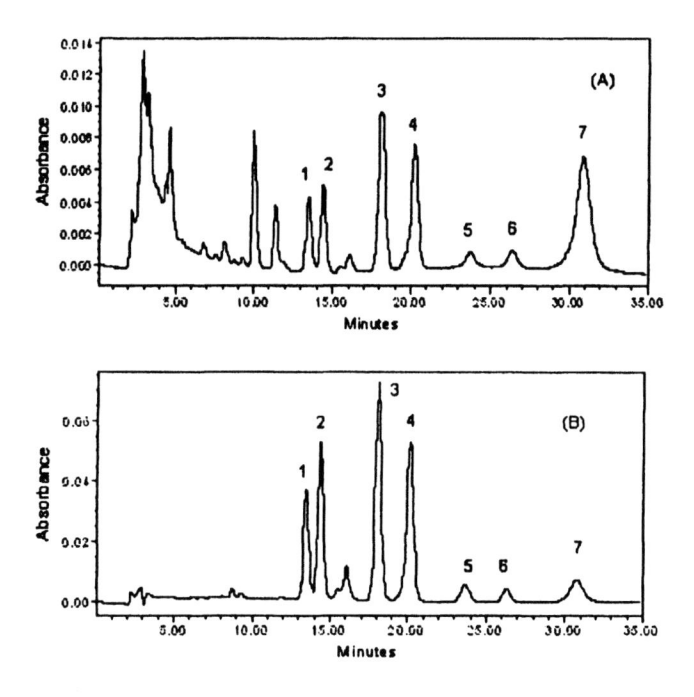

Figure 7. HPLC chromatograms of hot water infusion of tea-bag products. (A) tea-1 (B) tea-2. 1=M, 2=DHM, 3=K, 4=DHK, 5=Y, 6=DMY, 7=Internal standard

Table III. Kavalactone Content in Tea-bag (mg/g) Extracted by Methanol (MeOH) or Hot Water and Fruit-flavored Beverages (µg/mL).

Product[a]	Method	K	DHK	M	DHM	Y	DMY	Total
Tea-1	MeOH	1.38	4.92	0.85	1.23	0.76	0.51	9.65
	water	0.72	1.01	0.43	0.65	0.44	0.23	3.48
Tea-2	MeOH	7.82	10.87	7.24	10.68	5.51	2.25	44.37
	water	1.75	3.01	1.10	2.33	0.51	0.21	8.91
Beverage-1		36.13	51.92	13.67	30.31	11.47	10.80	154.30
Beverage-2		ND[b]	ND[b]	ND[b]	ND[b]	0.11	0.08	0.19

[a]Products from different manufacturers. Tea-l = 2.2 g/bag; Tea-2 = 2.0 g/bag.
[b]ND: not detectable, LOD 0.033 µg/mL for Beverage-2.

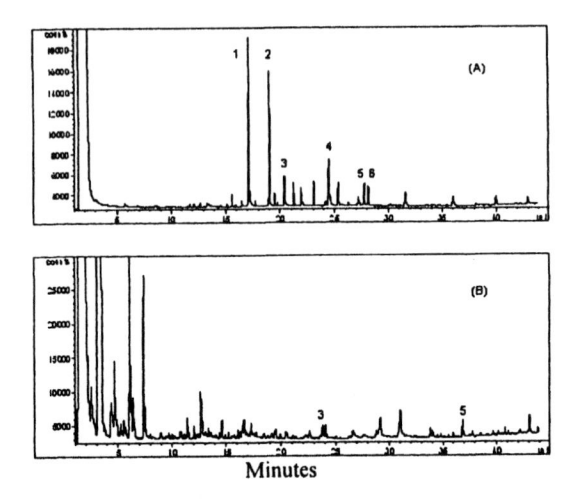

Figure 8. GC-FID chromatograms of fruit-flavored kava beverages under different GC conditions. (A), beverage-1; (B), beverage-2. 1 = DHK, 2=K, 3=DMY, 4=DHM, 5=Y, 6=M

or another form of kava. The amount of kavalactones as determined in the beverage-2 (Table III) was significantly lower than the label claim. One possible explanation for this low amount could be the instability of kavalactones during storage in acidic solution (beverage-1: pH 4.0; beverage-2: pH 2.7). The results suggest that further investigation is needed to better understand the stability of kavalactones in acidic aqueous solutions.

Conclusions

Botanicals and their products have complex matrices and contain many constituents with biological functions. When botanicals are added to a food or beverage, the matrix is more complex and the concentration of active ingredients is lower than in the original plant material. The analytical assessment of bioactive constituents in functional foods and beverages presents a new challenge. In a series of our studies, several analytical methods were developed for the determination of active constituents of SJW, ginkgo and kava in functional beverages. Composition of bioactive ingredients of the same type of product varies significantly between manufacturers. Regarding tea-bag products, actual levels of SJW and kava bioactive compounds determined in the water infusion (for consumption) were much lower than the values obtained from organic solvent extraction. Analytical results of aqueous buffer solutions and fruit-flavored beverages suggest that chemical changes may occur to the SJW constituents, such as degradation of hyperforin in acidic drink products, particularly under light exposure. Evaluation of the biological activity, quality, and safety implications of these chemical changes is needed.

References

1. American Herbal Pharmacopoeia. *St. John's wort. Hypericum perforatum. Quality control, analytical and therapeutic monograph;* Upton, R., Ed.; American Herbal Pharmacopoeia: Santa Cruz, CA, 1997; pp 1-32.
2. Laakmann, G.; Jahn, G.; Schule, C. *Nervenarzt.* **2002**, *73*, 600-6 12.
3. Barnes, J.; Anderson, L. A.; Phillipson, J. D. *J. Pharm. Pharmacol.* **2001**, *53*, 583-600.
4. Neary, J. T.; Bu, Y. *Brain Res.* **1999**, *816*, 358-363.
5. Hypericum Depression Trial Study Group. *JAMA.* **2002**, *287*, 1807-1814.
6. Greeson, J. M.; Sanford, B.; Monti, D. A. *Psychopharmacology.* **2001**, *153*, 402-414.
7. Nahrstedt, A.; Butterweck, V. *Pharmacopsychiat.* **1997**, *30*, 129-134.
8. Chen, Y.; Fergusson, S.S.; Negishi, M.; Goldstein, J.A. *J. Pharmacol. Exp. Ther.* **2004**, *308*, 495-501.
9. Moore, L.B.; Goodwin, B.; Jones, S.A.; Wisely, G.B.; Serabjit-Singh, C.J.; Willson, T.M.; Collins, J.L.; Kliewer, S.A. *Proc. Natl. Acad Sci. USA.* **2000**, *97*, 7500-7502.
10. Gray, D. E.; Rottingghaus, G. E.; Garrett H. E. G.; Pallardy, S. G. *J. AOAC Int.* **2000**, *83*, 944-949.
11. Piperopoulos, G.; Lotz, R.; Wixforth, A.; Schimerer, T.; Zeller, K-P. *J. Chromatogr. B.* **1997**, *695*, 309-316.
12. Ang, C. Y. W.; Cui, Y.; Chang, H. C.; Luo, W.; Heinze, T. M.; Lin, L. J.; Mattia, A. *J. AOAC Int.* **2000**, *85*, 1360-1369.
13. Jürgenliemk, G.; Nahrstedt, A. *Pharmazie.* **2003**, *58*, 200-203.
14. Ang, C. Y. W.; Hu, L.; Heinze, T. M.; Cui, Y.; Liu, F.; Freeman, J. P.; Kozak, K.; Luo, W.; Mattia, A.; DiNovi, M. *J Agric. Food Chem.* **2004**, (In press).
15. Gin, Y. L.; Wu, H. S. *Dictionary of Chinese Natural Supplement Sources* (In Chinese); Chung Zheng Publishing House: Beijing, China, 1993.
16. Wada, K.; Ishigaki, S.; Ueda, K.; Take, Y.; Sakata, K.; Haga, M. *Chem. Pharm. Bull.* **1988**, *36*, 1779-1782.
17. Scott, P. M.; Lau, B. P.; Lawrence G. A.; Lewis D. A. *J. AOAC Int.* **2000**, *83*, 1313-1320.
18. Gertz, H. J.; Kiefer, M. *Curr. Pharm. Des.* **2004**, *10*, 261-264.
19. Le Bars, P. L.; Kastelan, J. *Public Health Nutr.* **2000**, *3*, 495-499.
20. Clostre, F. *Ann. Pharm. Fr.* **1999**, *Suppl. 1*, 1S8-88.
21. van Beek, TA. *J. Chromatogr. A.* **2002**, *967*, 21-55.
22. Kressmann, S.; Muller, W. E.; Blume, H. H. *J. Pharm. Pharmacol.* **2002**, *54*, 661-669.
23. Glisson, J.; Crawford, R.; Street, S. *Nurse Practitioner.* **1999**, *24*, 28, 31, 35-36.
24. Rowin, J.; Lewis, S. L. *Neurology.* **1996**, *46*, 1775-1776.
25. Woerdenbag, H. J.; van Beek, T. A. *Adverse Effects of Herbal Drugs,* Springer-Verlag: Berlin, 1997; pp 51-60.

26. Maclennan, K. M.; Darlington, C. L.; Smith, P. F. *Prog. Neurobiol.* **2002**, *67*, 235-257.
27. DeFeudis, F.V.; Drieu, K. *Curr. Drug Targ.* **2000**, *1*, 25-58.
28. Deng, F.; Zito, S. W. *J. Chromatogr. A.* **2003**, *986*, 121-127.
29. Li, W.; Fitzloff, J. F. *J. Pharm. Biomed Anal.* **2002**, *30*, 67-75.
30. Lang, Q.; Wai, C. M.; Ang, C.Y.W.; Cui, Y.; Heinze, T.M.; Mattia, A.; Dinovi, M. *J. AOAC Int.* **2004**, *87*, 815-826.
31. Lebot, V.; Merlin, M.; Lindstrom, L. *Kava-The Pacific Elixir. the definitive guide to its ethnobotany, history, and chemistry;* Healing Art Press: Rochester, VT, 1997.
32. Singh, Y.N.; Singh, N.N. *CNS Drugs.* **2002**, *16*, 731-743.
33. Russmann, S.; Barguil, Y.; Cabalion, P.; Kritsanida, M.; Duhet, D.; Lauterburg, B.H. *Eur. J. Gastroenterol. Hepatol.* **2003**, *15*, 1033-1036.
34. U.S. Food and Drug Administration. Letter to health-care professionals. Available at http://www.cfsan.fda.gov/~dms/addskava.html.
35. Denham, A.; McIntyre, M.; Whitehouse, J. *J. Altern. Complement. Med.* **2002**, *8*, 237-263.
36. Unger, M.; Holzgrabe, U.; Jacobsen, W.; Cummins, C.; Benet, L. Z. *Planta Med.* **2002**, *68*, 1055-1058.
37. Stickel, F.; Baumuller, H.-M.; Seitz, K.; Vasilakis, D.; Seitz, G.; Seitz, H. K.; Schuppan, D. *J. Hepatol.* **2003**, *39*, 62-67.
38. Nerurkar, P. V; Dragull, K; Tang, C-S. *Toxicol. Sci.* **2004**, *79*, 106-111.
39. He, X.; Lin, L.; Lian, L. *Planta Med.* **1997**, *63*, 70-74.
40. Young, R.L.; Hylin, J.W.; Plucknett, D.L.; Kawano, Y.; Nakayama, R.T. *Phytochemistry.* **1966**, *5*, 795-798.
41. Smith, R. M. *Phytochemistry.* **1983**, *22*, 1055-1056.
42. Gracza, L.; Ruff, P. *J. Chromatogr. A.* **1980**, *193*, 486-490.
43. Smith, R.M.; Thakrar, H.; Arowolo, A.; Shafi, A.A. *J. Chromatogr. A.* **1984**, *283*, 303-308.
44. Boonen, G.; Beck, M.-A; Haberlein, H. *J. Chromatogr. B.* **1997**, *702*, 240-244.
45. Siméoni, P.; Lebot, V. *Biochem. Syst. Ecol.* **2002**, *20*, 413-424.
46. Shao, Y.; He, K.; Zheng, B.; Zheng, Q. *J. Chromatogr. A.* **1998**, *825*, 1-8.
47. Schmidt, A.H.; Molnar, I. *J. Chromatogr. A.* **2002**, *948*, 51-63.
48. Hu, L.; Jhoo, J.-W.; Ang, C.Y.W.; Dinovi, M.; Mattia, A. *J. AOAC Int.* **2004** (in press).

Chapter 6

Bioactive Natural Products from Chinese Tropical Marine Plants and Invertebrates

Y.-W. Guo, X.-C. Huang, W. Zhang, and Y.-Q. Sun

State Key Laboratory of Drug Research, Institute of Materia Medica, Shanghai Institutes for Biological Sciences, Chinese Academy of Sciences, Shanghai 201203, China

This paper discusses chemical and biological investigations of Chinese tropical marine plants and invertebrates 1) an unusual macrocyclic polydisulfide from the Chinese mangrove *Bruguiera gymnorrhiza*; 2) polyhydroxylpolyene compounds from marine dinoflagellate *Amphidinium* sp.; 3) two new series of uncommon steroids from marine invertebrates.

Introduction

Historically, China has been the leader in the use of marine medicinal agents. The use of marine products in Chinese medicine dates back to Pen Tsao Kang Mu (ca. 500 A.D.) in which a number of seaweeds, shellfish and fish used as medicinal agents are listed. Although worldwide competition during the last 40 years saw other nations emerging as leaders in the discovery of potent active natural marine products, China has once again assumed the lead. The Chinese government lunched a marine science and technology program (863 marine science program) in 1995 to support Chinese scientists carrying out research and development in all aspects of oceanology. Today China is vigorously exploring the ancient treasures of oceanic biodynamic deep, using modern analytical technology.

Our research on marine natural products is devoted mainly to discovering organic substances from marine organisms as a source of more effective and

© 2006 American Chemical Society

safer pharmaceuticals. Tropical mangrove of the genera *Bruguiera* (family Rhizophoraceae) interested us since it has been described to frequently afford cyclic disulfides, which are relatively uncommon in nature, but usually exhibit interesting biological activities (*1*). We recently investigated the mangrove *Bruguiera gymnorrhiza* collected at Guangxi Province, China. Our effort resulted in the isolation of a novel macrocyclic polydisulfide, gymnorrhizol (**1**), which possesses a previously unknown 3,4,8,9,13,14-hexathia-1α,6β,11β-triol-cyclopentadencane carbon-skeleton (*2*).

Marine dinoflagellates are of considerable current interest as a new and promising source of bioactive substance. Recently, marine dinoflagellates were proven to produce a variety of structurally and biologically intriguing secondary metabolites, some of which are expected to serve as lead compounds for drug development or pharmacological tools for basic studies of life sciences. Recently, we made a collection of dinoflagellates from Lingshui Bay, Hainan Province, China. On separation of methanol/toluene extract from the cultured dinoflagellates, three new polyhydroxylpolyene compounds, lingshuiol (**5**), lingshuiols A (**6**) and B (**7**) were isolated (*3,4*).

A large number of uncommon steroids have been described from marine sources. But steroids with C-24 spiroketal side chains and polyhydroxylated sterols bearing 9,11-epoxy moieties are rarely reported. Our recent chemical investigation on the ether extract of gorgonian *S. reticulata,* and an unidentified sponge of the genus *Dysidea* collected along the cost of South Chinese Sea, led to the isolation of two series of new polyoxygenated steroids, namely suberretisteroids (**9-13**) (*5*) and dysidesterols A-E (**14-18**) (*6*). Suberretisteroids A-E all bore rare C-24 spiroketal functionalities, while steroids **14-18** possessed unusual 9α,11α-epoxy nuclei.

An Unusual Macrocyclic Polydisulfide from Mangrove *Bruguiera gymnorrhiza*

While cyclic polysulfides are mainly found in the mushroom *Lentinus edodes* (*7,8*) and the red alga *Chondria californica* (*9*), tropical mangrove of the genera *Bruguiera* (family Rhizophoraceae) frequently afford cyclic disulfides (*10-16*). During our search for new bioactive natural products from Chinese marine organisms, we investigated the mangrove *Bruguiera gymnorrhiza* collected at Guangxi Province, China.

The chipped stems and leaves of *B. gymnorrhiza* were extracted with MeOH. The MeOH extract was subjected to solvent partitioning and repeated chromatographies on silica gel and Sephadex LH-20 to give gymnorrhizol (**1**) (as shown in Figure 1, 0.0005 % yield).

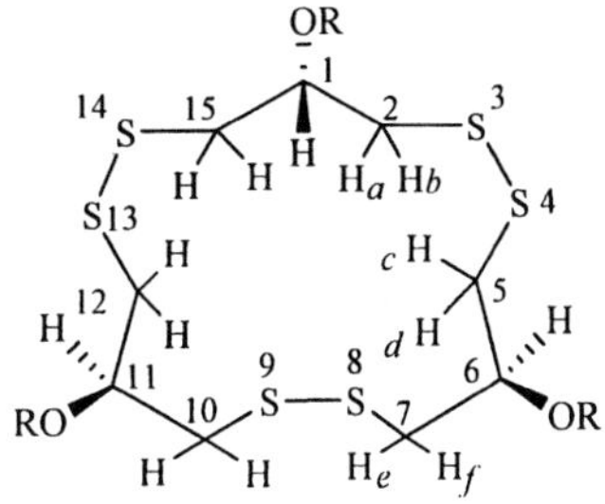

Figure 1. Chemical structures of compounds **1-4**.

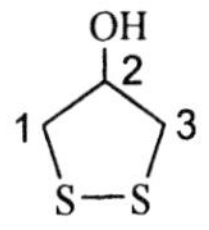

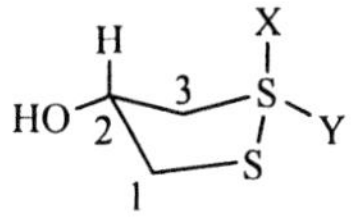

Gymnorrhizol (**1**) was obtained as a colorless and optically inactive crystal. Its molecular formula was deduced to be $C_9H_{18}O_3S_6$ by HRESIMS {m/z 388.9485, [M+Na]$^+$, Δ = +0.7 mmu}. Since only one unsaturation degree was accounted for from the molecular formula, (**1**) was inferred to be a macrocyclic polysulfide. IR spectrum of (**1**) contained signals at 3205, 1022 cm^{-1} indicating the presence of a hydroxyl group. The ^{13}C NMR spectrum of (**1**) was quite simple, showing only four lines for three sp^3 oxymethine carbons (δ_C 67.9) and six sp^3 sulfur-bearing methylene carbons (δ_C 46.0, 45.7 and 44.9), implying that the molecule is highly symmetric. The ^{1}H NMR spectrum of (**1**) simply showed two sets of signals corresponding to oxymethine protons [δ_H 3.97 (3H, m)] and methylene protons [δ_H centered at 2.90 (12H, m)].

The structural elucidation of (**1**) was not straightforward. In fact, although the NMR spectra of (**1**) was quite simple, the heavily overlapped signals in the ^{1}H NMR spectrum prevented us from interpreting them clearly. To determine the structure of gymnorrhizol, a decoupling experiment was performed. On irradiation at methine (δ 3.97), the upfield complex spectra were greatly simplified. Careful analysis of the splitting patterns and the coupling constants of the collapsed signals allowed recognition of three pairs of AB spin systems of ABX spectra for three methylenes, indicating the presence of partial structures **a** and **b**, as shown in Figure 2.

For partial structure **a**, evident ^{1}H-^{1}H COSY cross-peak between Ha (2.80, dd, J =13.4, 7.4 Hz) and Hb (2.95, dd, J = 13.9, 4.7 Hz) was observed, and both Ha and Hb, in turn, were correlated with H-1 (3.97). For partial structure **b**, COSY spectrum revealed H-6 (3.97) to be correlated to two pairs of AB type methylenes, [Hc (2.89, dd, J =13.3, 7.0 Hz), Hd (3.07, dd, J = 13.3, 4.7 Hz) and He (2.84, dd, J =13.6, 6.7 Hz), Hf (2.98, dd, J = 13.6, 6.0 Hz)]. This was further supported by HMBC cross-peaks (Hc, Hd/C-6, C-7 and He, Hf/C-5, C-6).

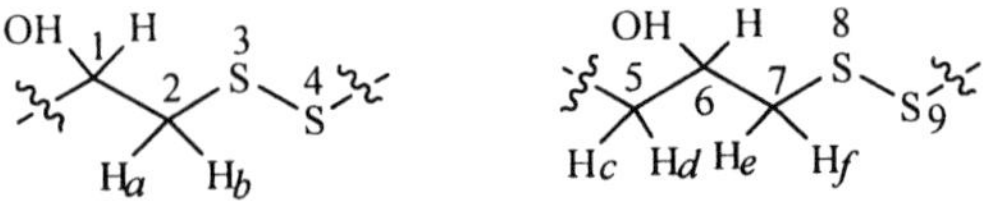

Figure 2. Partial structures of compound (1).

Subtraction of the atoms present in above elaborated partial structures from the molecular formula of gymnorrhizol (1) indicated that the rest of the molecule, $C_4H_8OS_2$, consisted of three methylenes, one –CHOH– and one –S-S–. Considering the highly symmetric nature of the molecule, these atoms have to be ascribed to the left half of the molecule (C-10-C-15), the same corresponding part (C-2-C-7) as depicted in (1) to complete the macrocycle.

The question remaining was how the three hydroxyl groups were oriented in the molecule. In fact, structurally, there were only two possibilities for the configuration of three hydroxyl groups: all *cis* or two *cis* one *trans*. The former could be ruled out because in case of all *cis*, all six methylene groups are symmetrically arranged and chemically equivalent in the molecule, and as a consequence, only one ABX signal pattern should be observed in the ^{1}H-NMR spectrum. While in case of (1), three pairs of methylenes, each chemically equivalent, were recognized. This fact could be explained only when 1-OH adopted a different configuration from that of 6-, and 11-OH. However, in order to ascertain the proposed structure of gymnorrhizol and, in particular, to confirm the suggested relative stereochemistry of the hydroxyl groups, (1) was acetylated with Ac_2O-pyridine to afford the expected triacetate derivative (1a). The presence of the same substructures **a** and **b** in both (1) in (1a) was easily recognized by detailed interpretation of the well resolved ^{1}H-^{1}H COSY, TOCSY, HMQC and HMBC spectra. Interestingly, the ^{1}H-NMR spectrum of (1a) showed well resolved resonances at δ 5.38 and 5.26, that were integrated for one and two protons, respectively, ascribable to three acetyl-bearing oxymethines. ^{13}C-NMR spectrum also exhibited corresponding signals at δ 70.6 and 71.2 (ratio is approximately 1:2), respectively. Furthermore, HMQC and HMBC experiments led to unequivocally localize three acetoxyl groups on C-1 ($δ_C$ 70.6 and $δ_H$ 5.38), C-6 and C-11 ($δ_C$ 71.2 and $δ_H$ 5.26). From these observations, the configurations of 6-OH and 11-OH were unambiguously determined to be *cis* (β), while that of 1-OH was *trans* (α) to 6-OH and 11-OH.

We can conclude that the structure of (1) is characterized by a symmetric fifteen-membered macrocyclic hexathiatriol composed of three repeated 1,3-dimercaptopropan-2-ol units. In the molecule the two hydroxyl groups at C-6 and C-11 are β-oriented while the 1-OH has an α configuration as depicted in Figure 1.

Gymnorrhizol (**1**) has an unprecedented skeleton related to dithiolanes **2** (*9*), **3** (*6*) and **4** (*6*) (Figure 1) previously isolated from different species of the genus *Bruguiera* although formally it appeared completely different from them. A plausible retrosynthetic pathway for gymnorrhizol (**1**) is proposed as shown in Scheme 1. The cleavage of the bonds between each pair of sulfur atoms (S3 to S4, S8 to S9 and S13 to S14) and successively recyclization between S3/S14, S4/S8, and S9/S13 gives 1,2-dithiolane (**2**), which, after oxidation on one of the sulfur atoms, yields brugierol (**3**) and isobrugierol (**4**).

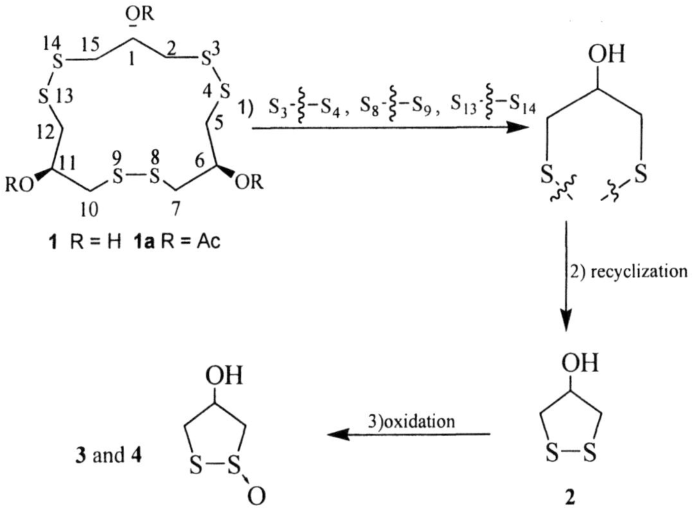

Scheme 1. Plausible biosynthetic correlation between gymnorrhizol (1) and compounds (2), (3) and (4).

Crude *n*-butanol extract of *B. gymnorrhiza* exhibited weak antitumor activity against A-549 and HL-60, but gymnorrhizol (**1**) proved inactive. Bioassays for the anti-fungal and antibiotic activity of gymnorrhizol are currently ongoing. Macrocyclic polythioethers are a group of molecules with potential extensive use as bioinorganic model systems, binucleating ligands, chelators for specific metal ions and phase-transfer catalysts (*17*). Further study should be conducted to test if the title compound also possess similar properties, as well as to understand its real biological role in the life cycle of the mangrove.

Polyhydroxylpolyene Compounds from Marine Dinoflagellate *Amphidinium* sp.

Microalgea play an important role in the marine biological system. With their photosynthetic ability, they are the major producer of biomass and organic compounds in the oceans. Many algal metabolites have unique structures and are formed by biosynthetic routes quite different from those known for terrestrial metabolites. Marine dinoflagellates are flagellated organisms, both photosynthetic and hetersynthetic. More than 4000 species are known to date. The living style of the organisms is diverse, and many of them occupy symbiotic environments (*18*). The chemistry of dinoflagellates has focused on their toxin production and pigment composition, and only a small number of the organisms have been investigated for secondary metabolites. As part of our ongoing research on the biologically active substances of Chinese marine organisms, we made a collection of dinoflagellates from Lingshui Bay, Hainan Province, China. On separation of the toluene soluble fraction of a methanol/toluene (3:1) extract of the cultured dinoflagellate, we isolated three novel polyhydroxylpolyene compounds, lingshuiol (**5**), lingshuiols A (**6**) and B (**7**) as shown in Figure 3.

From the surface wash of seaweeds collected at Lingshui Bay, Hainan Province, China, we isolated a strain of *Amphidinium* sp., which was deposited in the Herbarium of the South China Sea Institute of Oceanology, CAS for inspection (code-named as *Amphidinium* 2001-1). The strain was grown unialgally in sterilized seawater enriched with an ES-1 supplement at 24 °C for 3 to 4 weeks under illumination of a 12 h-light and 12 h-dark cycle. When the cell density reached 500,000-600,000 cells/mL, the algae were harvested by filtration and extracted with methanol /toluene (3:1, 500 mL×4). The combined extract was partitioned between toluene and 1N NaCl solvent, the organic layer (9.0 g) was subjected to separation by normal phase and reversed-phase silica gel column chromatography, followed by purification with C_{18} HPLC (MeCN/H_2O) to afford lingshuiol (**5**, 16.0 mg, 0.0008% wet weight), lingshuiols A (**6**, 25.4 mg, 0.0013% wet weight) and B (**7**, 22.5 mg, 0.0011% wet weight). All of the compounds were obtained as pale yellow amorphous solids. The IR spectra revealed the presence of hydroxyl group (3400 cm^{-1}), and carbon carbon double bonds (1624 cm^{-1}). Their molecular formulae were determined as $C_{69}H_{122}O_{25}$, $C_{66}H_{112}O_{23}$ and $C_{60}H_{99}O_{23}SNa$ by HRESIMS spectra, respectively.

All the compounds possessed two tetrahydropyran rings and the rest parts of the molecules are linear carbon chains with one exo-methelene and two (**6** and **7**) or three (**5**) methyl branches. The partial structure of the C-12-C-66 of lingshuiol A corresponded to that of the C-6-C-60 part of lingshuiol B, while the subunit of C-30-C-66 of lingshuiol was quite similar to the C-28-C-66 part of lingshuiol A and the C-22-C-60 fragment of lingshuiol B except that the former possessed a saturated chain bearing a terminal vinyl group instead of the hydrophobic polyene portions in one end of the molecule as in the other two compounds. The

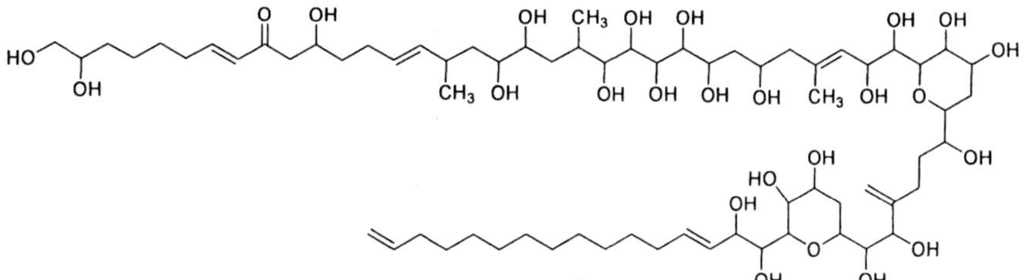

Figure 3. Chemical sturcutures of Compounds (5), (6) and (7).

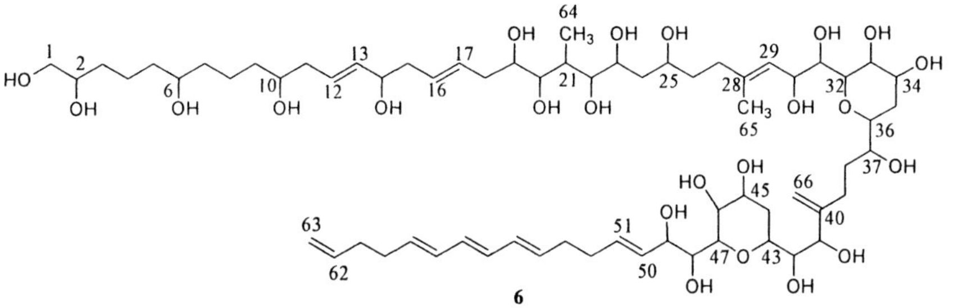

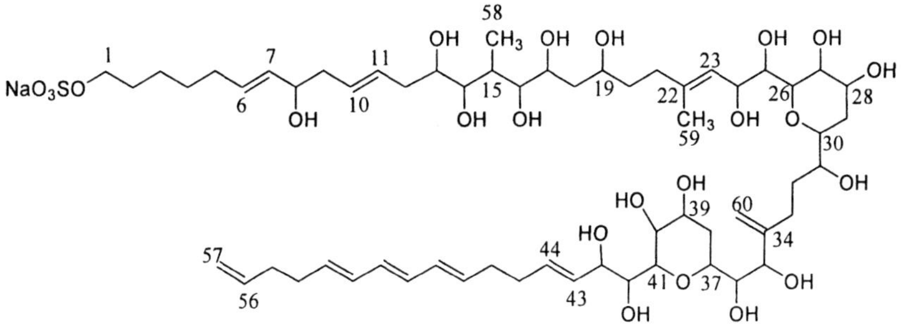

structural variations among the homologues are restricted to the start unit of the molecules and the terminal olefinic parts. Their structural variation may be accounted for by the recombination of polyketide synthase (PKS) domains; variable regions are responsible for the former part while the relatively constant regions are responsible for the latter part. We only describe the structure elucidation of lingshuiol here.

The ^{1}H and ^{13}C NMR data suggested that (5) contains one ketone carbon, two sp^2 quaternary carbons, eight sp^2 methines, two sp^2 methylenes, twenty-five oxymethines, one oxymethylene, two sp^3 methines, twenty-five sp^3 methlenes and three methyl groups. The molecular formula $C_{69}H_{122}O_{25}$ led to nine unsaturation equivalents, six of which were due to carbon-carbon double bonds, one due to the ketone carbon, the other two were ascribable to two rings.

Detailed analysis of DQF-COSY and TOCSY spectra of (5) led to the following four partial structures: from C-1 to C-8 (unit **a**), from C-10 to C-29 (unit **b**), from C-31 to C-41 (unit **c**) and from C-43 to C-69 (unit **d**). The proton connectivities for the two subunits from H$_2$-1 to H$_2$-3 (**a₁**) and from H$_2$-6 to H-8 (**a₂**) are evident from the DQF-COSY spectrum as shown in Figure 4a. The connectivities from C-3 to C-4 were assigned by HMBC spectrum of (5), which showed cross-peaks due to H-2/C-4 and H$_2$-3/C-4. The connectivities from C-5 to C-6 were revealed by HMBC correlations as H-7/C-5 and H$_2$-6/C-5. The assignment of C-4 connecting to C-5 was supported by the HMBC correlation for H$_2$-6/C-4. The geometry of the carbon-carbon double bond at Δ^7 was determined to be E since the ^{1}H-^{1}H coupling constant between H-7 and H-8 was found to be 16.2 Hz.

For partial structure **b**, connectivities from H$_2$-10 to H$_2$-17 (**b₁**), from H-18 to H-24 (**b₂**) andfrom H-25 to H$_2$-29 (**b₃**) were evident from DQF-COSY cross-peaks as shown in Figure 4b. These two subunits **b₁** and **b₂** were connected by HMBC spectrum, which showed cross-peaks due to H-18/C-16 and H$_2$-17/C-18. The TOCSY spectra data revealed H-18 to be correlated with H-16/H$_2$-17 and H-19/H$_2$-20, also supporting the connection of **b₁** to **b₂** through C-17/C-18. A secondary methyl group (C-66, δ 21.4; H$_3$-66 δ 0.84, d, J = 7.2 Hz) was located on C-16, which was shown by the DQF-COSY cross-peak for H$_3$-66 (δ 0.84)/H-16 (δ 2.34). The HMBC correlations for H$_3$-66/C-15, H$_3$-66/C-16, H$_3$-66/C-17, and H-15/C-66 further confirms this assignment. The E geometry of Δ^{14} double bond was assigned by the ^{1}H-^{1}H coupling constant ($J_{14,15}$ = 15.3 Hz). Observation of an evident NOE between H-14 and the allylic methine proton H-16 also supports this conclusion (5). Although there were no evident DQF-COSY cross-peaks observed from H-25 (δ 3.74) to any other protons, HMBC correlations for H-25/C-23 and H-25/C-24 undoubtedly determine the connection of **b₂** to **b₃**. The TOCSY spectrum confirms this determination, which reveals H-24 to be correlated to H-22, H-23, H-26, and H-27. A doublet (δ 0.93, 3H, d, J = 6.7 Hz, H$_3$-67) in the ^{1}H NMR spectrum of (5) suggests that there is

Figure 4a. partial structure **a**

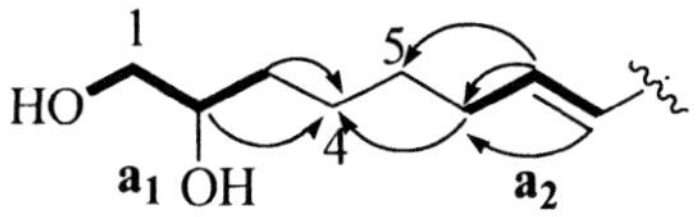

a

Figure 4b. partial structure **b**

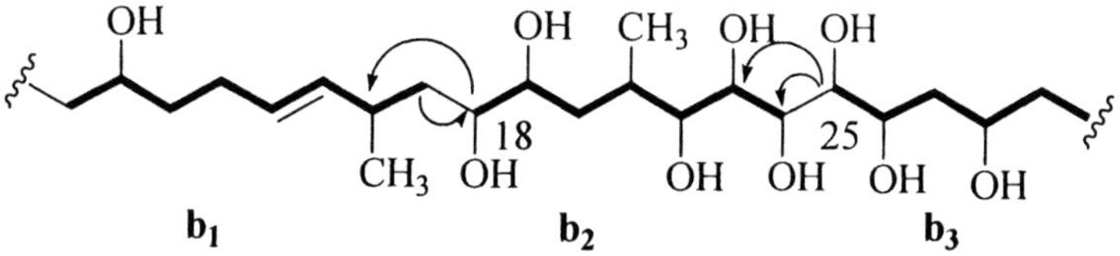

b

Figure 4c. partial structure **c**

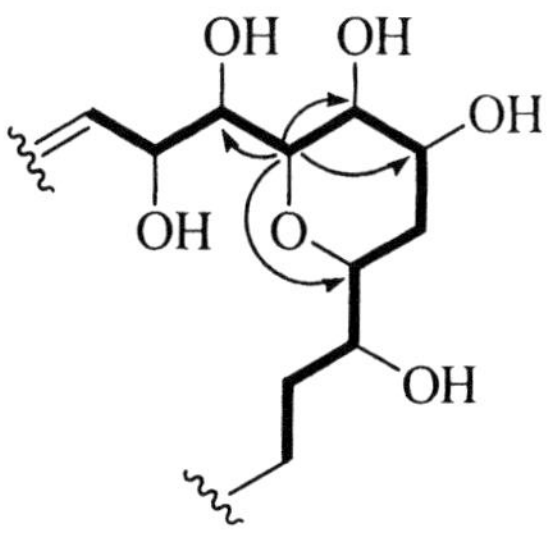

c

Figure 4d. partial structure **d**

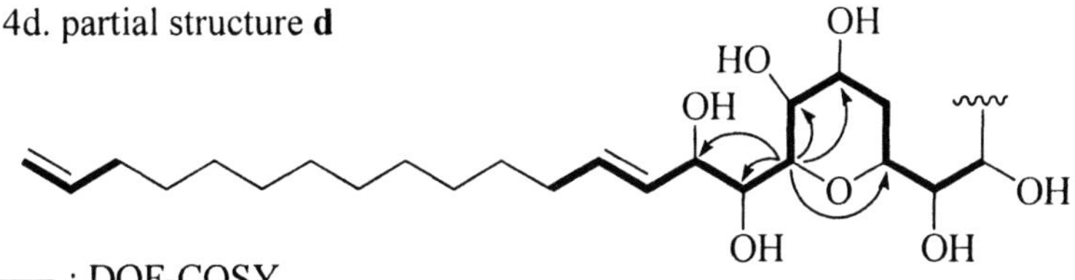

d

—— : DQF-COSY

⌣ : HMBC

*Figure 4. Partial structures of Compound (**5**).*

another, secondary, methyl group located in **5**. It was assigned on C-21 for that DQF-COSY cross-peak for H$_3$-67 (δ 0.93)/H-21 (δ 2.32) was evidently observed. This was further supported by HMBC spectrum (H$_3$-67/C-20, H$_3$-67/C-21, and H$_3$-67/C-22).

The ^{1}H and ^{13}C NMR data for C-31 to C-54 of (**5**), including the NOEs and ^{1}H-^{1}H coupling patterns derived from the NOESY and other 2D-NMR experiments, agreed quite well with those for C-31 to C-54 of amphidinol 2 (*19*) (**8**, as shown in Figure 5), indicating that they shared the same structure for that part as shown in Figure 4c and 4d. The relative stereochemistry of the tetrahydropyran ring (from C-34 to C-38) in partial structure C was established largely by NOESY spectrum as shown in Figure 6.

Figure 5. Chemical structure of compound (8).

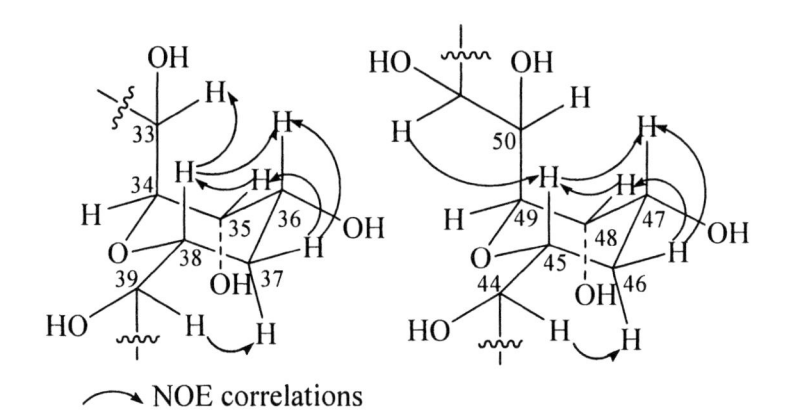

Figure 6. Relative stereochemistry of two tetrahydropyran rings of lingshuiol (5).

Major structural alterations in partial structure d from amphidinol 2 resided in the terminus C-56- C-61, where the conjugated triene moiety in amphidnol 2 was reduced to a saturated chain. Although the C_8-linear chain of C-55-C-62 could not be assigned because of heavily overlapped proton signals (δ 1.06-1.08), the existence of eight methylene carbons for unit d was confirmed by ^{13}C NMR data (δ 28.7, 28.9, 29.0, 29.1, 29.2, 29.3, 29.3, 29.4). The relative stereochemistry of the tetrahydropyran ring (from C-45 to C-49) in partial structure d was established by NOESY spectrum as shown in Figure 6.

Three partial structures **a-c** seemed to be connected through a ketone carbon (C-9) and a sp^2 quaternary carbon (C-30), as shown in Figure 7. The HMBC correlations for H-7/C-9, H-8/C-9, H-8/C-10, H_2-10/C-9 suggest that the partial structures **a** and **b** are connected through a ketone carbon (C-9 δ 200.1). On the other hand the HMBC spectrum showed cross-peaks due to H_3-68/C-29, H_3-68/C-30, H_3-68/C-31, H-31/C-29, H_2-29/ C-30 and H_2-29/C-31, indicating that units **b** and **c** are connected through a sp^2 quaternary carbon (C-30 δ 135.0). The geometry of the carbon-carbon double bond at Δ^{30} was determined to be *E* by the NOESY spectrum, which showed significant correlations of H-31/H-29a and H-31/H-29b. The conectivity of partial structures **c** and **d** was revealed by HMBC spectrum, which showed cross peaks due to H_2-40/C-42, H_2-41/C-42, H-43/C-42, H-43/C-41, H-44/C-42, H_2-41/C-69, and H_2-69/C-41.

From all of these observations, we conclude the planar structure of lingshuiol A as (**5**). Lingshuiol (**5**) possessed powerful cytotoxic activity against A-549 and HL-60 cell lines *in vitro* with an IC_{50} of 0.21 and 0.23 µM, respectively. Lingshuiols A and B also showed similar activity with IC_{50} of 0.87, 0.82 and 0.045, 0.49 µM, respectively.

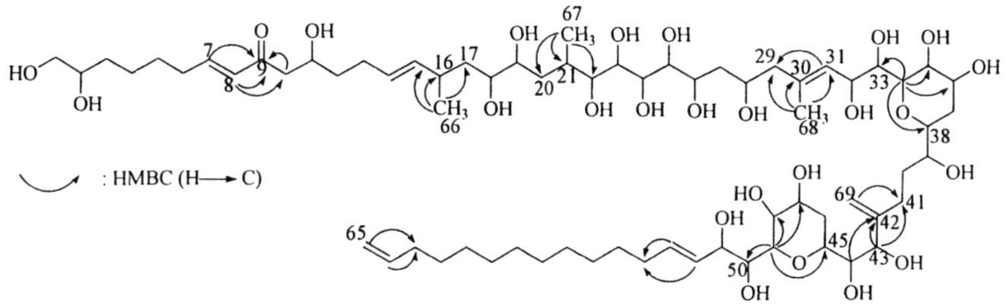

Figure 7. Key HMBC correlations of compound (5).

Two New Series of Sterols from Marine Invertebrates

Uncommon Steroids from Gorgonian *Subergorgia reticulata*

The subclass of Octocorallia comprises different orders among which Gorgonaceae and Alcyonacea are distinguished by their high content of polyhydroxylated sterols, often their major steroids. But only limited numbers of those steroids bearing a sprioketal ring structure on the side chains, such as C-22 spiroketal steroids hippurins (*20-24*), are reported. Steroids with C-24 spiroketal side chains are also infrequent. To the best of our knowledge, only two other compounds possessing the C-24 spiroketal functionalities have been described so far (*25,26*). Our chemical investigation on the ether extract of *Subergorgia reticulata*, collected along the coast of the South China Sea, led to the isolation of five new polyoxygenated steroids, named suberretisteroids A-E (**9-13**) as shown in Figure 8. All display the rare C-24 spiroketal functionalities.

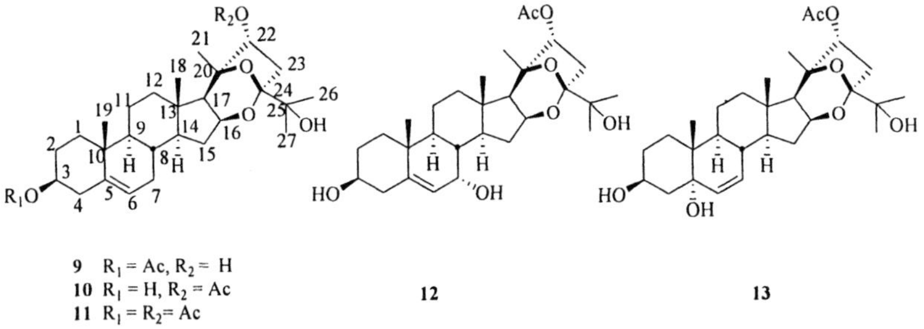

Figure 8. Chemical structures of Compounds (9)-(13).

Freshly collected animals from the South China Sea were frozen at once. The frozen animals were cut into pieces and subsequently extracted with acetone. After filtration the crude extract of *S. reticulata* was partitioned between ether and water. The ether extract was concentrated under rotation evaporation to give a dark green residue, a large portion of which was fractionated by gradient silica gel column chromatography (0-100% acetone in petroleum) followed by sephadex LH-20 gel column chromatography and repeated normal-phase and reversed–phase silica gel column chromatography. This afforded five pure compounds: (**9**) (suberretisteroid A, 11.9 mg), (**10**) (suberretisteroid B, 13.7 mg), (**11**) (suberretisteroid C, 7.5 mg), (**12**) (suberretisteroid D, 5.1 mg), (**13**) (suberretisteroid E, 1.1 mg).

The molecular formulae of suberretisteroids A-E were established as $C_{29}H_{44}O_6$, $C_{29}H_{44}O_6$, $C_{31}H_{46}O_7$, $C_{29}H_{44}O_7$ and $C_{29}H_{44}O_6$, respectively, by

HRESIMS on the pseudo-molecular ions at m/z 511.3031, 511.3041, 553.3139, 527.2999 and 527.2986, respectively. Their IR spectra indicated the presence of hydroxyl groups (3444 cm^{-1}), acetate functionalities (1736, 1259 cm^{-1}) and double bonds (1639 cm^{-1}). ^{1}H and ^{13}C NMR spectra of suberretisteroids showed that all the compounds belong to a homologous series containing a common tetracyclic nucleus and a C-24 spiroketal side chain. They all possessed five tertiary methyl groups typical for C-27 sterols, a 3β-oxygenated carbons and a carbon carbon double bond, suggesting that the compounds [except (13)] were 3β-oxygenated-Δ^5-cholostane derivatives. Spectroscopic investigations on suberretisteroid A are reported here.

Suberretisteroids A (9) was obtained as a colorless glass, [α]$_D^{20}$ − 22.5° (CDCl$_3$, c 0.25). The ^{1}H NMR showed one olefinic proton (δ 5.38, 1H, d, J = 4.9 Hz), one secondary acetoxyl group (δ 2.04, 3H, s) with correspond α-methine proton at δ 4.59 (1H, br m), one secondary hydroxyl group (δ 3.86, 1H, dd, J = 11.0, 4.9 Hz), one oxymethine proton (δ 4.51, 1H, br m), and five tertiary methyl groups (δ 1.04, 1.06, 1.15, 1.24, 1.37, each 3H, s). The ^{13}C NMR spectrum showed 29 carbon signals and demonstrated the same functionality: one olefinic proton (δ 139.7, s; 122.3, d), one secondary acetoxyl group (δ 170.7, s; 21.4, q; 73.9, d), one secondary hydroxyl group (δ 79.2, d), one oxymethine carbon (δ 72.6, d), one quaternary oxygen bearing carbon (δ 83.3, s), one quaternary hydroxyl group (δ 72.4, s), a ketal group (δ 107.6, s), and five tertiary methyl groups (δ 14.4, 19.3, 23.4, 23.4, 24.0, each s). These evidences suggest that compound (9) is a 3β-acetoxy-Δ^5-cholostane derivative, and the ketal group is in a side chain. These conclusions are supported by COSY (H-1/H-2, H-3/H-4, H-6/H-7, H-11/H-9, H-12, H-14/H-8, H-15, H-16/H-15, H-17) and HMBC (H-1/C-5, 10, H-2/C-5, 28,, H4/C-2, 3, 5, 6, 10, H-7/C-5, 6, 8, 9, H-18/C-12, 14, H-19/C-1, 5, 9) spectra. The structure of side chain was determined by HMBC (H-17/C-13, 16, 18, H21/C-17, 20, 22, H23/C-22, 24, H-26/C-24, 25, 27, H27/C-24, 25, 26), which suggests that compound 9 is a C-24 ketal connected to the C-16 and C-20 carbons with hydroxyl groups present at C-22 and C-25. The sterochemistry of the ketal and C-22 secondary hydroxyl group were inferred from the NOESY experiment. The NOE enhancement between H-16/H-14, 17, 23, H-18/H-21, 27 and H-21/H-12β, 22 indicate that H-16 and H-17 are in α-position while H21, H22 are in β-position. In this structure, the configuration of C-21β was uncommon since most steroids Me-21 are always in α-orientation. Because of this a single *X*-ray diffraction study was employed, as shown in Figure 9, the result of which strongly supported the previously conclusion. Compound (9) was then elucidated as 3,22,25-trihydroxy-16-24, 20-24-bisepoxy-(3β,16β,20S,22R,24S)-cholest-5-en-3-acetate.

Sanyaols A-E are a series of uncommon polyoxygenated C-24 spiroketal steroids isolated from gorgoiuan *Subergorgia reticulate*. Some of which were cytotoxic (IC$_{50}$, 0.6-1.5 μg/mL).

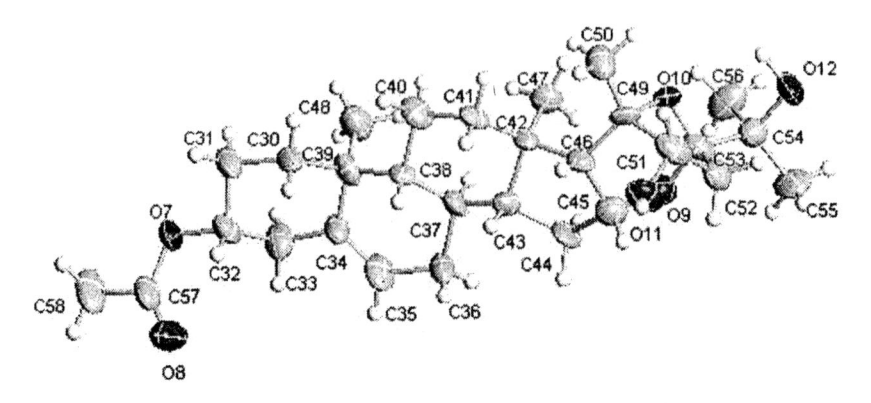

Figure 9. A computer generated perspective drawing of compound (9) showing the relative configuration.

Uncommon Steroids from Sponge *Dysidea* sp

The sponges have yielded the most varied and biogenetically unprecedented array of steroids found among the marine invertebrate phyla. Most of the 200 new monohydroxylated steroids found in marine organisms have been isolated from sponges (*27*). Recently, an increasing number of polyhydroxylated steroids have been isolated. Some authors have postulated that future studies would be likely to discover numerous examples of polar, polyhydroxylated steroids in marine organisms (*28*). Sponges of the genus *Dysidea* are widely distributed in tropical and subtropical waters around the world (*29*). It has been shown that highly fuctionalized steroids are widespread in this genus (*28,30-35*). In particular, C-9,11-oxygenated steroids, such as 5α-cholesta-7,22-diene-2α, 3β,5α,6β,9α,11α,19-heptaol and its analogue (*31*) are the most common whereas 9α,11α-epoxy steroids are infrequent. To the best of our knowledge, only one steroid possessing 9α,11α-epoxy functionality has been reported so far (*28*). We recently made a collection of the sponge *Dysidea* sp. off the Lingshui Bay Hainan Province, China. On separation of the Et$_2$O soluble fraction of an acetone extract of this sponge, five new uncommon polyhydroxylated steroids, named dysidesterols A-E (**14-18**), as shown in Figure 10, were isolated.

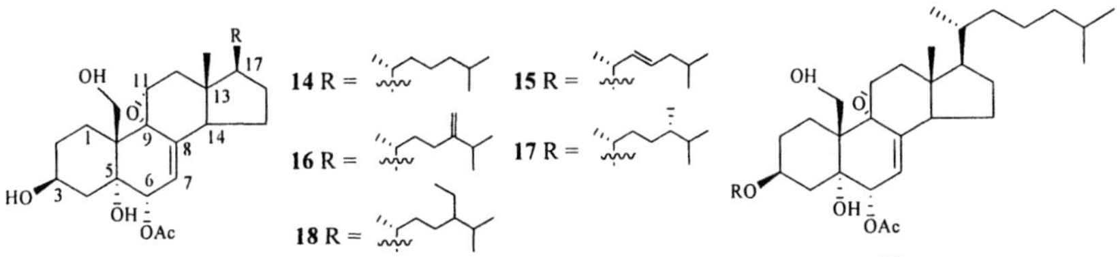

Figure 10. Chemical structure of Compounds (14)-(19).

Freshly collected animals from the South Chinese Sea were immediately put at -20 °C and kept frozen until used. Frozen material was extracted exhaustively with acetone. The acetone extract was then partitioned between diethyl ether and water. The organic-soluble extract was repeatedly subjected to silica gel column chromatography followed by purification on C-18 HPLC to afford compounds (14)-(18).

All of these compounds showed very similar spectroscopic properties. Their IR spectra indicated presence of hydroxyl groups (3498 cm^{-1}), an acetate (1710, 1261 cm^{-1}) and a double bond (1639 cm^{-1}). Their ^{13}C NMR spectra, as shown in Table I, were also very similar, showing six carbon signals between δ 55.0 and 75.0, indicating the presence of six sites of heteroatom functionality, along with a trisubstituted double bond [δ 122.7 (d), 141.1 (s)]. Moreover, careful analysis of their ^{1}H-^{1}H COSY, HMQC, HMBC spectra revealed that compounds (14)-(18) actually possessed identical 5α-cholest-7-ene-3β,5α,6α,19-tetraol-9α,11α-epoxy steroid framework and varied only in the side chains. Therefore, it appeared that we had only to establish the substitution pattern and stereochemistry of the most abundant steroid (14) to gain insight into the whole series.

Dysidesterols A (14) was obtained as colorless crystalline solid: mp 229-230°C. [α]^{20}$_D$ +60 (c 0.13 CHCl$_3$). The ^{1}H NMR spectrum of (14) exhibited signals for four of the five methyl groups typical of a cholesterol: δ 0.56 (s, 3H, Me-18), 0.816 (d, J = 6.5 Hz, 3H, Me-26), 0.821 (d, J = 6.5 Hz, 3H, Me-27), 0.85 (d, J = 6.1 Hz, 3H, Me-21). Although the methyl singlet typical for the Me-19 protons was absent, two doublets centered at δ 3.76 (d, J = 10.8 Hz, 1H, H-19a) and 3.95 (d, J = 10.8 Hz, 1H, H-19b) indicated an isolated methylene bearing a hydroxyl group in the place of C-19 methyl. The presence of an acetate group was reinforced by the ^{1}H NMR signal at δ 2.12 (s, 3H, OAc) and a ^{13}C NMR signal in the ester carbonyl region (δ 171.3).

The comparison of the ^{1}H NMR data, summarized in Table II, of the lead steroid (14) and model compound (19) [a polyhydroxylated steroid isolated from an unidentified sponge *Dysidea* sp (28).] showed a strong similarity between these two compounds. The only substantial difference being the coupling constants between H-6 and H-7 (the former $J_{6,7}$ = 1.9 Hz and the latter $J_{6,7}$ = 3.5 Hz) implying that (**14**) possessed the same 5α-cholest-7-ene-9α,11α-epoxy-3β,5α,19-triol steroid nucleus as (**19**) with a cholesterol-type side chain. Furthermore, the observable coupling constants ($J_{3, 4ax}$ = $J_{3, 2ax}$ = 12.7 Hz, $J_{3, 4eq}$ = $J_{3, 2eq}$ = 3.4 Hz, $J_{6, 7}$ = 1.9 Hz, $J_{11, 12eq}$ = 5.5 Hz in C_5D_5N) are also in full agreement with the 5α-cholest-7-ene-3β,5α,6α,19-tetraol-9α,11α-epoxy stereochemistry. Further support of the suggested configuration was achieved from NOE enhancements between: H-19a/H-6ax; H-6ax/H-4ax; Me-18/H-11ax, -12eq (β) and H-14ax/H-12ax, -17eq (α) as shown in Figure 11.

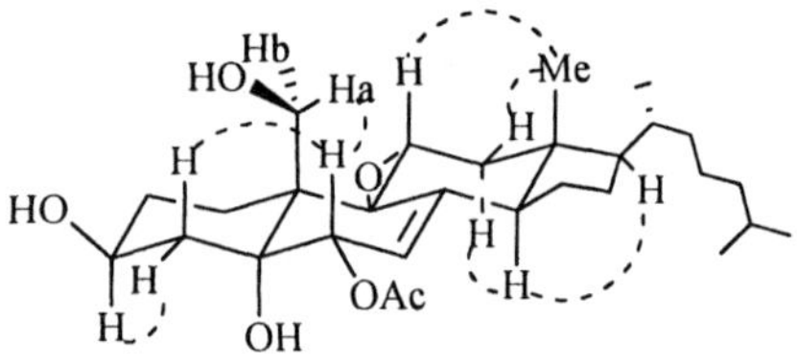

Figure 11. Selected NOESY corrections of compound (14).

The NOE between H-19a and H-6, together with the small coupling constant between H-6 and H-7 (1.9 Hz, agreeable with a dihedral angle $\Phi_{6\beta,7}$ ≈ 85°), determined the configuration of C-6, which was not self-evident because of the twisted conformation of ring B.

To verify our structural assignments for compound (**14**) were correct, we subjected (**14**) to single crystal *X*-ray diffraction analysis.

Crystal data:

$C_{29}H_{46}O_6$ FW = 490.66; monoclinic, $P2$ (1); a = 11.4638 (12), b = 6.2666 (7), c = 19.6140 (2) Å; V = 1402.5 (3) Å^3; Z = 2; D_{calcd} = 1.162 g cm^{-3}; λ = 0.71073 Å.

A total of 8637 reflections were collected in the range 1.78° < δ < 28.34° by using the φ/ω scan technique at 293(2) K. The structure was solved by direct method and refined by a full-matrix least-squares procedure. Sadabs corrections were applied to the data. All non-hydrogen atoms were given anisotropic thermal parameters. The refinement converged to the final R = 0.0628, Rw = 0.1461 for 6061 observed reflections with I > 2δ (I) and 536 variable parameters. The *X*-ray experiment did not define the absolute configuration. A computer generated perspective drawing of the final *X*-ray model is given in Figure 12.

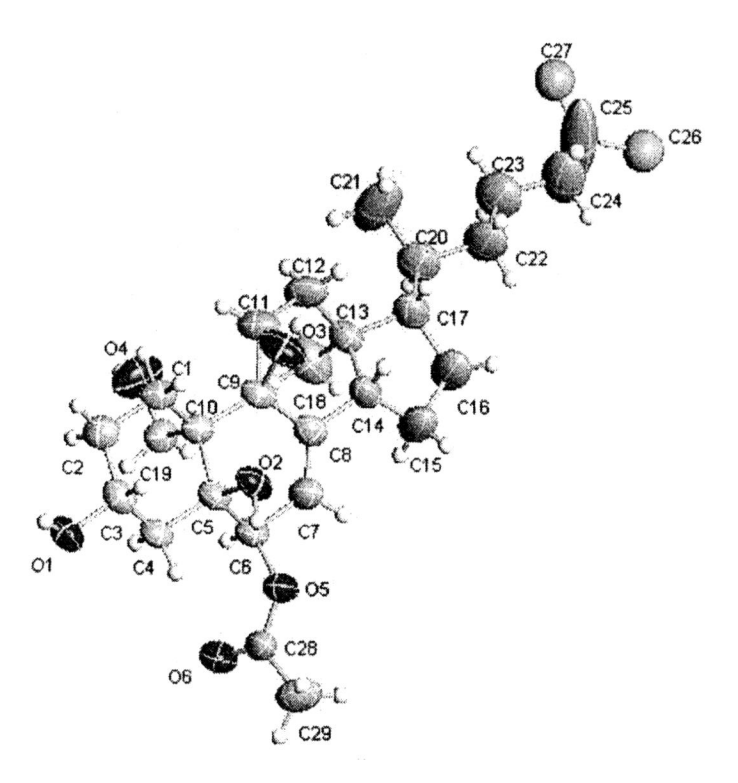

Figure 12. A computer generated perspective drawing of compound (14) showing the relative configuration.

On the basis of the knowledge of the relative relationships of the chiral centers around the steroid system, advanced stereochemical studies were performed. Compound (**14**) was esterified separately with (*R*)- and (*S*)- MTPA chloride in dry CH_2Cl_2 with pyridine at room temperature and the products purified by silica gel column chromatography to give two corresponding MTPA esters. Assignment of the ^{1}H NMR signals of the esters (**14R** = **14***R*-ester, **14S** = **14***S*-ester) was achieved by analyzing 2D-COSY spectra. The $\Delta\delta$ (δ_S-δ_R) values of the protons near the oxygenated carbon (C-3) are summarized in Figure 13. The negative values were recorded for the protons of H-2α, H-19a, H-19b, H-11β, H-12α, H-12β, Me-18; and the positive shifts were observed for the protons of H-4α, H-6β, H-7, H-14α and OAc. Interpretation of these data, according to the MTPA determination rule, allowed us to assign the absolute stereochemistry of the secondary alcohol at C-3 as *S*, which, in turn, led us to complete the absolute stereochemistry of (**14**) as (3S,5S,6S,9S,10R,11R,13S, 14R,17R,20R)-9,11-epoxycholest-7-ene-3,5,6,19-tetrol-6-acetate.

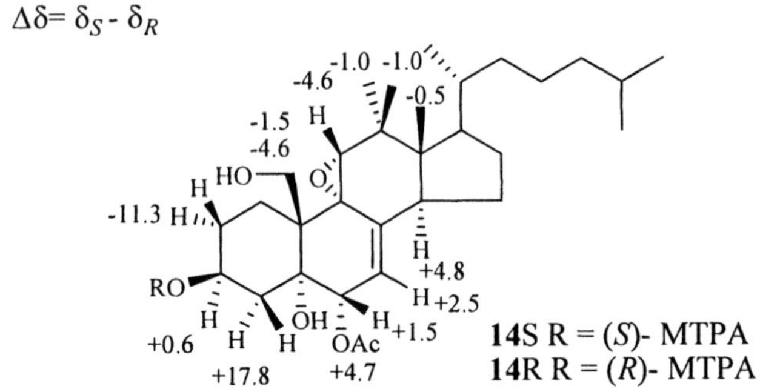

$\Delta\delta = \delta_S - \delta_R$

14S R = (S)- MTPA
14R R = (R)- MTPA

Figure 13. δ values (Hz) obtained for the protons near the C-3 OMTPA ester center.

It is clear from our work, and disclosures by the Faulkner (*30*) and Schmitz (*28*) groups, that highly functionalized steroids are widespread in the sponge genus *Dysidea*, but what role they play in the sponge remains undertermined. Their high level of functionality makes it unlikely that they contribute to the unique cell membrane structure in the sponge; similar steroids do not appear common in other members of the Demospongiae. The similarities between these compounds and the crustecdysones could imply a function of feeding deterrence to potential crustacean predators. An extension of this line of thought leads to consideration that they could be kairomones, which induce settling and metamorphosis in larvae of the dorid nudibranches, known to be associated with genus *Dysidea*.

Compound (**14**) was found to be slightly cytotoxic against A-549, HL-60 and P-388 cell lines.

References

1. Pandeya, S.N. *J. Sci. Ind. Res.* **1972**, *31*, 320-321.
2. Sun, Y.-Q.; Guo, Y.-W. *Tetrahedron Lett.* **2004**, *45*, 5533-5535.
3. Huang, X.-C.; Zhao, D.; Guo, Y.-W.; Wu, H.-M.; Enrico, T.; Guido, C. *Tetrahedron Lett.* **2004**, *45*, 5501-5504.
4. Huang, X.-C.; Zhao, D.; Guo, Y.-W.; Wu, H.-M.; Wang, Z-H.; Lin, Y.-S. *Bioorg&Med. Chem. Lett.*, **2004**, *14*, 3117-3120.

5. Zhang, W.; Guo, Y.-W.; Mollo, E.; Guido, C. *Helvetica Chimica Acta*, **2005**, *88*, 87-94.
6. Huang, X.-C.; Guo, Y.-W.; Mollo, E.; Cimino, G. *Helvetica Chimica Acta*, **2005**, *88*, 281-289.
7. Morite, K.; Kobayashi, S. *Tetrahedron Lett.* **1966**, *23*, 537-538.
8. Morite, K.; Kobayashi, S. *Chem. Pharm. Bull.* **1976**, *15*, 988-989.
9. Wratten, S.J.; Faulkner, D.J. *J. Org. Chem.* **1976**, *41*, 2465-2467.
10. Loder, J.W.; Russell, G.B. *Tetrahedron Lett.* **1966**, *51*, 6327-6329.
11. Loder, J.W.; Russell, G.B. *Aust. J. Chem.* **1969**, *22*, 1271-1275.
12. Kato, A.; Numata, M. *Tetrahedron Lett.* **1972**, *3*, 203-206.
13. Kato A.; Okutani, T. *Tetrahedron Lett.* **1972**, *29*, 2959-2960.
14. Cragg, R.H.; Weston, A.F. *Tetrahedron Lett.* **1973**, *9*, 655-656.
15. Kato, A. *Phytochemistry* **1975**, *14*, 1458.
16. Takahashi, A.J. *Phytochemistry* **1976**, *15*, 220-221.
17. Wolf, Jr. R.E.; Hartman, J.R.; Storey, J.M.E. *J. Am. Chem. Soc.* **1987**, *109*, 4328-4335.
18. Shimizu, Y. *Chem. Rev.* **1993**, *93*, 1685-1698.
19. Paul, G.K.; Matsumori, N.; Murata, M.; Tachibana, K. *Tetrahedron Lett.* **1995**, 36, 6279-6282.
20. Kazlauskas, R.; Murphy, P.Y.; Quinn, R.J.; Wells, R.J.; Schonholzer, P. *Tetrahedron Lett.* **1977**, 4439-4442.
21. Higa, T.; Tanaka, J.; Tachibana, K. *Tetrahedron Lett.* **1981**, 22, 2777-2780.
22. Rao, C.B.; Ramana, K.V.; Rao, D.V.; Fahy, E.; Faulkner, D.J. *J. Nat. Prod.* **1988**, 51, 954-958.
23. Anjaneyulu, A.S.R.; Murthy, M.V.R.K.; Rao, N.S.K. *J. Chem. Research (S)* **1997**, 450-451.
24. Tanaka, J.; Trianto, A.; Musman, M.; Issa, H.H.; Ohtani, I. I.; Ichiba, T.; Higa, T.; Yoshida, W.Y.; Scheuer, P. J. *Tetrahedron* **2002**, 58, 6259-6266.
25. Subrahmanyam, C.; Kuamr, S.R. *J. Chem. Research (S)* **2000**, 182-183.
26. Anjaneyulu, A.S.R.; Rao, V.L.; Sastry, V.G. *Nat. Prod. Res.* **2003**, 17, 149-152.
27. D'Auria, V.M.; Minale, L.; Riccio, R. *Chem. Rev.* **1993**, *93*, 1839-1895.
28. Gunasekera, S.P.; Schmitz, F.J. *J. Org. Chem.* **1983**, *48*, 885-886.
29. Gorde, S.H.; Cardellina II, J.H. *J. Nat. Prod.* **1984**, *47*, 76-83.
30. Capon, R.J.; Faulkner, D.J. *J. Org. Chem.* **1985**, *50*, 4771-4773.
31. West, R.R.; Cardellina, J.H. *J. Org. Chem.* **1988**, *53*, 2782-2787.
32. West, R.R.; Cardellina, J.H. *J. Org. Chem.* **1989**, *54*, 3234-3236.
33. Isaacs, S.; Berman, R.; Kashman, Y. *J. Nat. Prod.* **1991**, *54*, 83-91.
34. Milkova, T.S.; Mikhova, B.P.; Nikolov, N.M.; Popov, S.S. *J. Nat. Prod.* **1992**, *55*, 974-978.
35. Casapullo, A.; Minale, L.; Zollo, F. *Tetrahedron Lett.* **1995**, 36, 2669-2672.

Chapter 7

Curcumin: Potential Health Benefits, Molecular Mechanism of Action, and Its Anticancer Properties In Vitro and In Vivo

Resmi Ann Matchanickal and Mohamed M. Rafi

Department of Food Science, The New Jersey Agricultural Experiment Station, Rutgers, The State University of New Jersey, New Brunswick, NJ 08901

In recent years the food industry has seen an exponential rise in the interest of companies and consumers alike towards the development of foods that are not only nutritious and pleasing to the senses, but also provide the body with important health benefits. In 1989 after much debate, Dr Stephen DeFelice defined nutraceuticals as any substance classified as a food or part of a food with medical or health benefits that range from the prevention to the treatment and/or cure of a disease. Soon after a new discipline was born: nutragenomics, dedicated to the study of how natural molecules present in foods alter the molecular expression of genes in each individual. The nutraceutical curcumin is a polyphenolic yellow pigment found in turmeric, the ground form of the *Curcuma longa* species. The interest that the medical world has shown towards this molecule stems from the fact that several studies have shown promising results in the use of this nutraceutical as a potential agent to prevent and fight cancer. Researchers are not yet clear on the exact mechanism of action, but they believe that curcumin's anticancer potential has to do with its ability to suppress the proliferation of cells. Curcumin has been shown to down-regulate the activity of two major transcriptor factors NF-κB and AP-1, scavenge reactive oxygen species (ROS), suppress mitogen-activated protein kinases (MAPKs) generated by inflammatory stimuli, suppress the expression of pro-inflammatory enzymes cyclooxygenases (COX-2) and lipoxygenases (LOX), induce apoptosis in the cancer cells by up regulating p53 protein and induce Phase II detoxification enzymes. In addition to its anticancer potential, curcumin has also been found to be a good antioxidant, anti-inflammatory and anti-mutagenic agent as a result of the different moieties in its chemical structure. So far turmeric has been found to be safe at doses up to 2200 mg daily that is equivalent to 180 mg of curcumin. However, its low bioavailability due to extensive intestinal metabolism has put into question its efficacy in the body.

Cancer Chemprevention and Diet

Cancer continues to be one of the most difficult and complicated diseases to tackle despite the major advances in medical technology for its diagnosis and treatment. As a result, the medical community is looking to find new ways to decrease the burden of the disease. Nowadays, the focus has shifted from treatment to prevention, in light of some promising research that has shown that of all the environmental factors, dietary components appear to play a major role in the initiation/progression of the disease. One of the countries that has inspired researchers to approach the cancer problem through a dietary perspective is India, a rapidly developing country characterized for having one of the most diverse populations and diets in the world.

Spices have been used for many years to add exotic flavors, aromas and colors to foods. In Asian cultures (eg. India and China), spices also play a special role as "healing tools" for a variety of diseases (*1*). It was not until recently that researchers discovered that spices contain as many beneficial phytochemicals as fruits and vegetables. Thus, there has been a recent interest by the scientific world to study and understand how these phytochemicals transform themselves into "natural healers" when they are ingested in adequate quantities. Scientists have discovered that common herbs/spices such as parsley, ginger, turmeric, fenugreek, clove, cardamon, cinnamon, mint, garlic, mustard and chili peppers have various hidden therapeutical properties such as: production of enzymes that detoxify carcinogens, inhibition of cholesterol synthesis, estrogen blockers, lowering of blood pressure etc. In many cultures, spices are viewed as capable of boosting energy, relieving stress, protecting the nervous system, aiding in digestion and relieving symptoms of the common cold (*1*). Turmeric is one of the spices that has caught most of the attention because its main component, curcumin, has shown significant potential as an anti-cancer agent, anti-inflammatory, anti-mutagenic agent in both *in vitro* and *in vivo* studies (*2-3*). However, the efficacy and bioavailability of curcumin in the human body is under investigation, due to its extensive intestinal metabolism (*4*).

One of thee major hurdles in the earlier stages of the nutraceutical movement was establishing a complete and descriptive definition for the term nutraceutical. Finally in 1989, Dr Stephen DeFelice put to rest the feud by defining nutraceuticals as "any substance considered a food, or part of a food, with medical or health benefits, including the prevention, treatment or cure of disease"(*5*). There are over 35 plant based nutraceuticals each with a distinctive bioactive molecule. For example: garlic contains diallyl sulfide, soy contains genestein, grapes contain resveratrol and turmeric contains curcumin (*6*).

The completion of the Human Genome Project in 2003 (*7*) was an important milestone for researchers around the world because it confirmed the existence of genetic differences amongst individuals in response to external factors such as diet. Thus, the science of nutragenomics was born to study how nutraceuticals such as curcumin affect the molecular expression of genetic information in each individual. It is a discipline that combines nutrition, molecular biology and genomics (*8*). The key difference between nutrition and nutrigenomics is that the former treats individuals as genetically identical while the latter takes into consideration genetic differences also known as single nucleotide polymorphisms or SNP's (*8*).

Curcumin is a natural yellow, orange dye derived from the rhizome of *Curcuma longa*. It is not water soluble, but it does solubilize in ethanol and other organic solvents (*9*). There is one study that reports having increased the water solubility of curcumin by a factor of 10^4 as well as improving its photochemical stability by complexing curcumin with a cyclodextrin (*10*). Curcumin's melting point is 183° C and its molecular weight is 368.37. There are three main molecules present in curcumin, which are collectively known as curcuminoids (Figure 1). Curcumin has a brilliant yellow hue at pH ranges of 2.5-7.0 and becomes reddish at pH >7 (*2*).

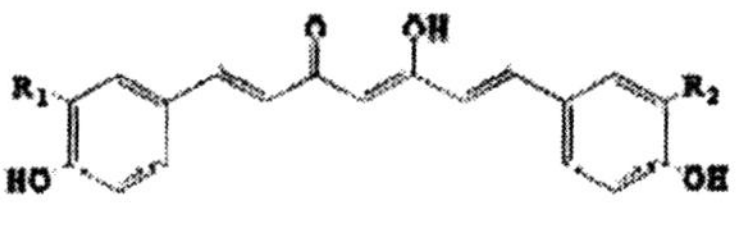

Curcumin   : $R_1=R_2=OCH_3$
Demethoxycurcumin : $R_1=OCH_3$, $R_2=H$
Bisdemethoxycurcumin: $R_1=R_2=H$

Figure 1. Structures of curcuminoids)

Chemoprevention encourages the use of special diets and chemicals of natural or synthetic origin to reverse, suppress or prevent cancer from progressing and becoming invasive (*11*). Chemopreventative agents can be divided into two subgroups: blocking agents that interfere with initiation and tumor suppressor agents that affect promotion/progression stage of carcinogenesis. Most nutraceuticals (including curcumin) are considered to be tumor suppressor agents because they cannot prevent initiation from taking place but they can reverse the promotion and progression stages. A disease as complicated as cancer needs a good chemotherapeutic agent that has a variety of mechanistically distinct but complementary properties that will aid in regulating cell growth, differentiation and survival (*11*). Chemoprevention of cancer through the consumption of edible phytochemicals is the cheapest, most readily applicable and accessible approach to cancer control known so far. With healthcare costs being so high, chemoprevention is a cost effective approach to promote the awareness and encourage the consumption of food containing phytochemicals as a cancer preventative strategy for the general public (*11*).

Phytochemicals are defined as non-nutritive components in plants that have significant anti-carcinogenic and anti-mutagenic properties (*6*). Plants not only provide us with essential vitamins and minerals, but they are also responsible for 80,000-100,000 of the secondary metabolites on the planet. Unfortunately for us, some of the best phytochemicals (eg. curcumin) are unique to certain plant species (turmeric), which are often either absent or present at very low levels in the typical American diet (*12*). Thus, due to poor dietary choices the average consumer is ignoring many of these beneficial phytochemicals. Although it helps to classify phytochemicals as blocking or suppressing agents, we have to acknowledge that the ability of a single phytochemical to prevent tumor development is a combination of several sets of intracellular effects rather than a single biological response. The key is to first figure out how nutraceuticals function in plants in order to further understand their possible mechanism within the human body. Ironically, we use spices as food additives to provide added flavor, while some plants produce these phytochemicals as a self defense mechanism in order to discourage consumption by their predators (*13*).

Mechanisms of Action of Curumin *In Vitro*

Suppression of NF-κB and AP-1

NF-κB and AP-1 are ubiquitous eukaryotic transcription factors that act as mediators between external and internal stimuli in the cellular signaling cascades and are therefore prime targets of a diverse class of chemopreventative phytochemicals (*6*). NF-κB is a transcription factor that is inducible and ubiquitously expressed in genes involved in cell survival, cell adhesion, inflammation, differentiation and growth (*14*). It also plays a fundamental role in the immune system where it controls the expression of cytokines and the major histocompatibility complex genes (2).

In resting cells, NF-κB can be found in the cytoplasm bound to the inhibitory IkB proteins, which prevent it from relocating to the nucleus. NF-κB is activated by various carcinogens and tumor promoters such as benzopyrene, UV radiation and phorbol esters. Once activated, the NF-κB is released into the nucleus as a result of the phosphorylation, ubiquitination and degradation of IkB. Activation of NF-κB is tightly regulated by IkB which complexes and sequesters NF-κB in the cytoplasm. There are many kinases involved in the phosphorylation of IkB, which are all linked to cytokine specific receptors and proteins and eventually converge on NF-κB inducing kinase (NIK). Once this kinase gets activated, it then phosphorylates and activates the IkB kinase complex (IKK). The activation of the IKK complex leads to IkB phosphorylation and degradation followed by the release of NF-κB. A study using intestinal epithelial cells (IEC), found that curcumin blocked NF-κB activation by blocking the signal leading to IKK activity (15). The activation of NF-κB promotes cell survival, due to the expression of several genes such as bcl-2, bcl-xl, cyclin D1, COX-2 and others, which are regulated by NF-κB and are involved in the survival of cancerous cells. Curcumin suppresses many inflammatory genes that are regulated by NF-κB such as TNF, COX-2 and NOS (2).

Activating protein-1 (AP-1) activity is induced by a wide range of stimuli and environmental insults such as growth factors, cytokines, neurotransmitters, polypeptide hormones, bacterial and viral infections and other physical and chemical stresses. The ability for this transcription factor to control an eclectic collection of biological processes is based on its structural and regulatory complexity (16). AP-1 is a complex between dimers of Jun proto-oncogene family with the Fos proto-oncogene family. The Fos proteins form stable heterodimers with Jun proteins thereby enhancing the DNA binding capacity. We have yet to find a clear explanation on how curcumin inhibits AP-1, but there are three potential mechanisms which are:

1) Alteration of the redox status of cells,

2) Inhibition of Jun-N-terminal kinase (JNK) which is needed for AP-1 activation

3) Inhibition of the fos-jun-DNA complex.

Thus down–regulation of AP-1 by curcumin may explain its ability to suppress chemical carcinogenesis (2).

p53, BAX, BCL-2, P21, Apotosis, Cell Cycle

Typically cells die either by necrosis or by programmed cell death, also known as apoptosis. The difference between both is that necrosis is characterized by the loss of plasma membrane integrity without damage to the nucleus, while apoptosis is an organized degradation of the cellular components (17). Curcumin induces apoptosis in cancer cells taken from the colon, liver, and breast.(18).

One of the major properties of curcumin is the inhibition of apoptosis, which takes place through either a mitochondria-dependent or mitochondria-independent pathway depending on the cell type involved. Curcumin activates caspase 8, which leads to cleavage of BID, loss of mitochondrial potential, opening of transition pores, release of cytochrome C, activation of caspase 9 and 3, activation and cleavage of PARP and finally DNA fragmentation and apoptosis. Curcumin also exerts apoptosis by down-regulating anti-apoptotic proteins such as bcl-2 and bcl-xL. A study found that curcumin induced Bax through p53, which caused the release of cytochrome c from the mitochondria, thus leading to apoptosis (19). The tumor suppressor gene p53 is mutated, deleted or rearranged in more than 50% of all human tumors. This gene is also known as "the guardian of the genome" because it triggers apoptosis or cell cycle arrest in response to DNA damage. In normal cells p53 is in charge of preventing the replication of damaged DNA, reducing genetic instability and allowing the cells to perform critical repair functions before going through the cell cycle (19). Bax is a pro-apoptotic protein (member of the Bcl-2 family) that is up-regulated in a number of cells during p53-mediated apoptosis. It has been well established that the fate of cells depends upon the balance between proteins that mediate cell death, such as Bax (pro-apoptotic) and Bcl-2/Bcl-xl (anti-apoptotic). The ratio of proapoptotic vs. antiapoptotic proteins is critical in defining the cell's threshold for undergoing apoptosis.

Curcumin is known to suppress the proliferation of a wide variety of tumor cells from breast, colon, renal and prostate induced by growth factors such as IL-2, PDGF and PHA. This attribute is due in part to curcumin's effect on the cell cycle (2). Treatment of cells with curcumin arrests the cell cycle at the G_2-M stage in several cell types, which consequently causes a reduction in the expression of genes such as p53, p21, Bcl-xl and up-regulation of Bax (20). The possible molecular targets and mechanism of action of curcumin is shown in Figure 2.

Molecular targets of Curcumin

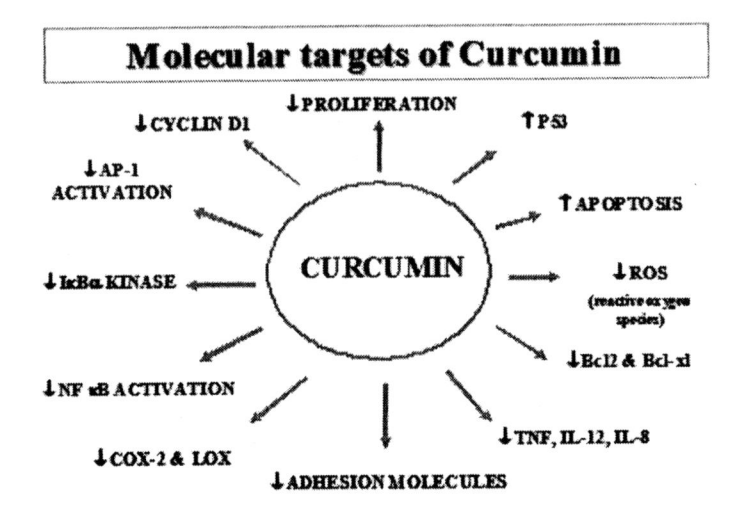

- ↓ CYCLIN D1
- ↓ PROLIFERATION
- ↑ P53
- ↓ AP-1 ACTIVATION
- ↑ APOPTOSIS
- ↓ IκBα KINASE
- ↓ ROS (reactive oxygen species)
- ↓ Bcl2 & Bcl-xl
- ↓ NF-κB ACTIVATION
- ↓ TNF, IL-12, IL-8
- ↓ COX-2 & LOX
- ↓ ADHESION MOLECULES

(CURCUMIN)

Figure 2. Molecular targets of curcumin

Suppression of Mitogen Activated Protein Kinase (MAPK)

Many molecular alterations that occur during carcinogenesis affect cell-signaling pathways that regulate cell proliferation and differentiation. One of the central components of this signaling network is the mitogen-activated protein kinases (MAPK's). MAPK cascades are present in all eukaryotes and are activated by a diverse group of extracellular signals, which are then transduced into the nuclei. They are also avid participants in the regulation of various cellular programs such as cell differentiation, cell division, cell movement and cell death. A typical MAPK cascade is made up of three kinases mitogen acitivated protein kinase kinase kinase (MAPKKK), mitogen acitivated protein kinase kinase (MAPKK) and mitogen acitivated protein kinase MAPK (21). These kinases are activated through phosphorylation mechanisms. Many of these pathways converge and activate NF-κB as well as AP-1, which are the prime targets of many chemopreventative phytochemicals (6). Any signaling that involves MAPK's, AP-1 and NF-κB will favor cell proliferation and survival (20).

Abnormalities or improper activation or silencing of the MAPK pathway or its downstream transcription factors could lead to uncontrolled cell growth and thus malignant cell growth. Some phytochemicals such as curcumin can switch on or turn off specific signaling molecules which prevent abnormal cell proliferation. Apart from MAPK's, there are other cell signaling kinases such as protein kinase C (PKC) and phosphatidylinositol 3-kinase (PI3K) that are also important targets of many chemopreventative phytochemicals (6). In a study curcumin suppressed PMA induced activation of cellular PKC by 26 % to 60% (22).

Most inflammatory stimuli activates three independent mitogen activated protein kinase (MAPK) pathways which lead to the activation of: 1) epidermal growth factor receptor/extracellular regulated kinase (EGFR/ERK) which activates transcription factors such as elk-1 and c-fos, 2) stress activated pathways (JNK/SAPK) and 3) p38 MAPK pathway which activates c-jun (20). Curcumin was found to inhibit the JNK pathway in an indirect manner, by interfering with the signaling molecule at the same level or close to the MAPKKK level (23). The inhibition of this important pathway, also explains curcumin's ability to inactivate AP-1 and NF-κB conferring it potent anti-inflammatory and anticarcinogenic properties (2).

Suppression of COX-2

COX-2 (cyclooxygenase) and LOX (lipoxygenase) are enzymes that promote inflammation by producing prostaglandins and leukotrienes respectively. Several studies have demonstrated curcumin's ability to suppress protein expression and enzymatic activity of both of these enzymes (2).

COX-2 is an important modulator protein that helps catalyze the conversion of arachidonic acid to prostaglandins. This enzyme has been the focus of many researchers because it has been found to be over-expressed in various types of cancer. COX-2 is over expressed in 80-90% of all colorectal cancers (24). In the late 1980's the common belief was that the formation of PG's was limited by the amount of arachidonic acid available, which is the substrate for cyclooxygenase. But in 1988, a scientist known as Needleman suggested that the amount of COX protein in inflamed tissues was much higher than in normal tissues. This type of COX was induced by external factors such as LPS and IL-1. He also concluded that there is a non-inducible form of cyclooxygenase, also known as COX-1, or a "house keeping enzyme" responsible for basic levels of PG's. This form is expressed in all types of tissue cells like colon, kidney, spleen, stomach, liver, lung, heart and brain. The inducible form, COX-2, is activated by different stimuli that mediate inflammatory reactions such as pro-inflammatory cytokines, LPS, mitogens and oncogenes, growth factors etc. This isoform can be induced and constitutively expressed and remains undetected unless induced by inflammatory stimuli. Both of these isoenzymes are membrane bound in the endoplasmatic reticulum (25). One suggested mechanism by which COX-2 may lead to cancer is over expression of COX-2 which inhibits apoptosis and leads to tumorigenesis (26).

Scientists believe that non-steroidal anti-inflammatory drugs (NSAID) inhibit colon carcinogenesis by suppressing arachidonic acid metabolism via COX enzymes, which in turn affects cell proliferation, tumor growth and immune responsiveness by modulating eicosanoid production. But frequent use of NSAID's comes with a high price that leads to gastrointestinal ulceration, bleeding and renal toxicity due to the fact that NSAID's lack selectivity for inhibition of COX-2. This is a common problem with most therapeutic agents in use today; they do not have the capability to distinguish between malignant and normal cells. Curcumin, on the other hand, can selectively inhibit COX-2 mRNA and protein expression without disturbing COX-1 expression. Most of curcumin's biological activity is achieved at concentrations ranging in the 10-100 µM (27). One major difference between curcumin and NSAID's lies in the fact that curcumin inhibits both the cyclooxygenase and peroxidase activities of COX enzyme, while NSAID's only inhibit the cyclooxygenase function of the COX enzyme. Thus, curcumin has more potential as an anticancer agent than regular NSAID's.

Activation of Phase I/II Enzymes

Phase II detoxifying enzymes include: NAD(P)H, quinine oxidoreductase /DT-diaphorase (QR), glutathione S-transferases (GST's), UDP-glucuronosyl transferases and epoxide hydrolases. These enzymes are responsible for converting reactive or toxic compounds into less toxic and more readily excretable products in order to protect the cells against various chemical stresses and carcinogens (21).

A study looked upon the effect that turmeric/curcumin had on the activity of rat liver cytochrome P450 enzymes involved in the metabolism of B(a)P and NNK, two carcinogenic compounds present in tobacco. B(a)P is activated by CYP 1A1 and 1A2 isoenzymes and NNK is activated by CYP 2B1. The experiments were designed to study the effect of curcumin (C), demethoxycurcumin (dmC) and bisdemethoxycurcumin (bdmC) on CYP 1A1, 1A2 and 2B1 *in vitro* while using phenyl-isothiocyanates (PITC) and phenethyl-isothiocyanates (PEITC) as positive controls for inhibition of these isoenzymes. The experiments showed that curcumin had a higher inhibitory potency than demethoxycurcumin and bisdemethoxycurcumin, while PEITC had higher potency than phenylisotiocyanate(PITC). Overall, the curcuminoids had a higher inhibitory potency for CYP 1A1 and 1A2 as compared to both PITC and PEITC. In the case of CYP 2B1, PEITC seemed to have a higher potency than the curcuminoids or PITC (28). The isoenzymes 1A1/1A2 and 2B are closely involved in the detoxification process for PAH's and nitrosamines (carcinogens present in tobacco smoke). An agent capable of inhibiting these enzymes, will ultimately block the conversion of procarcinogens into ultimate carcinogens. The study demonstrated that curcumin can effectively inhibit the conversion of B(a)P and NNK into their ultimate carcinogenic form (*28*).

Another study found that curcumin is a potent inhibitor of rat liver cytochrome P450 enzyme subgroups 1A1/1A2, less potent inhibitor of 2B1/2B2 and a weak inhibitor of 2E1. The author suggests adding glutathione GSH or adding an antioxidant such as ascorbic acid to lower the pH and overcome curcumin's instability in neutral buffers (*30*).

Antioxidant/Prooxidant Activity

The most effective way of preventing cancer is to prevent DNA damage by cancer initiating molecules such as ROS (reactive oxygen species). ROS modifies the redox status of cells and thus the cysteine residues of proteins. This in turn affects the conformation and function of proteins as well as binding of transcription factors to DNA, all of which lead to altered gene expression. All aerobic organisms produce ROS, but they are maintained on check by the presence of low molecular weight antioxidants such as glutathione, and with enzymes such as catalase and superoxide dismutase (*20*).

Many spices and herbs contain phenolic substances that make them powerful antioxidative and chemopreventative agents, and the phenol moiety is thought to be responsible for the antioxidative properties. Any compound capable of preventing free radical formation, removing harmful radicals, repairing oxidative damage, eliminating damaged molecules or preventing mutations is a good candidate for cancer chemoprevention. Curcumin is an excellent antioxidative and antimutagenic compound because of its chemical structure. Its two alpha, beta unsaturated carbonyl groups can react with nucleophiles such as glutathione (GSH) through a Michael reaction, which inhibits lipid peroxidation and intercepts and neutralizes reactive oxygen species (*31*). Oxidation of biological molecules induces a number of harmful pathological events that include carcinogenesis and aging. These events result from the generation of reactive oxygen species (ROS). Studies *in vitro* have shown that curcumin can act as a potential scavenger of a variety of ROS, thus protecting lipids, hemoglobin and DNA against oxidative degradation (32). Cellular membranes are more vulnerable to oxidative damage due to their high content of unsaturated fatty acids. Lipid peroxidation (LPO) is initiated by ROS produced by the cells as a result of partial reduction of oxygen during metabolism. In addition to ROS, ionizing radiation and transition metals such as iron and copper can also lead to LPO. A study designed to compare the protective effect of ginger derived phenolic 1,3-diketone compounds vs.

curcumin against lipid peroxidation using different biological models found that curcumin showed excellent antioxidant activity compared to the ginger compounds. This is probably due to the additional phenolic hydroxyl group, which lend it better iron chelating and radical scavenging properties. The ginger compounds possessed a partially methylated cathecol moiety, a 1,3- diketone group and extended conjugation similar to that of curcumin. But unlike curcumin they only contained one phenolic hydroxyl group (*32*).

An interesting and important finding, which contradicts curcumin's antioxidant capabilities, is that curcumin causes oxidative stress, which leads to the release of cytochrome c and the activation of caspases. Researchers showed that the presence of an antioxidant known as N-acetylcysteine (NAC) prevented the apoptosis induced by curcumin. Thus, perhaps curcumin can play the dual role of acting as an antioxidant and in some cases as a pro-oxidant (*27*).

Metabolism, Pharmacokinetics, Tissue Distribution and Excretion of Curcumin (Animals/Humans)

An extensive study was conducted to re-evaluate the pharmacokinetic properties of curcumin through the use of a more sensitive HPLC method, as well as examine the degradation kinetics of curcumin under various pH conditions and its stability in a physiological matrix (*29*). The study found that curcumin is very sensitive to light. Approximately 5% of its absorbance is lost during the time for typical sample preparation when using clear glassware rather than amber containers. In addition, more than 90% of the curcumin decomposes in buffers at neutral conditions. It is believed that curcumin's stability in acidic media is due to its conjugated diene system. When the pH approaches neutral conditions, a proton is removed from its phenolic group causing the disruption of the conjugated diene system. It is not yet clear why curcumin degrades rapidly at neutral pH, but it appears to occur through an oxidative mechanism because the presence of an antioxidant such as ascorbic acid or NAC completely blocks the degradation of curcumin at a pH of 7.4. In the stomach and intestine, curcumin should be relatively stable or degrade very slowly since the pH in this environment remains between 1 and 6 (*33*).

Trans-6-(4-Hydroxy-3-methoxyphenyl)-2,4-dioxo-5-hexenal is the major degradation product of curcumin, while vanillin, ferulic acid and feruloyl methane are minor degradation products of short-term reactions. Vanillin is a naturally occurring flavoring agent that shares many of the biological properties that have been conferred to curcumin such as: inhibition of mutagenesis, scavenging of ROS and inhibition of lipid peroxidation. Therefore, careful consideration must be taken during sample preparation, in order to avoid the formation of vanillin, which may skew the results of the experiment due to the similar biological properties as curcumin. It would be interesting to conduct future studies comparing the potency of vanillin vs. curcumin on these aspects (29).

In most organisms, glucuronidation is the most common conjugation pathway for converting xenobiotics into water-soluble metabolites (33). A general diagram of the conversion undergone by plant secondary metabolites such as curcumin in the body is presented in (34). Once curcumin reaches the gut, it is activated by microsomal reactions such as reductase and glucuronidation which eventually lead to the formation of curcumin glucuronide and curcumin sulfate which are then further bioreduced to tetrahydrocurcumin (THC), hexahydrocurcumin and hexahydrocurcuminol. THC is the major metabolite of curcumin and it also has anti-inflammatory and antioxidant activities comparable to those of curcumin. Interestingly enough, THC remains quite stable at neutral pH while almost 90% of the curcumin gets degraded. Curcumin becomes stable when it is placed in 10% serum rather than phosphate buffer. The biotransformation of curcumin and the stability of its major metabolite THC are very important factors for the overall biological effect of curcumin (24). The fact that some studies have reported THC to exhibit similar physiological and pharmacological properties as curcumin, leads to the possibility that perhaps THC serves as the available form of curcumin *in vivo*. However, further studies have to be conducted to test this hypothesis (33).

A study aimed at corroborating earlier findings about curcumin's biotransformation in the intestinal tract of humans and rodents, revealed that curcumin is metabolized to curcumin glucuronide, curcumin sulfate, THC and hexahydrocurcumin in the intestine and liver of both humans and rats. There was however important quantitative differences in metabolite generation between the human and rat tissue and between gut and liver. In general, sulfation of curcumin was four times higher in human than in rat intestine, whereas liver sulfation in humans was only one fifth of what was found in rat liver. Gut metabolism was found to be the major contributor to the overall yield of curcumin metabolites *in vivo*. The enzymes SULT1A1 and 1A3 are part of the human phenol xenobiotic-metabolizing SULT family of isoenzymes, which are expressed in the gastrointestinal tract. Curcumin is a substrate of both enzymes, the latter being more efficient at sulfating curcumin (35). UDP-glucuronosyltransferase activity is higher in the liver than in the intestinal mucosa. Therefore, orally administered curcumin is conjugated to the glucuronide form in the intestine and enters the portal vein to be further conjugated with sulfate and form glucuronide/sulfate conjugates in the liver (36). Variations were also observed amongst human cells in the amount of curcumin glucuronide formed in the intestine probably due to differences in the individual's UDP-glucuronosyl-transferase enzyme, which depends on the donor's genotype, disease state and lifestyle (35). Although the study was not designed to specifically compare curcumin metabolism between humans and rats, it did serve to examine the suitability of the rat model to study curcumin metabolism in humans. Eventhough the pattern of metabolites was qualitative similar, there were considerable quantitative differences. The author concluded that in quantitative terms, the rat model might severely underestimate the extent of intestinal metabolism of curcumin that takes place in humans (*35*).

All of these findings lead us to conclude that extensive intestinal sulfation, glucuronidation and reduction in humans coupled with extremely low aqueous solubility may be the reasons for curcumin's poor bioavailability in the body. Because of the highly lipophilic character of curcumin, one would assume that the body fat would contain high amounts of curcumin bound to it. But because of its poor absorption in the intestine and its avid metabolism in the liver as well as rapid elimination in the bile, it makes it highly unlikely that high concentrations of this molecule would be found in the body long after ingestion (*33*).Thus curcumin's biotransformations in the intestine constitutes a pharmacological deactivation process because the metabolites that are generated are either devoid of biological activity or are less potent than their progenitor in terms of their anti-cancer properties (*35*).

Dietary constituents with poor systemic bioavailability, such as curcumin, are currently being considered for the chemoprevention of cancers in the gastrointestinal tract (such as colon), but not in other tissues (*37*). Most of the promising effects of agents such as curcumin are achieved with concentrations above 10^5 mol/L, which are unlikely to be achieved when they are consumed as part of the normal diet of human beings. In the case of curcumin, the dietary level that has proven to be the most effective in rodent models ranges from 0.1% to 2%, which is a daily dose of 150-3000 mg/kg.

Black pepper (*Piper nigrum*) like turmeric is another well-known spice that has been used since ancient times throughout the world. One of its major components is the alkaloid piperine (1-piperoylpiperidine). This compound has been found to increase the bioavailability of drugs by inhibiting glucuronidation that takes place in the liver and the intestine. Given that one of the major drawbacks in the pharmacological properties of curcumin is its poor systemic bioavailability, it seems logical to study whether piperine could improve the bioavailability of curcumin as it does with other drugs (*38*). A study found that administration of piperine along with curcumin significantly increased the bioavailability of curcumin in both humans and rats without any adverse side-effects. Again there were some differences in the overall serum concentrations between rats and humans. In the rats, a dose of 2 g/kg of curcumin alone reached moderate serum concentrations in about 4 hours, and the addition of piperine increased the serum concentration of curcumin for a period of 1-2 hours. The overall bioavailability increased 154%. But in humans a dose of 2 g of curcumin alone showed either undetectable or very low serum concentration levels. The addition of piperine increased the concentration levels rapidly and the overall bioavailability was increased 200%. Thus, piperine enhances the serum concentration and bioavailability of curcumin probably due to increased absorption and reduced metabolism (*38*).

Conclusions

There is no doubt that curcumin has gained a spot in the medical world as a potentially effective tool in the fight against cancer. Proof is the fact that it is currently undergoing clinical trials to test its effectiveness against colon cancer. Its powerful antioxidant property, ability to alter gene expression, regulation of the cell cycle, specifically induce apoptosis in cancer cells without affecting healthy ones have certainly caught the attention of oncologists and cancer patients around the world. Finding a drug that attacks multiple pathways and is pharmacologically safe is very difficult. However, curcumin fits both criterions quite well. The inability of big pharmaceutical companies to patent this naturally derived molecule is preventing their involvement in the exploration of this molecule as a therapeutic or preventative agent (*2*).

We have to understand that foods are not completely safe, and although there are anti-carcinogenic agents present such as indoles, flavonoids and polyphenols, there are also tumor promoters, carcinogens and co-carcinogens such as nitrates, heterocyclic amines and aflatoxins just to name a few. One future challenge for the food industry will be to educate the public to make wise choices in food consumption, especially promoting the inclusion of larger amounts of plant based foods containing phytochemicals. The challenge for food technologists will be to find ways of including phytochemicals into foods taking into account issues such as: the nutritional activity of the phytochemical in the food, off flavors and color problems, solubility and secondary effects on functionality, safety, regulations and health claims (*39*).

With the recent completion of the human genome project, we can expect many scientists to begin studying and understanding how substances such as curcumin affect our body at the genetic level, which is the basis of nutragenomics research.

Acknowledgement

This work was supported in part by New Jersey Commission on Cancer Research and New Jersey Agricultural Experimentation Station project to Dr. Rafi.

References

1. Uhl, S. *Food Technol.* **2000**, *54(5)*, 61-65
2. Aggarwal, B.B.; Kumar, A.; Bharti, A.C. *Anticancer Res.* **2003**, *23*, 363-398
3. Majeed, M.; Badmaev, V.; Shivakumar, U.; Rajendran, R. 1995. Curcuminoids: Antioxidant Phytonutrients. Nutriscience Publishers, Inc.: Piscataway, NJ.
4. Sharma, R.A.; McLelland, H.R.; Hill, K.A.; Ireson, C.R.; Euden, S.A.; Manson, M.M.; Pirmohamed, M.; Marnett, L.J.; Gescher, A.J.; Steward, W.P. *Clinical Cancer Res.* **2001**, *7*, 1894-1900.
5. http://www.foodproductdesign.com/archive/1993/1293CS.html
6. Surh, Y.J. *Nature Rev.* **2003**, *3,* 768-780
7. http://www.ornl.gov/sci/techresources/Human_Genome/home.shtml
8. Fogg-Johnson, N.; Kaput, J. *Food Technology.* **2003**, *57(4)*, 60-67
9. Sinha, R.; Anderson, D.E.; McDonald, S.S.; Greenwald, P. *J. Postgraduate Med.* **2003**, *49*, 222-228.
10. Tonessen, H.; Masson, M.; Loftsson, T. *Int. J. Pharm.* **2002**, *244*, 127
11. Gesher, A.; Patorino, U.; Plummer, S.M.; Manson, M.M. *Brit. J. Clin. Pharma.* **1998**, *45*, 1-12
12. DellaPenna, D.; Grusak, M.A. *Ann. Rev. Plant Physiol. Plant Mol. Biol.* **1999**, *50*, 133-161
13. Lampe, J.W. *Amer. J. Clin. Nutr.* **2003**, *78*, 579S-583S
14. Bharti, A.C.; Aggarwal, B.B. *Biochem. Pharmacol.* **2002**, *64*, 883-888
15. Jobin, C.; Bradham, C.A.; Russo, M.P.; Juma, B.; Narula, A.S.; Brenner, D.A.; Sartor, R.B. *J. Inmmunol.* **1999**, *163*, 3474-3483
16. Shaulian, E.; Karin, M. *Nature Cell Biol.* **2002**, *4*, E131-E136
17. Aggarwal, B.B. *Biochem. Pharmacol.* **2000**, *60*, 1033-1039
18. Rashmi, R.; Santosh, T.R.; Karunagaran, D. *FEBS Lett.* **2003**, *538*, 19-24
19. Choudhuri, T.; Pal, S.; Aggarwal, M.; Das, T. *FEBS Lett.* **2002**, *512*, 334-340
20. Manson, M.M. *Trends Mol. Med.* **2003**, *9(1)*, 11-18
21. Yu, R.; Lei, W.; Mandlekar, S.; Weber, M.J.; Channing, J.D.; Wu, J.; Kong, A.N. *J. Biol. Chem.* **1999**, *39(24)*, 27545-2755
22. Liu, J.Y.; Lin, S.J.; Lin, J.K. *Carcinogenesis* **1993**, *14*, 857-861
23. Chen, Y.R.; Zhou, G.; Tan, T.H. *Mol. Pharmacol.* **1999**, *56*, 1271-1279
24. Chauhan, D.P. *Current Pharmaceutical Design* **2002**, *8*, 1695-1706
25. Dannhardt, G.; Kiefer,W. *Eur.J.Med.Chem.* **2001**, *36*, 109-126
26. Levi, M.S.; Borne, R.F.; Williamson, J.S. *Current Med. Chem.* **2001**, *8*, 1349-1362

27. Woo, J.H .; Kim, Y.H.; Choi, Y.J.; Kim, D.G.; Lee, K.S.; Bae, J.H.; Min, D.S.; Chang, J.S.; Jeong, Y.J.; Lee, Y.H.; Park, J.W.; Kwon, T.K. *Carcinogenesis* **2003**, *24*, 1199-1208
28. Thapliyal, R.; Maru, G.B. *Food Chem. Toxicol.* **2001**, *39*, 541-547
29. Wang, Y.J.; Pan, M.; Cheng, A.L.; Lin, L.I.; Ho, Y.S.; Hsieh, C.Y.; Lin, J.K. *J. Pharma. Biomed. Anal.* **1997**, *15*, 1867-1876
30. Oetari, S.; Sudibyo, M.; Commandeur, J.N.; Samhoedi, R.; Vermeulen, N.P. *Biochem. Pharmacol.* **1996**, *51*, 39-45
31. Calabrese, V.; Scapagnini, C.; Colombrita, C.; Ravagna, A.; Pennisi, G.; Stella, A.M.; Galli, F.; Butterfield, D.A. *Amino Acids* **2003**, *25*, 437-444
32. Patro, B.S.; Rele, S.; Chintalwar, G.J.; Chattopadhyay, S.; Adhikari, S.; Mukherjee, T. *ChemBiochem* **2003**, *3*, 364-370
33. Pan, M.H.; Huang, T.M.; Lin, J.K. *Drug Metabolism and Disposition* **1999**, *27(1)*, 486-494
34. Singh, B.; Bhat, T.J.; Singh, B. *J. Agric. Food Chem.* **2003**, *51*, 5579-5597
35. Ireson, CR.; Jones, D.J.; Orr, S.; Coughtrie, M.W.; Boocock, D.J.; Williams, M.L.; Farmer, P.B.; Steward, W.P.; Gescher, A.J. *Cancer Epidemiology, Biomarkers and Prevention* **2002**, *11*, 105-111
36. Asai, A.; Miyazawa, T. *Life Sci.* 2000, 67, 2785-2793
37. Gescher, A.J.; Sharma, R.A.; Steward, W.P. *Lancet Oncology* **2001**, *2*, 371-379
38. Shoba, G.; Joy, D.; Joseph, T.; Majeed, M.; Rajendran, R.; Srinivas, P.S. *Planta Med.* **1998**, *64*, 353-356
39. Dillard, C.J.; German, J.B. *J. Sci. Food Agric.* **2000**, *80*, 1744-1756

Natural Product Chemistry and Analysis

Chapter 8

Need for Analytical Methods and Fingerprinting: Total Quality Control of Phytomedicine *Echinacea*

Jatinder Rana and Amitabh Chandra

Department of Analytical Services, Access Business Group, 7575 Fulton Street East, Ada, MI 49355

Increased popularity of phytoceutical products has generated interest in the benefits of products such as Echinacea as a dietary supplement. Echinacea phytomedicines and herbal preparations are widely used for the treatement of colds and influenza. Echinacea contains a variety of phytochemicals with demonstrated biological activities, which are suitable "markers" for quality assurance. The need for analytical methods for qualitative and quantitative analysis of these markers in herbal and botanical products has increased and is very challenging. Challenges in analytical method development, isolation of markers, and their identification using fingerprinting to identify different species has been used as a quality control tool for many herbal preparations.

Echinacea, one of the most popular herbs in the United States marketplace is the Native American medicinal plant. In recent years, Echinacea products have been the best-selling herbal products in natural food stores in the US (*1-4*). The term refers to several plants in the genus *Echinacea,* derived from the aboveground parts and roots of *Echinacea purpurea* (L.) Moench, *E. angustifolia* D.C., and *E. pallida* (Nutt.) Nutt. [Fam. Asteraceae]. Herbalists and pharmacognosists point out the irony that almost all of the scientific research on this medicinal plant has been conducted not in the United States but in Germany. Echinacea preparations have become increasingly popular in Germany since the early 1900s. The herb was first analyzed and tested for homeopathic purposes in Germany and its medical use was later investigated by Dr. Gerhard Madaus in 1938 (*5*). Echinacea was formerly used in the United States by Native Americans and by Eclectic physicians in the late 1800s and early 1900s. Preparations made from various plants and plant parts of the genus *Echinacea* constituted the top-selling herbal medicine in health food stores in the United States (*4*).

Echinacea is used for preventing and treating the common cold, flu, and upper respiratory tract infections (URIs) (*5*). It is also used to increase general immune system function and to treat vaginal candidiasis. The clinical literature tends to support the *treatment* for symptoms of colds, the flus, and URIs (*6*). Recent studies do not support its use to *prevent* URI (*7*).

The genus *Echinacea* is assigned to the Heliantheae that is the largest tribe within the Compositae (Asteracea) family. The current taxonomy of the genus *Echinacea,* also used in the present National list of Scientific Plant Names, is based on a comparative morphological and anatomical study, according to which the genus comprises nine species and two varieties. But, only *E. purpurea* (L.) Moench *E. angustifolia* DC. and *E. pallida* (Nutt) is widely used as medicinal plants (*7,8*).

The phytochemical constituents of *Echinacea* species range over a wide range of polarity, from the polar polysaccharides and glycoproteins, via the medium polar, caffeic acid (phenolic acid) derivatives and flavonoids to the lipophilic polyacetylenes and isobutyl amides commonly known as alkamides. The pungent property of the Echinacea roots is due to the presence of alkamides.

E. purpurea herb contains caffeic acid derivatives, mainly cichoric acid (1.2–3.1% in the flowers), caftaric acid and chlorogenic acid (Figure 1); 0.001–0.03% alkamides, mainly isomeric dodeca-2E,4E,8Z,10E/Z-tetraenoic acid isobutylamides (Figure 2); water soluble polysaccharides, including PS I (a 4-0-methylglucoronylarabinoxylan) and PS II (an acidic rhamnoarabinogalactan), fructans.

Echinacea angustifolia root contains caffeic acid derivatives, mainly echinacoside (0.3–1.7%) followed by chlorogenic acid, isochlorogenic acid, and its characteristic constituent cynarin (1,5-*O*-dicaffeoyl-quinic acid); flavonoids of the quercetin and kaempferol type in free and glycoside forms, including rutoside, luteolin, kaempferol, quercetin, apigenin and Isorhamnetin; polysaccharides, including inulin (5.9%) and fructans; glycoproteins comprised of approximately 3% protein of which the dominant sugars are arabinose (64–84%), galactose (1.9–5.3%) and glucosamines (6%); 0.01–0.15% alkamides, mainly derived from undeca- and dodeca-noic acid, primarily the isomeric dodeca-2E,4E,8Z,10E/Z-tetraenoic acid isobutylamides.

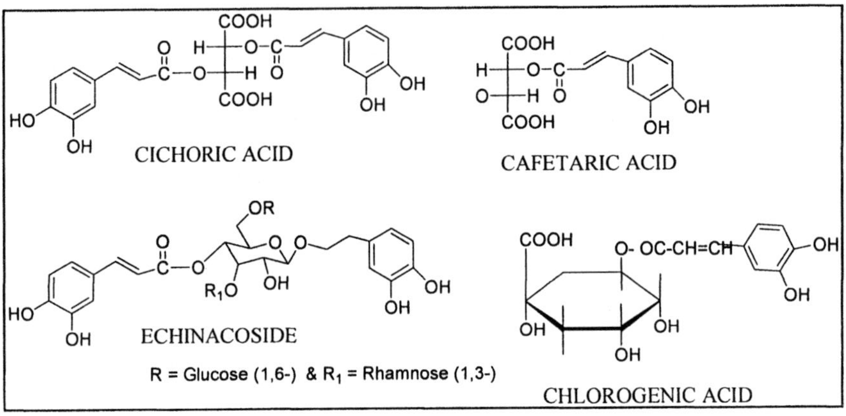

Figure 1.: Chemical structures of Echinacea polyphenols

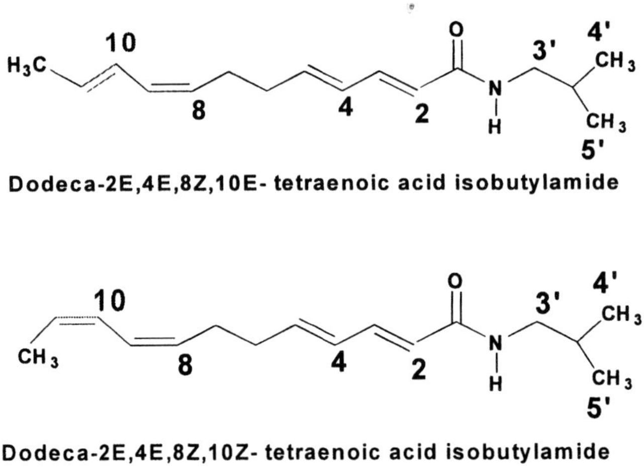

Figure 2. Chemical structures of major isobutylamides in Echinacea spp.

Echinacea pallida herb contains caffeic acid derivatives, including cichoric acid, caftaric acid, echinacoside, verbascoside, chlorogenic acid, and isochlorogenic acid; flavonoids mainly rutoside; alkamides, mainly of the 2,4-

diene type with the isomeric mixture of dodeca-2E,4E,8Z,10E/Z-tetraenoic acid isobutylamides; and <0.1% essential oil (9,10)

There are striking differences in the constituents between the alkamide constitution in roots of E. angustifolia, E. purpurea and E. pallida (1). Mainly of the undeca-2,4-diene type with the isomeric mixture of dodeca-2E,4E,8Z,10E/Z-tetraenoic acid isobutylamides; polysaccharides; and < 0.1% essential oil (8); trideca-1-en-3,5,7,9,11-pentayne and ponticaepoxide have been detected in the flowerbuds.

Commission E approved the internal use of E. purpurea herb as supportive therapy for colds and chronic infections of the respiratory tract and lower urinary tract.

In vitro tests using alcoholic root extracts showed an increase in phagocytic elements of 23% when tested in granulocyte smears (11,12).

Identification of Echinacea Species by HPLC

Phenolic / Caffeic acids

- Weigh approximately 0.3g of Echinacea powdered sample in a Pyrex heating tube. Add 25 mL of 80/20 methanol/water extraction solution to the Pyrex tube along with a stir bar. The solution is heated in a 60 °C stirring block digester for 1 hour. (Note: It is important that stirrer is working and the top of the solution is spinning along with the motion of the rest of the solution for a period of 1 hour.)
- Remove the tubes from the heating block, cool and sonicate for 15 minutes and vortex for 1 to 2 minutes.
- Filter into a 50 mL volumetric flask using funnel and GFA filter paper, rinse tube and its contents with 80/20 extraction solution, filter and dilute to volume.
- Filter the solvent through 0.45 micron syringe filters into labeled HPLC autosampler vials.
- HPLC Conditions:
  Mobile phase: A: 0.2% Phosphoric acid in water
                         B: Acetonitrile
  Solvent Program: t = 0   A = 92%   B = 8%
                         t = 13min   A = 79%   B = 21%
                         t = 20min   A = 72%   B = 28%
                         t = 28min   A = 72%   B = 28%
                         t = 30min   A = 92%   B = 8%

Flow rate: 1.2 mL/min.
Injection volume: 10 µL
Column temperature: 35 °C
Detection wavelength: 200 nm-450 nm, extract spectrum at 330 nm.
Integration: Peak area
Run time: 30 minutes

HPLC profiles of polyphenols present in *Echinacea angustifol* and *Echinacea purpurea* are shown in Figures 3 and 4, respectively.

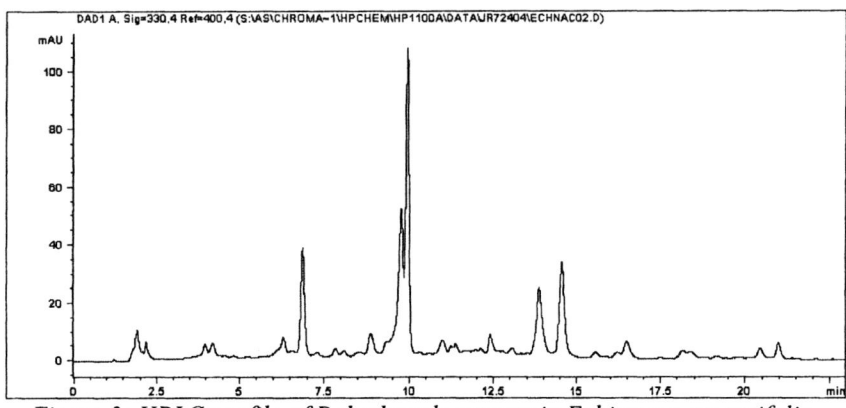

Figure 3: HPLC profile of Polyphenols present in Echinacea angustifolia.

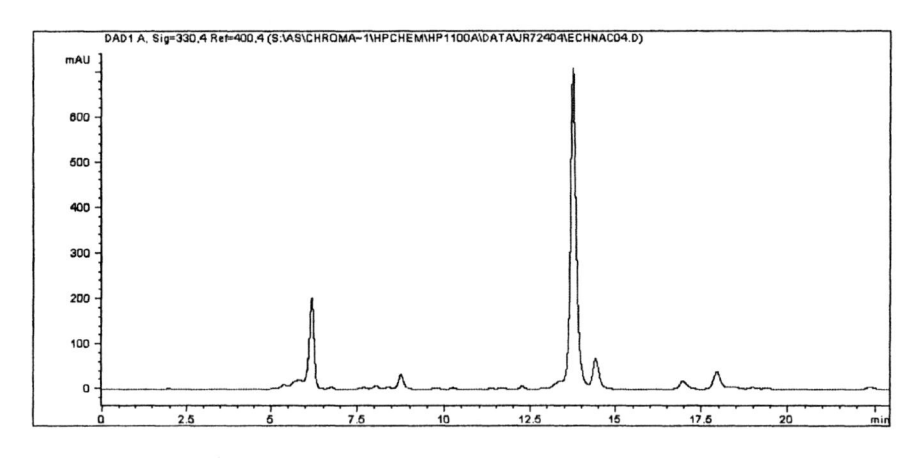

Figure 4: HPLC profile of Polyphenols present in Echinacea purpurea.

Alkamides

Weigh approximately 0.5 g of Echinacea powdered sample in a 50 mL volumetric flask. Add 25 mL of 80/20 methanol/water extraction solution. Mix and slightly warm on steam for 1 minute to quick boil.
- Cool and sonicate for 15 minutes and vortex for 1 to 2 minutes.
- Filter into a 50 mL volumetric flask using funnel and GFA filter paper, rinse tube and its contents with 80/20 extraction solution, filter and dilute to volume.
- Filter the solvent through 0.45 micron syringe filters into labeled HPLC autosampler vials.
- HPLC conditions:

Mobile phase:	A: Deionized water		
	B: Acetonitrile		
Solvent program:	t = 0	A = 40%	B = 60%
	t = 30min	A = 80%	B = 20%
Flow rate:	1.0 mL/min.		
Injection volume:	10 µL		
Column temperature:	Ambient		
Detection wavelength:	200 nm-450 nm, extract spectrum at 260 nm.		
Integration:	Peak area		
Run time:	30 minutes		

HPLC profiles of alkamides present in *Echinacea angustifolia* and *Echinacea purpurea* are shown in Figures 5 and 6, respectively.

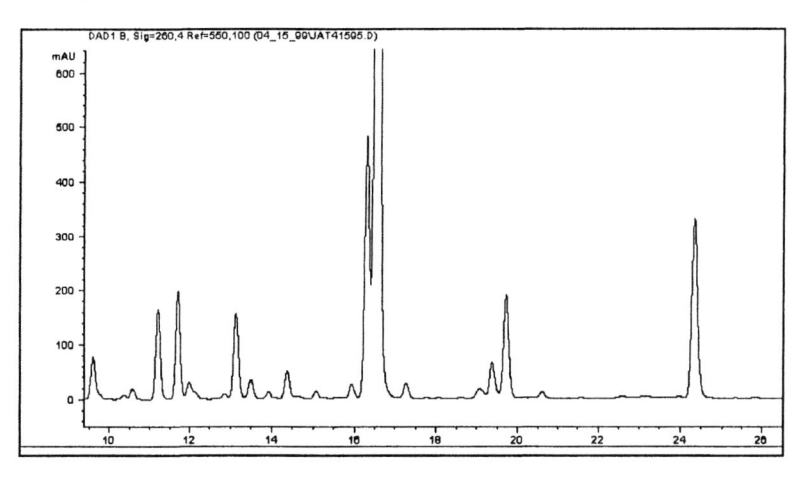

Figure 5: HPLC profile of Alkamides present in Echinacea angustifolia.

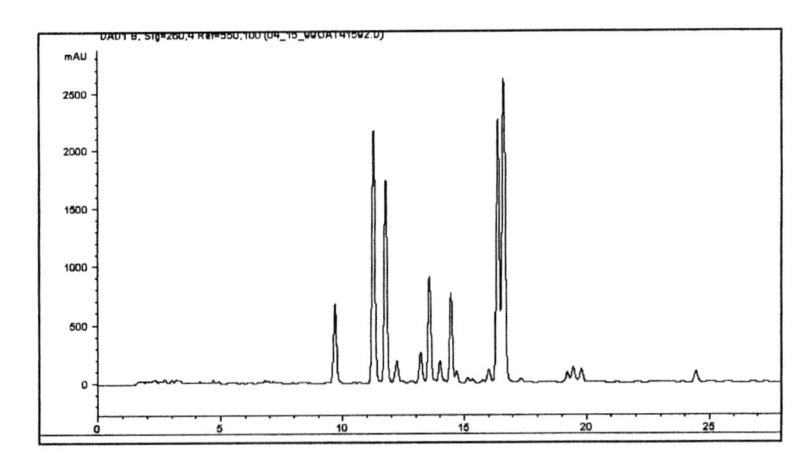

Figure 6: HPLC profile of Alkamides present in Echinacea purpurea.

References

1. Bergeron, C.; Livesey, J.F.; Awang, D.V.C.; Arnoson, J.T.; Rana, J.; Baum, B.R.; Wudeneh L. *Phytochemical Analysis*, **2000**, *11*, 207-215.
2. Bauer, R., Wagner, H. in, *Economics and Medicinal Plant Research,* Wagner, H., Farnsworth, N.R. Eds. Academic Press, London, New York, 1991, Vol. 5/8, pp 253-321.
3. Brevoot, P. *Herbal Gram*, **1995**, 36, 49-57.
4. Bradley, P.R (Ed). *British Herbal Compendium*, 1992, Vol. 1. Bournemouth: British Herbal Medicine Association. Pp 81–83.
5. Barrett, B, M.; Vohmann, C. C;. *J. Fam. Pract.* **1999**, *48(8)*, 628–635.
6. Hobbs, C. *Herbal Gram*, **1994**, *30,* 33-47.
7. Bauer, R.; Remiger, P.; Jurcic, K.; Wagner, H.Z. *Phytotherpy.* **1989**, *10*, 43-48.
8. Bauer, R.; Remiger, P.; Wagner, H. *Phytochemistry,* **1998**, *27*, 2239-2342.
9. Nigel B. P.; John W. Van K.; Elaine J. B.; Graeme A. P. *Planta Medica* **1997**, *63*, 58-62.
10. Bauer, R.; Remiger, P.; Wagner, H. *Phytochemistry* **1989**, *28*, 505-508.
11. Bauer, R., Wagner. H. In *Economic and Medicinal Plants Research,* Wagner, H., Farnsworth N. R (Eds.). Echinacea species as potential immunostimulatory drugs. New York: Academic Press. 1991, Vol. 5. pp. 253–321.
12. Snow, J.M. *Protocol. J. Botany Med.* **1997**, *2(2)*, 18–24.

Chapter 9

Bioassay-Guided Isolation, Identification, and Quantification of the Estrogen-Like Constituent from PC SPES

Shengmin Sang[1], Zhihua Liu[2], Robert T. Rosen[2], and Chi-Tang Ho[2]

[1]Department of Chemical Biology, Ernest Mario School of Pharmacy, Rutgers, The State University of New Jersey, 164 Frelinghuysen Road, Piscataway, NJ 08854-8020
[2]Department of Food Science and Center for Advanced Food Technology, Rutgers, The State University of New Jersey, 65 Dudley Road, New Brunswick, NJ 08901-8520

> The herbal mixture PC SPES, a dietary supplement for patients with prostate cancer, has been proved to have estrogenic activity in clinical trials. To determine the active components in PC SPES for estrogenic activity, a bioassay-guided fractionation of PC SPES using a functional estrogen receptor (ER) assay led to the isolation and identification of the most active constituent, ethinyl estrodiol, together with two flavonones, neobaicalein and tenaxin I. The isolation, identification and quantification of ethinyl estrodiol are discussed.

Introduction

The herbal mixture PC SPES has been widely used by patients diagnosed with prostate cancer. PC is short for prostate cancer and SPES is the Latin word for hope. This herbal extract is a combination of seven Chinese herbs (*Dendranthema morifolium* Tzvel; *Ganoderma lucidium* Karst; *Glycyrrhiza uraensis* Fisch; *Isatis indigotica* Fort; *Panax pseudo-ginseng* Wall; *Rabdosia rubescens* Hara; *Scutellaria baicalensis* Georgi) and one American herb (*Serenoa repens*, saw palmetto). Its efficacy as an antiprostatic dietary regimen has been supported by numerous studies in cell lines, animal models and some clinical trials (*1-4*). PC SPES inhibited cell proliferation and induced apoptosis in all tested cell lines including androgen-sensitive (LNCaP) and androgen-independent prostate cancer cells (*5,6*). Animal studies using two different models further confirmed the antiprostatic activity of PC SPES. The growth of DU-145 prostate tumors in immunodeficient mice was suppressed by more than 50% by the treatment of PC SPES (*7*). In a separated study, PC SPES significantly reduced the incidence and growth rate of androgen-independent MAT-LyLu prostate tumors in Copenhagen rats (*1*). No detectable side effects were observed in these studies. Clinically, most patients with androgen-dependent or androgen-independent prostate cancer had a biochemical antitumor response to PC SPES treatment, with a decrease rate of 50 to 80% of the prostate specific antigen (PSA) level (*1*). PC SPES has been associated with a variable incidence of breast tenderness or enlargement, loss of libido and sexual potency, and a reduction in circulating testosterone levels (*1*). To characterize the most active components in PC SPES responsible for these effects, Sovak, *et al*. (*8*) found synthetic compounds indomethacin (between 1.07-13.19 mg/g) and diethylstilbestrol (DES) (between 107.28-159.27 µg/g) from PC SPES lots manufactured between 1996 and mid-1999; in lots manufactured after mid-1999, the concentrations of indomethacin (from 1.56 to 0.70 µg/g) and DES (from 46.36 to 0.00 µg/g) were gradually decreased. The PC SPES used in this study showed potent estrogenic activity by an estrogen receptor assay, but no DES, estrone, and estrodiol were detectable by high-performance liquid chromatography (HPLC) and gas chromatography (GC) (*3*). To determine the active estrogenic components from PC SPES, we systematically fractionated the methanol extract of PC SPES guided by a functional estrogen receptor (ER) assay. The isolation, identification and quantification of the active component are discussed herein.

Materials and Methods

General Experimental Procedures

PC-SPES lots 5430265, 5431106, 5411164, and 5431219 were purchased from the manufacturer (BotanicLab, Brea, CA) in sealed bottles. ^{1}H (600 MHz), and ^{13}C (150 MHz) NMR spectra were acquired on a Varian ^{Unity}INOVA 600 NMR spectrometer (Palo Alto, CA) equipped with a z-gradient inverse-detection triple resonance probe, with TMS as internal standard. Mass spectrometry was performed on a VG Platform II single quadrupole mass spectrometer (Micromass Co., MA) equipped with an atmospheric pressure ion source and atmospheric pressure chemical ionization (ApCI) interface. The effluent from the LC column was delivered to the ion source (150°C) through a heated nebulizer probe (450 °C) using nitrogen as drying gas (300 L/hr) and sheath gas (150 L/hr). Positive ions were acquired in full scan (m/z 450-1300 for quantification and m/z 200-1000 for identification, 1s scan time). RP C-18 silica gel and Sephadex LH-20 gel were purchased from Sigma Chemical Co. (St. Louis, MO). Thin-layer chromatography was performed on Sigma-Aldrich TLC plates (250 μm thickness, 2-25 μm particle size), with compounds visualized by spraying with 5% (v/v) H_2SO_4 in ethanol solution. HPLC-grade solvents and other agents were obtained from VWR Scientific (South Plainfield, NJ). Ethinyl estradiol (EE), β-estradiol (ED) and CD_3OD were purchased from Sigma (St. Louis, MO). BSTFA (Bis-silyltrifluoroacetamide) was ordered from Sulpelco Inc (Bellefonte, PA).

Bioassay-guided Extraction and Isolation.

The fractionation of PC-SPES extracts was guided by a functional estrogen receptor (ER) assay. The growth-based assay used Saccharomyces cerevisiae strain PL3, which carries a URA3 gene under the control of the human ER alpha and beta (3,4). 10 bottles of PC SPES (Lot: 5430265, 60 capsules each bottle and 320 mg each capsule, total 192 g powder) were extracted with 1000 ml methanol at 50 °C for three times (each time four hours). After evaporation of methanol *in vacuo*, the residue was suspended in water and then extracted successively with hexane (3×300 mL), EtOAc (3×300 mL), and *n*-BuOH (3×300 mL). Butanol fraction (7.0 g), the fraction with greatest ER activity was subjected to Dial HP-20 column chromatography using a EtOH-H_2O gradient system (0%-100%) to give 4 fractions. Subfraction 3 (2.6 g) was then subjected to a silica gel column eluted with $CHCl_3$-MeOH-H_2O solvent system (4:1:0.15, 3:1:0.15 and MeOH) to give 7 fractions (fractions 3-1 to 3-7). Among them, fraction 3-1 was the most active fraction. However, the amount of fraction 3 is small and also has the overlap with the ethyl acetate fraction and subfraction 4 of the butanol fraction. So we combined these three fractions and loaded it to a silica gel column eluted with $CHCl_3$- MeOH solvent system (1:0-0:1) to get 10 fractions (fractions 1-10). Fraction 3, the strongest active fraction, was subjected to Sephadex LH-20 column eluted with 95% ethanol to give four subfractions (fractions 3-1 to 3-4). Fraction 3-2 (60 mg), which showed the strongest effect at activating both the alpha and beta receptors, was subjected to RP-C18 column eluted with MeOH-H_2O solvent system (70% -80%) to give 8 fractions (fractions 3-2-1 to 3-2-8). Fractions 3-2-3, 3-2-5 and 3-2-6 are three pure compounds. Among them, fraction 3-2-5 (**1**, 3 mg) was the strongest active fraction. **Compound 1:** amorphous solid, ^{1}H NMR (CD_3OD) of **1**: δ 7.09 (H-1, d, J=8.4

Hz), 6.55 (H-2, dd, J=3.0, 8.4 Hz), 6.48 (H-4, d, J=3.0 Hz), 1.2-3.0 (16H, m), 0.85 (H-18, s); EI-GC-MS (m/z): 296 (M$^+$), 270, 228, 213, 172, 160, 145, 133, 107, 91, 77, 55, 41.

Quantitative Determination of Ethinyl Estrodiol by Gas Chromatography-Mass Spectrometry (GC-MS) Analysis

Four samples designed as S-1 (Lot No 5431106), S-2 (Lot No. 5431219), S-3 (Lot No. 5431219 duplicate), and S-4 (Lot No. 5431164) were used in this study. Sixty capsules in each bottle were emptied and the contents weighed as 19.56 g, 19.31 g, 19.15 g, and 19.41 g respectively for these samples. Soxhlet extraction was performed on the samples in the following sequence: Blank 1 (empty thimble without sample), S-1, S-2, Blank 2, S-3, and S-4. For each extraction, 250 ml Chloroform was added into the system and kept refluxing for 4 hrs. The final extracts had the weights of 34.9 mg (Blank 1), 578.3 mg (S-1), 552.3 mg (S-2), 7.9 mg (Blank 2), 494.8 mg (S-3), and 508.5 mg (S-4).

The compound of interest in the extract was further concentrate through normal phase silica gel column chromatography. Five columns packed with 50g silica gel (100-200 Å, Selecto Scientific, Georgia) each were used for S-1, S-2, Blank 2, S-3 and S-4, respectively. Each column was eluted first with 500 ml chloroform and then followed by 50:1 chloroform/acetone (1000 mL × 3). HPLC analysis indicated that the elution from 50:1 chloroform/acetone contained the compound of interest. The weight of the 50:1 elution for the above five samples were 259.1 mg (S-1), 225.3 mg (S-2), 3.4 mg (Blank 2), 199.5 mg (S-3), and 182.6mg (S-4), respectively.

In order to quantify the level of ethinyl estradiol (EE) in the above samples, electron ionization – mass spectrometry (EI-MS) was used, with β-estradiol (ED) as the internal standard. Each sample went through silylation by reacting with BSTFA (Bis-silyltrifluoroacetamide). In brief, each sample was dissolved in 100 μl methylene chloride and then reacted with 400 μl BSTFA under 80°C for one hour. 1 μL of the above sample was injected into GC (GC column: DB5MS 30 m × 0.32 mm I.D). A program of 3 minutes at 150°C, followed by a 17-min gradient of 10°C/min up to 320°C was used.

The calibration curve of this method was conducted by injecting the standard solution of EE with four different concentrations (50, 100, 200, and 400 ppm) using 100 μg ED as internal standard. Following the same procedure, the levels of EE in four different samples (25.9 ppm S-1, 22.5 ppm S-2, 19.9 ppm S-3, and 18.2 ppm S-4) were measured.

Results and Discussions

The methanol extract of PC SPES was systematically fractioned through different column chromatography, such as Diaon HP-20, silica gel, Sephadex LH-20, and RP C-18 column, guided by a functional estrogen receptor (ER) assay (Figure 1). Three compounds were purified from the most active subfraction. The compound with the greatest activation of alpha and beta ER in the yeast system was identified as ethinyl estradiol (Figure 2) by comparing its ^{1}H NMR, EI-GC-MS, and the retention time of GC with those of the standard sample. The other two compounds were identified as neobaicalein and tenaxin I (Figure 2) by comparing their NMR data with those reported in the literatures (*9,10*).

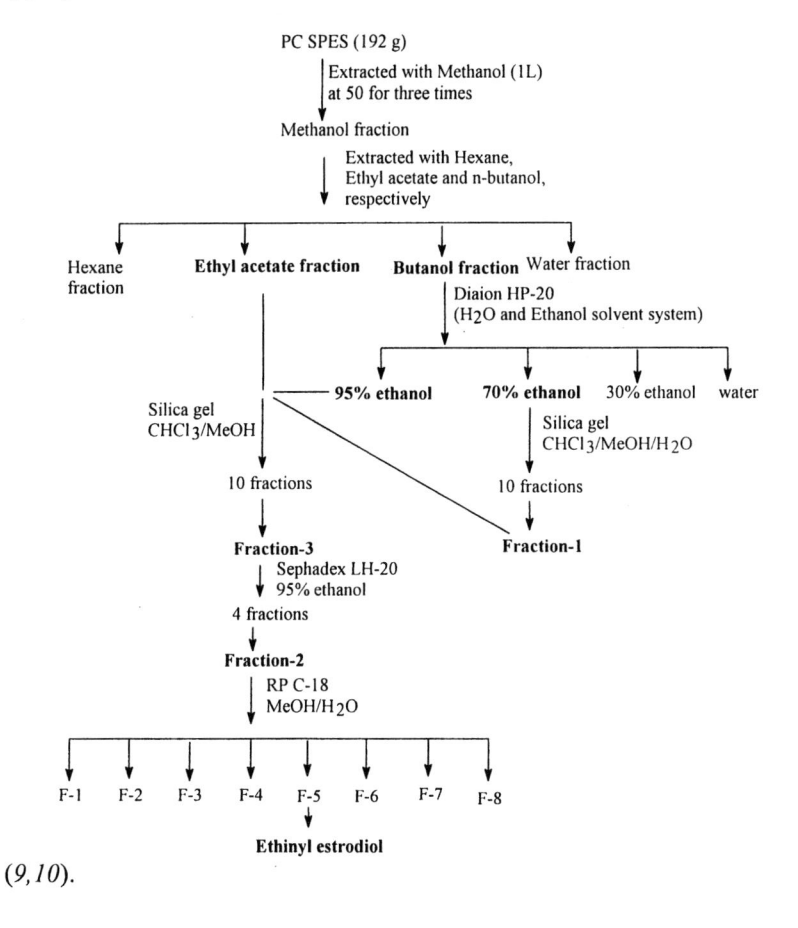

(*9,10*).

Figure 1. Extraction and isolation procedure

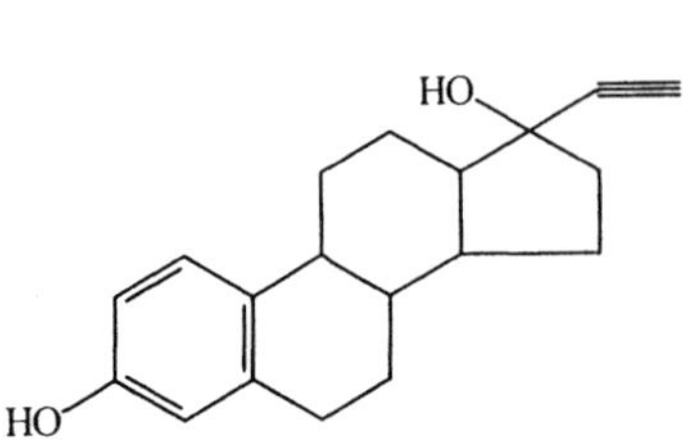

Figure 2. Structure of compounds 1-3

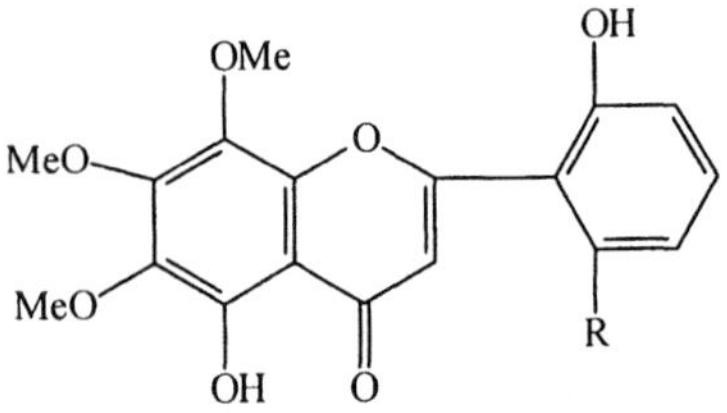

EE was detected in three different lots of PC SPES, in which one sample was using for a phase II trial in patients with androgen-independent prostate cancer (4). To quantify the level of EE in these products, we developed an electron ionization GC/MS method with estradiol as internal standard. The retention time of estrodiol di-TMS ether (m/z 416) and ethinyl estradiol mon-TMS ether (m/z 368) was 15.22 and 15.51 minutes, respectively (Figure 3). EE was detectable in all three different lots with the range from 8 to 19 ppm (Table I).

Due to the limited treatment options in androgen-independent prostate cancer (AIPC), the herbal mixture PC SPES had become one of the best prospects for an alternative medicine cancer treatment from 1996 to 2002. The

Table I. Area of the Peaks for the Samples and ED.

No. Of sample	Area of sample	Conc. of ED (ppm)	Area of ED	Area ratio of sample/ED	Conc. of EE in sample (ppm)
1 (Lot No. 5431106)	478	200	24841	0.0192	8.0
2 (Lot No. 5431219)*	1302	200	20115	0.0647	18.4
3 (Lot No. 5431219)*	1490	200	25351	0.0587	17.2
4 (Lot No. 5431164)	598	200	26842	0.0222	9.5

* This lot product was using in clinical trial.

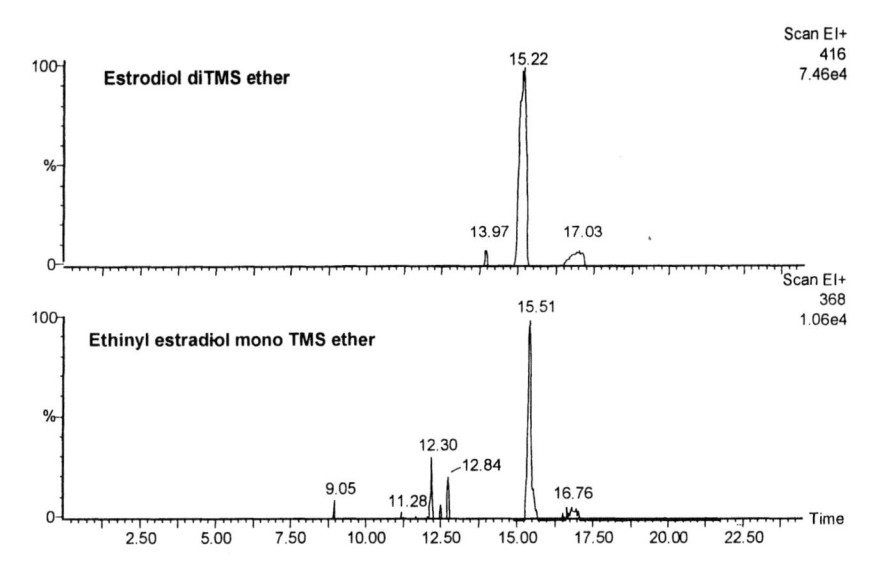

Figure 3. Gas chromatogram of Estradiol-diTMS (top) at 15.22 minutes (m/z 416) and Ethinylestradiol-monoTMS (bottom) at 15.51 minutes (m/z 368).

pioneering work of Dipaola and coworkers demonstrated *in vitro*, in mice, and in patients, that PC SPES induced potent estrogenic activity and androgen ablation (*3*). Later, three other clinical studies further confirmed the efficacy of PC SPES on androgen-dependent and androgen independent prostate cancer of patients (*4,11,12*). However, the estrogenic side effects in PC SPES trials made researchers try to identify the active estrogenic components in it and to determine whether these components account for all, or part, of the direct cytotoxic effects. Dipaola *et al* suspected the presence of a synthetic estrogen in PC SPES, but no diethylstilbestrol (DES), estrone, and estradiol were detected by HPLC and GC/MS analysis in their study (*3*). Then, we collaborated to further isolate and identify the estrogenic components in PC SPES guided by a functional estrogen receptor (ER) yeast assay. After a systematic fractionation and isolation, the most active component was purified and identified as ethinyl estrodiol, a synthetic estrogen. Independently, the California Department of

Health Services found contamination with synthetic compounds DES and warfarin (*13*); Sovak and coworkers (*8*) found synthetic indomethacin and DES from lots manufactured from 1996 through mid-1999. Interestingly, in lots manufactured after mid-1999, the concentration of these two compounds dramatically decreased. In some lots, DES was not detectable. This could explain why Dipaola *et al* could not detect DES in their samples. Our results indicated the presence of ethinyl estrodiol in the lots lacking of DES. On February 8, 2002, the California Department of Health Services released a warning of the adulteration of PC SPES (*13*). In the meantime, the manufacturer, BotanicLab, voluntarily recalled PC SPES nationwide and went out of business on June 1, 2002.

Herbal medicine research is complicated enough without having the problem of adulteration of potential product. The variation of the concentration of the active components for herbal extracts research is a big problem due to 1) inconsistent and inaccurate identification of plant species; 2) known and unknown genetic and environmental factors; 3) contamination by molds, mycotoxins, or pesticides; and 4) insufficient regulatory oversight of the production facility (*14*). Perhaps, the story of PC SPES will raise the issue regarding the quality control of herbal medicine, which should meet the standards of quality control under the Good Manufacturing Practice system, and should serve as a lesson to help researchers resolve such confounding factors earlier in their research.

Acknowledgement

We thank Dr. George Lambert of the University of Medicine and Dentistry of New Jersey for his help in functional estrogen receptor (ER) assay analysis of samples.

References

1. Marks, L.S.; Dipaola, R.S.; Nelson, P.; Chen, S.; Heber, D.; Belldegrun, A.S.; Lowe, F.C.; Fan, J.; Leaders, F.E.; Pantuck, A.J.; Tyler, V.E. *Urology*, **2002**, *60*, 369-377.
2. Hsieh, T.; Lu, X.; Chea, J.; Wu, J.M. *J. Nutr.* **2002**, *132*, 3513s-3517s.
3. DiPaola, R.S.; Zhang, H.; Lambert, G.H. Meeker, R.; Licitra, E.; Rafi, M.M.; Zhu, B.T.; Spaulding, H.; Goodin, S.; Toledano, M.B.; Hait, W.N.; Gallo, M.A. *N. Engl. J. Med.* **1998**, *339*, 785-791.

4. Oh, W.K.; Kantoff, P.w.; Weinberg, V.; Jones, G.; Rini, B.I.; Derynck, M.K.; Bok, R.; Smith, M.r.; Bubley, G.J.; Rosen, R.T.; DiPaola, R.S.; Small, E.J. *J. Clin. Oncol.* **2004**, *22*, 3705-3712.
5. Che, S. *Urology*, **2001**, *58*, 28-35.
6. Kubota, T.; Hisatake, J.; Hisatake, Y.; Said, J.W.; Chen, S.; Holden, S.; Taguchi, H.; Koeffler, H.P. *Prostate*, **2000**, *42*, 163-171.
7. Bennett, M.W.; O'Connell, J.; O'Sullivan, g.C.; Roche, D.; Brady, C.; Kelly, J.; Collins, J.K.; Shanahan, F. *Gut*, **1999**, *44*, 156-162.
8. Sovak, M.; Seligson, A.L.; Konas, M.; Hajduch, M.; Dolezal, M.; Machala, M.; Nagourney, R. *J. Natl. Cancer Inst.* **2002**, *94*, 1275-1281.
9. Iinuma, M.; Matsuura, S.; Kusuda, K. Chem. Pharm. Bull. **1980**, *28*, 708-716.
10. Liu, M.; Li, M.; Wang, F. *Yaoxue Xuebao* **1984**, *19*, 545-546.
11. Small, E.J.; Frohlich, M.W.; Bok, r.; Grossfeld, G.; Kelly, W.K.; Reese, D.M. *J. Clin. Oncol.* **2000**, *18*, 3595-3603.
12. de la Taile, A.; Hayek, O.R.; Buttyan, R.; Bagiella, e.; Burchardt, M.; Katz, A.E. *BJU. Int.* **2001**, *58*, 28-35.
13. Available at: http://www.applications.dhs.ca.gov/pressreleases/store/Press Release/02-03.html.
14. White, J. J. Natl. Cancer Inst. **2002**, *94*, 1261-1263.

Chapter 10

Intraspecific Variation in Quality Control Parameters, Polyphenol Profile, and Antioxidant Activity in Wild Populations of *Lippia multiflora* from Ghana

H. Rodolfo Juliani[1], Mingfu Wang[1], Hisham Moharram[1], Julie Asante-Dartey[2], Dan Acquaye[2], Adolfina R. Koroch[1], and James E. Simon[1]

[1]New Use Agriculture and Natural Plant Products Program, Cook College, Rutgers and the New Jersey Agricultural Experimental Station (NJAES), Rutgers, The State University of New Jersey, 59 Dudley Road, New Brunswick, NJ 08901–8520
[2]Agribusiness in Sustainable Natural African Plant Products, Accra, Ghana

This study evaluates the variation of quality, polyphenol profile and antioxidant activity in wild varieties of *Lippia multiflora* from different regions of Ghana and reviews the pharmacological properties of this species. Our results showed that while the essential oil composition is highly variable, the polyphenols profile remain similar among the discrete populations with verbascoside being the main component followed by nuomiside. Our study also demonstrated that the antioxidant activity of one gram of *Lippia multiflora* leaves is equivalent to 2.7-12.4 g of Trolox (a Vitamin E analogue). Since the amount of total phenols, antioxidant activities and phenylpropanoid glycosides in some *Lippia* populations were found to be comparatively low, and these are the compounds largely responsible for many of the plants medicinal attributes, those populations with higher contents of bioactive constituents and those processing techniques that can maximize their content should be used in the commercialization of this species to improve its nutraceutical properties. Our study also showed the importance of the evaluation of bioactive components from wild varieties of *Lippia multiflora*, and that such natural variation can be used as a source of new components and bioactivities to develop new or improved products by selection and/or breeding to improve the genetic materials that are commercialized. Because *Lippia multiflora* is not well know in the international markets, this plant species may serve as a unique 'African bush tea' in the western marketplace, and as a possible source of aroma chemicals.

The *Verbanaceae* is an important family of aromatic plants widely used in popular medicine (*1-4*). Within this family, the genus *Lippia,* that comprises approximately 200 species of aromatic plants, is used in folk medicine in Africa, Central and South America (*5-11*). *Lippia multiflora* (syn. *L. adoensis*), a common savannah plant, is distributed in subtropical Africa, from Senegal to Cameroon and South Africa (*12,13*). *L. multiflora*, also known as Gambian bush tea, has been used by indigenous people in many West African countries for a variety of medicinal purposes. In Ghana, the infusion of sun-dried leaves is consumed like tea with sugar for stomach pains, used as laxative and appreciated by children (*12*). This species has also been used as a febrifuge, antihypertensive, muscle relaxant, laxative, antimicrobial and sudorific (*14*). The leaves are used in the treatment of xerostomia, abdominal pains, conjunctivitis, fever, hypertension, insomnia and lactation failure. The whole plant is also used to treat fever while the roots are traditionally used to treat malaria, and placental retention (*15*). In Cote d'Ivoire, the most important beverage, next to water, is the tea prepared from *L. multiflora* leaves that is usually prepared for breakfast (*16*), and a decoction of roots and leaves is consumed for gastro-intestinal problems and enteritis (*12*).

Herbal teas are becoming important because of their functional properties and consumer interest in the health promoting properties of such beverages (*17*). The chemistry and pharmacology of other teas such as green tea (*Camelia sinensis*) and hibiscus (*Hibiscus sabdariffa*) has been extensively studied and their health benefits are mainly due to their polyphenol content (*18-20*).

However, the phenolic composition and pharmacological properties of L. multiflora have not been well researched even though the plant is used on a large scale in West African folk medicine and is commercially available as a herbal tea in Ghana (21), has recently entered the US marketplace (22), Limited clinical observations have demonstrated that the aqueous extract prepared from the leaves possess antihypertensive properties and a calming/sedative effect (13).

Volatile oils represent the class of components that partially impart the aroma and flavor to the teas (21). The essential oils and their components showed a wide range of bioactivities including antiviral, antimicrobial, anticancer, antioxidant, analgesic, and digestive, among others (2,23-25).

The essential oils of wild populations of *Lippia* species are typically recognized by a high degree of genetic variation (22,27-29). Although the phenolic composition in *L. multiflora* has been studied (30), the variability of the phenolic components in wild populations is still unknown.

Since this species is wild-gathered by local communities and in view of the increasing commercial importance in the local and international markets, the introduction into cultivation of improved germplasm and the development of new or improved products can generate new commercial opportunities for the local communities. However, the lack of scientific information on the chemical composition, quality parameters and health benefits of *L. multiflora*, may limit the access to markets. Studies that can identify new applications and uses of exotic products can potentially lead to new bioactivities that improve health and ultimately assist growers and rural communities by increasing interest in their products (24).

The development of a consistent Lippia herbal tea will thus need a consistent genetic population, consistent aroma and flavor that resonate favorably with the intended consumer and marketplace. This study evaluates the variation of quality, polyphenol profile and antioxidant activity in wild varieties of *L. multiflora* from different regions of Ghana and reviews the pharmacological properties of this species.

Materials and Methods

Plant Material

In the spring 2004, wild plants of *L. multiflora* were collected from the region of Brong-Ahafo, (Nsawkaw and Amantin), Ashanti (Aframso Bridge), Volta region (Buem Nsuta, Ho) and Eastern region (Apeguso). The ground sun-dried leaves were used to conduct all the quality control and chemical analysis. A commercial sample was included for comparison purposes. Each procedure was run twice and at least in duplicate.

Quality Control Analysis

The moisture, total ashes, total insoluble acids and volatile oil content were assessed for each sample using methods described by the Food Chemical Codex (*31*).

Tea Tasting

Each sample was placed in tea bags and brewed (3 g/200 mL) in hot water during 5 min, the tea was stirred a couple of times each minute. The effective sensory evaluations of the seven *L. multiflora* teas were performed by a taste panel that was composed of 10 people that were previously trained in tasting other African herbal teas. The panel evaluated the intensity and attractiveness of color, aroma and flavor and the hedonic (magnitude of like/dislike), each character was ranked from 0 to 9 (0 dislike, 5 neutral, 9 like).

Chemical and Antioxidant Activity Analysis

The ground leaves (100 mg) were extracted by 25 mL of methanol 60% through sonication for half hour and subjected to HPLC, total phenol content and antioxidant activity analyses. The phenolic components were recovered with methanol 60% for accurate analysis. The total phenols were measured using the Folin Ciocalteu's regent (*32*) the results were expressed as gram of gallic acid equivalents on a dry basis (g gallic acid/100 g DW). The antioxidant activity was evaluated using the ABTS method and the results were expressed as g of Trolox (a water soluble analog of vitamin E) on a dry weight basis (g Trolox/100 g DW) (*33*).

Qualitative HPLC-MS Analyses

The analyses were performed on a Hewlett-Pachard 1100 modular system with MSD Trap (*17*). Negative and positive ESI-mass spectra were measured on Agilent 1100 LC-MSD system (Agilent Technologies, Germany) equipped with an electrospray source, Bruker Daltonics 4.0 and Data analysis 4.0 software. The electrospray mass spectrometer (ESI-MS) was operated under a positive ion mode or negative mode and optimized collision energy level of 60%, scanned from *m/z* 100 to 800. ESI was conducted using a needle voltage of 3.5 kV. High-purity nitrogen (99.999%) was used as dry gas and flow rate at 9 L/min, capillary temperature at 325 °C. Helium was used as Nebulizer at 45 psi. The ESI interface and mass spectrometer parameters were optimized to obtain maximum sensitivity. and the mobile phase included water (containing 0.1% formic acid, solvent A), acetonitrile (solvent B) and isopropanol (Solvent C) in the following gradient system, 0-5 min (15% B, 2% C), 20 min (25% B and 2% C), 30 min (50% B and 2% C), 40 min (80% B and 2% C). The total running time was 40 minutes and post running time is 15 minutes and the flow rate was 0.8 mL/min. Two major phenolic compounds (nuomioside and verbascoside) were tentatively identified by comparing the UV and MS spectra with the reference standards and by their $[M+1]^+$ and $[M+Na]^+$ ions.

Quantitative HPLC Analysis

The HPLC was run on a Phenomenex Phenyl-hexyl column (3 μm, 150 x 4.6 mm) and the detection wavelength was 280 nm, the injection volume was 20 μL and flow rate was 0.8mL/min. The mobile phase was the same as that in the HPLC/MS but using phosphoric acid. Rosmarinic acid was used for the quantification of these phenylpropanoid glycosides and the results were expressed as percent (g/100 g of leaf dry weight).

Results and Discussion

Quality Standards

The moisture content ranged from 8.4% in the sample from Aframso Bridge to 11% in the commercial variety (Table I), the required maximum values for moisture is usually 12% according to international standards (*34*). The total ashes ranged from 8.5% (Buemnsuta) to 12.7% (Amantin), the value for total ashes for dried botanicals usually ranges from 3.5% to 16% (Table I) (*34*). The total insoluble ashes, measure the sand content and is a classical determination of the cleanliness of herbs, in *L. multiflora* ranged from 0.8 (Apeguso) to 4.1% (commercial) (Table 1), while the range for other botanicals is 0.4 (pimento) to 4% (thyme) (*34*). The volatile oil content showed also a great variation within the different populations, the sample from Nsawkaw showed the highest content (1.8%), while the commercial sample showed the lowest amount (0.3%) (Table I). The level of volatile oils in leaf botanicals according to international standards also showed great variations according to species, basil showed the lowest minimum content (0.5%) while higher minimum contents were described for oregano (1.5%) *(34)*.

Table I. Selected Quality Control Parameters of Different *L. multiflora* Varieties

Site	Co[1]	Ns[1]	Af[1]	Am[1]	Bu[1]	Ho[1]	Ap[1]
Moisture content % (m/m)	11.1 ± 0.3	10.7[1] ± 0.5	8.4 ± 0.1	8.6 ±0.2	9.5 ±0.7	11.0 ±0.5	10.1 ±0.2
Total ashes % (m/m)	10.6[a]	10.2 ± 1.5	12.5 ±0.1	12.7 ±0.1	8.5 ±0.3	10.3 ±0.1	9.2[a]
Total insoluble ashes % (m/m)	4.1 ± 1	1.8 ± 0.2	4.1 ±0.4	3.8[a]	1.0[a]	1.8 ±0.5	0.4 ± 0.2
Essential oil content % (m/m)	0.3[a]	1.8[a]	0.6[a]	0.6[a]	0.7[a]	1.1[a]	1.0[a]

[1] Co: Commercial; Ns: Nsawkaw; Af: Aframso bridge; Am: Amantin; Bu: Buemnsuta; Ho, Ho; Ap: Apegu
1-mean ± standard deviation (STD). a-STD less than 0.1

Total Phenols and Antioxidant Activities

Total phenols also varied among the populations. Lower levels were observed on the commercial (1.8%), Aframso bridge and Amantin (3.4%) samples. The samples from Nsawkaw, Ho and Apeguso showed intermediate levels (4.8-6%, respectively), while the maximum value was observed in the sample from Buemnsuta (8.3%) (Table II). The antioxidant activity measured as Trolox (Vitamin E analogue) equivalents, showed a similar trend, the lowest activity was observed in the commercial sample (2.7%) while the highest activity in the sample from Buemnsuta (Table II).

Contents of Major Phenols, Verbascoside and Nuomiside

Nuomiside and verbascoside were the two main phenylpropanoids glycosides detected in the methanolic extract of *L. multiflora* (Figures 1 and 2, Table II). The content of both polyphenols, that showed also a great variation among the populations, was in accordance with the total phenol and antioxidant activities (Table II). The lower percentages of nuomiside and verbascoside were found in the commercial sample (0.13, 0.16), Aframso bridge and Amantin (0.35-0.39, 0.52-0.55%, respectively) samples, intermediates values were observed in the Nsawkaw (0.7-1.4%), Ho (0.9-1.7%) and Apeguso (0.9-1.5%)

and the highest content was observed in the sample from Buemnsuta (1.4-3.1%) (Table II). The profile of polyphenols are similar, and in all samples verbascoside (0.16-3.8%) was the main component followed by nuomiside (0.1-1.4%) and the ratio nuomiside:verbascoside showed a low variation in all the samples (0.46-0.79) (Table II).

Table II. Total Phenols, Antioxidant Activity and Phenolic Composition of Different *L. multiflora* Varieties.

	Co[1]	Ns[1]	Af[1]	Am[1]	Bu[1]	Ho[1]	Ap[1]
Total phenols[2]	1.8[4] ± 0.1	4.8 ± 0.1	3.4 ± 0.4	3.4 ± 0.1	8.3 ± 0.3	5.5 ± 0.1	6.0 ± 0.3
Antioxidant activity[3]	2.7 ± 0.3	6.6 ± 0.3	4.5 ± 0.5	5.0 ± 0.2	12.4 ± 0.3	7.8 ± 0.7	9.0 ± 1.7
Nuomiside[4]	0.13[a]	0.69[b]	0.35[a]	0.39[b]	1.42[b]	0.90[b]	0.90[c]
Verbascoside 4	0.16[a]	1.36[c]	0.52[c]	0.55[b]	3.08[b]	1.73[b]	1.47[c]
Nuomiside + verbascoside	0.29	2.05	0.88	0.93	4.50	2.64	2.37

[1] Co: Commercial; Ns: Nsawkaw; Af: Aframso bridge; Am: Amantin; Bu: Buemnsuta; Ho, Ho; Ap: Apegu

[2] 1-g gallic acid equivalents/100g dry weight; [3] 2-g Trolox equivalents/100g dry weight; [4] 3- g/100g dry weight; a-Standard deviation less than 0.001, b-less than 0.02, c-less than 0.06. 4 - 1-mean ± standard deviation.

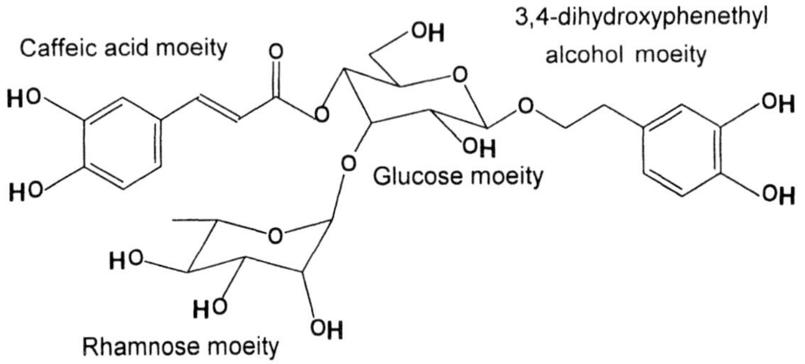

Figure 1. Chemical structure of verbascoside, the main polyphenol occurring in L. multiflora.

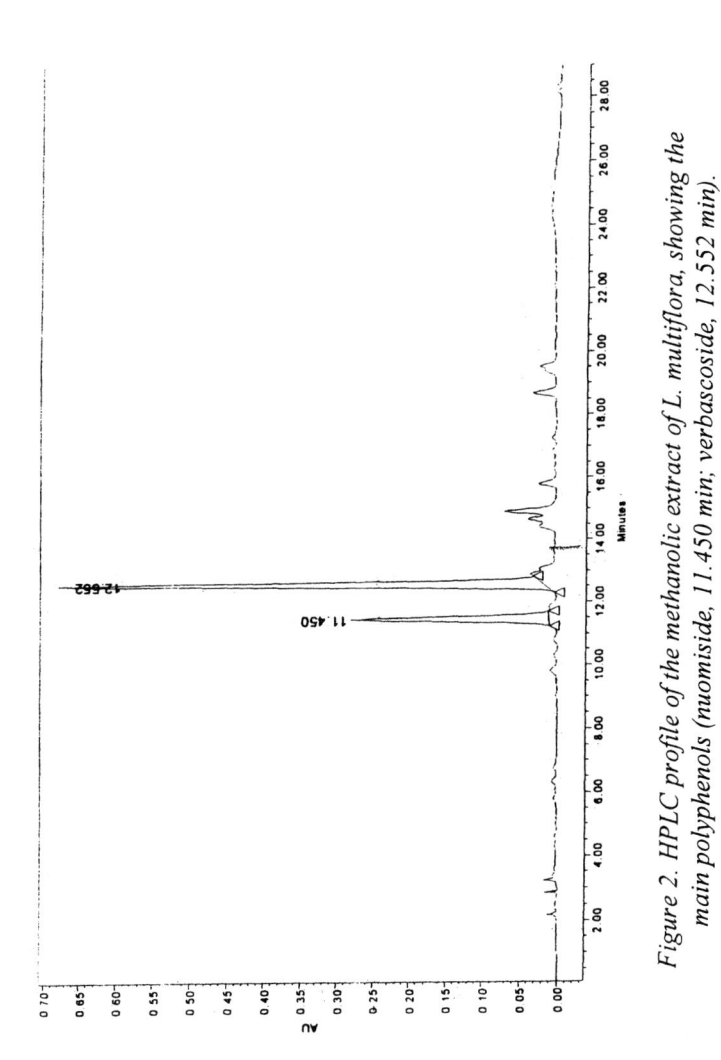

Figure 2. HPLC profile of the methanolic extract of L. multiflora, showing the main polyphenols (nuomiside, 11.450 min; verbascoside, 12.552 min).

Essential Oils

Our previous study on the essential oil composition from the commercial sample (farnesene, 21%, caryophyllene oxide, 36%, farnesol, 20%), Nsakaw (sabinene 12%, 1,8-cineole 43%, linalool 10%, α-terpineol 13%), Aframso bridge (linalool 29%, germacrene D 28%), Amantin (para cymene 16%, thymol 30%, thymyl acetate 17%), Buemnsuta (two unknown sesquiterpenes 43%) and Ho (para cymene 23%, farnesol 24%), and Apeguso (limonene) showed that *L. multiflora* exhibits a great intraspecific variability. This important intraspecific variability has also been observed in *L. junelliana* and *L. alba*, and seems to be common feature for the genus *(6,22,26-27)*. Our results showed that the polyphenol profile remain with minor variations; in all samples verbascoside was the main component followed by minor amounts of nuomiside (Table II), while the essential oil composition is highly variable *(22)*.

Review of Pharmacological Activities

Pharmacological studies have shown that the tea made from *L. multiflora* was found to have antihypertensive effect *(21)*. The aqueous extract of *L. multiflora* caused a lowering of blood pressure in normal cats and rats. The decrease in the blood pressure is due to the direct skeletal muscle relaxation which leads to the reduction of tension on the blood vessels and also vasodilatation resulting from a combination of alpha-adrenoceptor antogonism and beta-adrenoceptor stimulant *(13)*.

Lippia multiflora extracts exhibited most of the peripheral nervous system pharmacological action of the minor tranquilizers such as diazepam. The observed muscle relaxation which might contribute very significantly to a generalized reduction in tension may be responsible for the calming effect of *L. multiflora (21)*. The fractionation of the *L. multiflora* methanolic extract yielded a fraction containing polyphenols, flavonoids and phenolic acids, that showed the highest hypotensive effects. The subsequent fractioning of this phenol fraction led to the isolation of two pure components, verbascoside and an unknown caffeic acid ester. Verbascoside produce a weak tension fall while the caffeic acid ester was responsible for a marked and long lasting tension fall *(35)*. Since the thromboxane A2 is involved in hypertension, this caffeic acid derivative of *L. mulitflora* leaves showed anti-tromboxane synthase activity showing that this unknown component is responsible for the hypotensive activity of *L. multiflora* leaves *(36)*.

Verbascoside (acteoside) and other phenylethanoid glycosides represent a group of natural products mainly spread in several botanical families of Tubiflorae (Verbenacea, Lamiacea, Scrophulariacea) *(37)*.

In a cardiovascular test with male Wistar rats, a mixture of verbascoside and orobanchoside extracted from *Orobanche hederae* had no significant effect on arterial blood pressure (*38*).

Versbascoside isolated from *Aloysia citriodora* (syn *Lippia trypylla*) showed analgesic and weak sedation activities. The modification of the chemical structure of verbascoside showed that the methylation of the hydroxyl groups did not decrease the activity, while the modification of the sugar moiety such as the deletion or migration of the rhamnose decreased the activity, and the removal of the caffeoyl and phenethyl moieties completely reduced the potency (*39*).

After intraperitoneal administration of gradual doses of verbascoside containing aqueous extract obtained from leaves of *Stachytarpheta jamaicensis*, sedation and analgesia were observed in rats, suggesting the depression of the central nervous system (*40*). Verbascoside extracted from *O. hederae* showed also analgesic activities (*41*).

Some phenylpropanoids inhibited the Ca^{++}/phospholipid dependent protein kinase C (PKC) that plays an important role in signal transduction, cellular proliferation and differentiation. The PKC is thus involved in tumor proliferation. Molecules that inhibit PKC showed antitumoral activities. Verbascoside isolated from *Lantana camara*, showed inhibitory effect against the PKC by acting directly on the catalytic site of the enzyme (*42*).

Verbascoside and other phenylpropanoid glycosides, isolated from members of the Lamiaceae family, showed a bifasic effect on cancer cells, both cytostatic (cell growth inhibition without cell death) and cytotoxic (cell death) activities. While normal cells (rat hepatocytes) were not affected (*43*). The antitumor activity of verbascoside was demonstrated also by the inhibition of telomerase activity in tumor cells (*44*). These activities were found to be mainly dependant on the ortho-dihidroxy aromatic (phenolic) system, since methylation of at least one of the phenolic hydroxy groups abolished the activity (*43*). It was demonstrated that either the reactive oxygen species scavenging and antioxidative activities, or the anti-tumor activities of verbascoside was dependent on the number of phenol hydroxyl groups at conjugating positions (*44*).

Our studies have demonstrated that the antioxidant activity of one gram of *L. multiflora* leaves is equivalent to 2.7-12.4 g of Trolox (a Vitamin E analogue). Our observation is in accordance with the fact that verbascoside isolated from *Carypteris incana* exhibited potent radical scavenging activity against DPPH, hydroxyl and superoxide anion radicals (*45*), while this component extracted from *Ballota nigra* showed higher inhibition of LDL oxidation when compared with quercetin and caffeoyl malic acid (*46*), and the extract containing verbascoside from *Pedicularis striata* also inhibited of lipid peroxidation (*47*).

Verbascoside has been reported to exhibit antibacterial and antiviral activities. Verbascoside exhibited moderate activity against *Proteus mirabilis* and *Staphylococcus aureus* (*48*). The ethanolic leaf extracts of *Budleja globosa* showed antibacterial activity against *Staphylococcus aureus* and *Escherichia coli*, and TLC analyses led to the identification of verbascoside (*49*). *In vitro* studies showed that increasing the doses of verbascoside extracted from *Nepeta ucrainica* reduced the intracellular antibacterial activity of neutrophils, suggesting a possible immunosuppressive and antioxidant effect. However, since the immune system is a complex network, further *in vivo* studies are needed to give a definite conclusion (*50*).

Verbascoside isolated from *Sideritis lycia* and extracts from *Verbena officinalis* showed anti-inflammatory activities in mouse paw oedema models (*51-52*). This component also reduced significantly body temperature in rats (*40*), while in other reports this glycoside showed no effects on body temperature (*39*). The presence of verbascoside in *L. multiflora* may explain the uses of this plant to treat fever or as febrifuge (*14-15*), however, pharmacological studies are needed to confirm this activity for *L. multiflora*.

Tea Tasting

The affective sensory evaluation of *L. multiflora* tea was performed to evaluate consumer acceptance using *L. multiflora* leaves from the different populations. This preliminary study showed that the magnitude of like or dislike (hedonic) in the commercial and the Nsawkaw samples received a positive feedback (value higher than 6) (Table III). The aroma and flavor attractiveness also received positive feedback. The commercial sample although containing low levels of essential oils received a positive feedback regarding aroma intensity and attractiveness. The sample from Nsawkaw received a positive feedback and this is probable the best sample for herbal teas since it contains high levels of essential oils and also higher levels of total phenols, antioxidant activity and phenylpropanoid glycosides (Tables I and II). The sample from Amantin was included as a negative control because it contained spicy essential oils (Thymol) (*22*), usually not suitable for teas, thus the tea tasting showed a neutral response for this tea. However, the sample from Nsawkaw contained essential oils rich in 1,8-cineole, a component with refreshing notes, thus being suitable as herbal tea. That observation was also supported by the tea tasting panel. Since the phenol profile was more stable among the varieties, our work suggests that the essential oils is largely responsible for the aroma and flavor of *L. multifora* teas.

The essential oil from the Nsawkaw sample was rich in 1,8 cineole (*22*), the essential oils containing this component exhibited a choleretic effect in rats increasing the bile flux, thus helping in the process of digestion (*53*). Phenolic

Table III. Affective Sensory Evaluation of Different Varieties of *L. multiflora* Teas.

		Ns[1]	Af[1]	Am[1]	Bu[1]	Ho[1]	Ap[1]	Co[1]
Color	Intensity	6.4[2]	5.8	5.9	3.4	5.4	7.3	7.1
	Attractiveness	7.3	6.1	7.0	5.6	6.6	6.6	6.9
Aroma	Intensity	6.0	4.3	6.2	5.0	6.8	6.5	7.0
	Attractiveness	5.8	5.4	6.0	5.4	6.4	5.6	8.0
Flavor	Intensity	5.2	4.8	6.7	5.5	6.3	6.3	6.7
	Attractiveness	5.8	4.8	5.1	5.4	4.2	4.5	7.6
Hedonic[3]		6.2	4.7	5.3	5.1	4.1	4.8	7.3

[1] Ns: Nsawkaw; Af: Aframso bridge; Am: Amantin; Bu: Buemnsuta; Ho, Ho; Ap: Apegu; Co: Commercial.

[2] each character was ranked from 0 to 9 (0 dislike, 5 neutral, 9 like).

[3] Hedonic=magnitude of like or dislike of a herbal tea.

components usually showed hepatoprotective activities because of their antioxidant properties (*19,54*). The hepatoprotective action may have an indirect effect over digestion, through the production of bile, however this activity also need to be confirmed for *L. multiflora*.

Since the amount of total phenols, antioxidant activities and phenylpropanoid glycosides in the commercial samples were low, and in view that the pharmacological activities are largely due to the essential oils and polyphenols, those populations with higher contents of bioactive components and with the most agreeable taste should be selected for commercialization of *L. multiflora* in order to have a product with better functional properties. In addition, a better understanding of the post-harvest handling and processing techniques that optimize the content of active principles are needed.

We propose the following initial standards for *L. multiflora*, thus providing to the users and international community with consistent and defined products (Table IV). The development of clear grades and standards for *L. multiflora* should provide a foundation upon which processors, producers as well as buyers and users can objectively define this product.

Although the phenolic composition of *L. multiflora* has been described (*30,35-36*), this is the first study of the HPLC analysis of polyphenol, antioxidant activity and quality control parameters in different wild varieties of *L. mulitflora*.

Table IV. Quality Standards for *L. multiflora* Dried Leaves

Characteristic	Requirement
Moisture content % (m/m) maximum	12
Total ashes % (m/m) maximum	15
Total insoluble ashes % (m/m) maximum	4
Essential oil content % (m/m) minimum	0.6
Total phenols % (m/m) minimum	4
Antioxidant activity % (m/m) minimum	6
Verbascoside % (m/m) minimum	1

Conclusions

Many reports have shown the hypotensive activity of *L. multiflora* that is due to the muscle relaxation and vasodilatation properties of the extracts. The muscle relaxation properties also support the calming effects of *L. multiflora*. Further studies are needed to identify the constituent(s) responsible for this activity, and it appears that verbascoside has no effect on blood pressure. However, other authors have shown that the isolated verbascoside exhibited analgesic and weak sedation activities, the presence of this component in *L. multiflora* leaves may explain the calming/sedative effects of this plant. This component could be responsible for other activities as well, such as anti-inflammatory and microbial properties against certain virus and bacteria.

L. multiflora is a rich source of bioactive and aromatic components, is devoid of alkaloids (*13*), and is a very promising caffeine-free tea with potential hypotensive and calming effects. *L. multiflora* has been consumed since ancient times as herbal tea and without any reported adverse effects, thus, this species could be considered safe for human consumption, however, further pharmacological studies are needed to confirm this assumption.

This species could also have other additional and important health benefits due to the antioxidant properties of their constituents, these activities include, antitumor, hepatoprotective, and digestive. However, further studies are needed to confirm all these activities in humans.

This study also showed the importance of the evaluation of aromatic and bioactive components from wild populations of *L. multiflora*, and that the natural variation can be used as a source of new components and bioactivities to develop new or improved products as well as the identification of the most promising populations for tea relative to both flavor, aroma and bioactive profiles.

Because *Lippia multiflora* is not well know in the international markets, this plant species may serve as a unique 'African bush tea' in the western marketplace, and as a possible source of aroma chemicals. Such commercialization could generate new commercial opportunities for the West African local communities.

Acknowledgements

We wish to express our thanks and appreciation to, Drs. Charles Quansah and MLK Mensah, for their collaboration in the population collection studies and domestication of Lippia. We thank the Ghanaian communities involved in these domestication and commercialization studies, who, with assistance from the ASNAPP project, were among the first to actually begin to export this specialized tea collected from the wild. We thank Jerry Brown, USAID project officer, for his support and encouragement. This work originally began as part of the Agri-Business in Sustainable African Natural Plant Products Program (ASNAPP) with funding from the USAID (Contract Award No. HFM-O-00-01-00116). We also thank Carol Wilson, USAID Chief Technical Officer of our Partnership for Food and Industry in Natural Products (PFID/NP) project supported by the Office of Economic Growth, Agriculture and Trade (EGAT/AG) of the USAID (Contract Award No. AEG-A-00-04-00012-00) in support of their global economic development programs; and finally to the New

Use Agriculture and Natural Plant Products Program (NUANPP) and the New Jersey Agricultural Experiment Station, Rutgers University.

References

1. Juliani, H.R.; Biurrun, F.; Koroch, A.R.; Oliva, M.M.; Demo, M.S.; Trippi, V.S.; Zygadlo. J.A. *Planta Med.* **2002**, *68*, 756-762.
2. Zygadlo, J.A.; Juliani, H.R. In *Recent Progress in Medicinal Plants. Phytochemistry and Pharmacology II.* Vol 8, Majundar, D.K.; Govil, J.N.; Singh, V.K., Eds. Studium Press LLC: Houston, TX, 2003, pp. 273-291.
3. Juliani, H.R.; Koroch, A.R.; Simon, J.E.; Biurrun, F.N.;Castellano, V.; Zygadlo J.A. *Acta Horticulturae (ISHS)* **2004**, *629*, 491-498.
4. Zunino, M.P.; Lopez, M.L.; Zygadlo, J.A. *Advances in Phytochemistry* **2003**, *29*, 209-245.
5. Iwu, M.A. *African Medicinal Plants.* CRC Press: Boca Raton, FL. 1993, p. 69.
6. Retamar, J.A. In *On Essential Oils.* James, V., Ed., Synthite: India, 1986, pp.124–279.
7. Juliani, H.R. Jr., Juliani, H.R.; Ariza-Espinar, V.S.T.L. *Anales de la Asociación Quimica Argentina,* **1994**, *82(1)*, 53-55.
8. Terblanché, F.C.; Kornelius, G. *J. Essent. Oil Res.* **1996**, *8*, 471-485.
9. Juliani, H.R. Jr.; Koroch, A.R.; Juliani, H.R.; Trippi, V.S. *Anales de la Sociedad Quimica Argentina* **1998**, *86(3/6)*, 193-196.
10. Juliani, H.R. Jr.; Koroch, A.R.; Juliani, H.R.; Trippi, V.S. *Plant Cell Tissue and Organ Culture* **1999**, *59*, 175-179.
11. Juliani, H.R. Jr.; Biurrun, F.; Koroch, A.R.; Juliani, H.R.; Zygadlo, J.A. *Planta Med.* **2000**, *66*, 567-568.
12. Irvine, F.R. *Woody Plants of Ghana.* Oxford University Press: London. UK. 1961, pp. 758-759.
13. Noamesi, B.K.; Adebayo, G.I.; Bamgbose, S.O. *Planta Med.* **1985**, *3*, 253-255.
14. Ayiku, M.N.B. *Ghana Herbal Pharmacopeia.* Technology Transfer Center: Accra, Ghana, 1992, pp. 86-88.
15. Mshana, N.R.; Abbiw, D.K.; Addae-Mensah, I.; Adjannouhoun, E.; Ahyi, M.R.A.; Ekpere, J.A.; Enow-Orock, E.G.; Gbile, Z.O.; Noamesi, G.K.; Odei, M.A.; Odunlami, H.; Yeboah, A.A.O.; Sarprong K.; Sofowora, A.; Tackie, A.N. *Traditional Medicine and Pharmacopoeia. Contribution to the Revision of Ethno botanical and Floristic Studies in Ghana.* Organization of African Unity. Institute for Scientific and Technological Information: Accra, Ghana, 2000, p. 593.

16. Herzog, F.; Gautier-Béguin, D.; Müller, K. *Uncultivated plants for human nutrition in Côte d'Ivoire.* Non-Wood Forest Products 9, Food and Agriculture Organization of the United Nations: Rome, Italy, 1998.
17. Wang, M.; Juliani, H.R.; Simon, J.E.; Ekanem, A.; Liang, C.P.; Ho, C.-T. 2004. In *Phenolics in Foods and Natural Health Products*, Shahidi, F.; C.-T. Ho, Eds.; ACS Symp. Ser., American Chemical Society: Washington, DC., 2005, in press.
18. Duthie, G.G.; Duthie, S.J.; Kyle, J.A.M. *Nutr. Res. Rev.* **2000**, *13*, 79 - 106.
19. Chau-Jong, W.; Wang, J.M.; Lin, W.L.; Chu, C.Y.; Chou, F.P.; Tseng, T.H. *Food Chem. Toxicol.* **2000**, *38*, 411-416
20. Chen, C.C.; Chou, F.P.; Ho, Y.C.; Lin, W.L.; Wang, C.P.; Huang, A.C.; Wang, C.J. J. Sci. Food Agric. 2004, in press.
21. Noamesi, B.K.; Adebayo, G.I.; Bamgbose, S.O.A. *Planta Med.* **1985**, *3*, 256-258.
22. Juliani, H.R.; Acquaye, D.; Moharram, H.; Renaud, E.; Wang, M.; Simon, J.; Mensah, M.L.K.; Fleischer, T.C.; Dickson, R.; Mensah, A.Y.; Quansah, C.; Asare, E.; Akromah, R. 2005, in preparation.
23. Zygadlo, J.A.; Juliani, H.R. Jr. *Curr. Top. Phytochem.* **2000**, *3*, 203-214.
24. Juliani H.R.; Simon, J.E.; Ramboatiana, M.M.R.; Behra, O.; Garvey, A.; Raskin, I. *Acta Horticulturae*, **2004**, *629*, 77-81.
25. Juliani, H.R.; Simon, J.E. In *New Crops and New Uses: Strength in Diversity.* Janick J. Eds., ASHS Press: Virginia, 2002, pp 575-579.
26. Juliani, HR. Ph.D. Dissertation, Facultad de Ciencias Exactas, Fisicas y Naturales, Cordoba, Argentina, 1997.
27. Juliani, H.R. Jr.; Koroch, A.R.; Juliani, H.R.; Trippi, V.S.; Zygadlo, J.A. *Biochem. Systematics Ecol.* **2002**, *30*, 163-170.
28. Koumaglo, K.H.; Akpagana, K.; Glitho, K.; Garneau, A.I.; Gagnon, F.X.; Jean, H.; Moudachirou, F.I.; Addae-Mensah, M. *J. Essen. Oil Res.* **1996**, *8*, 237-240.
29. Kanko, C.; Koukoua, G.; N'Guessan, Y.T.; Lota, M.L.; Tomi, F; Casanova, J. *J. Essen. Oil Res.* **1999**, *11*, 153-158.
30. Taoubi, K.; Fauvel, M.T.; Gleye, J.; Fouraste, C.M.A.I.. *Planta Med.* **1997**, *63*, 192-193
31. Committee on Food Chemical Codex. Food Chemical Codex: National Academy Press: Washington, D.C., 1996.
32. Gao, X.; Bjork, L.; Trajkovski, V.; Uggla, M. *J. Sci. Food Agric.* **2000**, *80*, 2021-2027.
33. Re, R.; Pellegrini, N.; Protegente, A.; Pannala, A.; Yang, M.; Rice-Evans, C. *Free Rad. Biol. Med.* **1999**, *26*, 1231-1237.
34. Muggeridge, M.; Foods, L.; Clay, M. In *Handbook of Herbs and Spices*, Peter, K.V., Ed., CRC Press: Boca Raton. FL, 2001.
35. Chanh, P.H.; Koffi, Y.; Chanh, A.P.H. *Planta Med.* **1988**, *54*, 294-296.

36. Chanh, P.H.; Yao, K.; Chanh, A.P.H. *Prostaglandins, Leukotrienes Essential Fatty Acids* **1988**, *34*, 83-88.
37. Cometa, F.; Tomassini, I.; Nicoletti, M.; Pierretti, S. *Fitoterapia* **1993**, *64*, 195-271.
38. Capasso, A.; Pieretti, S.; Di Giannuario, A.; Nicoletti, M. *Phytotherapy Res.* **1993**, *7*, 81-83.
39. Nakamura, T.; Okuyama, E.; Tsukada, A.; Yamazaki, M.; Satake, M.; Nishibe, S.; Deyama, T.; Moriya, A.; Maruno, M.; Nishimura, H. *Chem. Pharm. Bull.* **1997**, *45*, 499-504.
40. Rodriguez, S.M.; Castro, O. *Revista de Biologia Tropical* **1996**, *44*, 353-359.
41. Pieretti S.; di Giannuario, A.; Capasso, A.; Nicoletti, M. *Phytotherapy Res.* **1992**, *6*, 89-93.
42. Herbert, J.M.; Maffrand, J.P.; Taoubi, K.; Augereau, J.M.; Fouraste, I.; Gleye, J. *J. Nat. Prod.* **1991**, *54*, 1595-1600.
43. Saracoglu, I.; Inoue, M.; Calis, I.; Ogihara, Y. *Biol. Pharm. Bull.* **1995**, *18*, 1396-1400.
44. Zhang, F.; Jia, Z.; Deng, Z.; Wei, Y.; Zheng, R.; Yu, L. *Planta Med.* **2002**, *68*, 115-118.
45. Gao, J.J.; Igalashi, K.; Nukina, M. *Biosci. Biotechnol. Biochem.* **1999**, *63*, 983-988.
46. Seidel, V.; Verholle, M.; Malard, Y.; Tillequin, F.; Fruchart, J.C.; Duriez, P.; Bailleul, F.; Teissier, E. *Phytotherapy Res.* **2000**, *14*, 93-98.
47. LiJi, G.R.C.; Liang, Z.R.; ZiMin, L.; ZhongJian, J. *Acta Pharmacologica Sinica* **1997**, *18*, 77-80.
48. Didry N.; Seidel,V.; Dubreuil, L.; Tillequin, F.; Bailleul, F. *J. Ethnopharmacol.* **1999**, *67*, 197-202.
49. Pardo, F.; Perich, F.; Villarroel, L.; Torres, R. *J. Ethnopharmacol.* **1993**, *39*, 221-222.
50. Akbay, P.; Calis, I.; Ündeger, Ü.; Basaran, N.; Basaran, A.A. *Phytotherapy Res.* **2002**, *16*, 593-595.
51. Akcos, Y.; Ezer, N.; Cals, I.; Demirdamar, R.; Tel, B.C. *Pharma. Biol.* **1999**, *37*, 118-122.
52. Deepak, M.; Handa, S.S. *Phytotherapy Res.* **2000**, *14*, 463-465.
53. Peana, A.; Satta, M.; Moretti, M.D.L.; Orecchioni, M. *Planta Med.* **1994**, *60*, 478-479.
54. Alam, K.; Nagi, M.N.; Badary, O.A.; Al-Shabanah, O.A.; Al-Rikabi, A.C.; Al-Bekairi, A.M. *Pharmacology Res.* **1999**, *40*, 156-163.

Chapter 11

Protein Tyrosine Phosphatases 1B Inhibitors from Traditional Chinese Medicine

Tianying An, Di Hong, Lihong Hu, and Jia Li

National Center for Drug Screening, Shanghai Institute of Materia Medica, Shanghai Institutes for Biological Sciences, Chinese Academy of Sciences, Shanghai 200031, People's Republic of China

In our research for natural products with antidiabetes activity, we screened our extract bank for inhibitors of PTP1B enzyme and found that some fractions from ethanol extract of Traditional Chinese Medicine showed strong inhibitory bioactivity against PTP1B enzyme. Using the PTP1B enzyme bioassay as a guide, chromatography of the fractions afforded several potential PTP1B inhibitors. Modification of oleanolic acid yielded novel more potent and higher selective PTP1B inhibitors.

Introduction

Diabetes is an endemic in industrialized society. A lucrative market is waiting for companies that can find an effective remedy. Diabetes is a common disorder in which blood sugar is ineffectively processed. Type 2 diabetes is the most prolific form, representing ~90% of cases. The primary cause is either insufficient secretion or action of insulin, whose role is to facilitate the passage of glucose into cells. When insulin is scarce or ineffective, extracellular glucose accumulates, starving cells of their energy source and causing a diversity of symptoms that can include blindness and heart disease. Type 2 diabetes can often be controlled by careful lifestyle management, but insulin injections or medication often become necessary. Unfortunately, diabetes drugs often suffer

© 2006 American Chemical Society

from low efficacy as monotherapy and can cause side-effects; therefore, there is a large medical need for an effective treatment.

Protein tyrosine phosphatases 1B (PTP1B) is a well-validated target in type 2 diabetes drug research. This enzyme modulates the signalling cascade that is activated upon insulin binding to its cell-surface receptor. Specifically, PTP1B dephosphorylates tyrosine residues on the receptor, stopping insulin action. Studies with PTP1B knockout mice have demonstrated that the lack of PTP1B activity resulted in an increased activity to insulin and obesity-resistance (1,2). PTP1B appears to be exemplary target for obesity and diabetes. Today, the obesity epidemic is an important factor in the rise of diabetes. Unfortunately, most of today's diabetes drugs, such as peroxisome proliferator-activated receptor (PPAR)-γ agonist and Metformin^{TM} tend to have the side effect of increasing body weight. Therefore, a drug aiming to target both obesity and hyperglycemia would have enormous advantage. Several companies have been pursuing the development of PTP1B inhibitors as drugs (Figure 1) (3). Wyeth began testing Ertiprotafib/PTP112 in Phase II trials in March 2001, but discontinued the trials in June 2002 because of unsatisfactory efficacy, and the occurrence of dose-limiting side effects among several trial participants. Whether these problems stem from this particular chemical serials, species-related differences in PTP1B function, or limited selectivity of the inhibitor for different PTPs is not yet clear. The PTP family of enzymes is large and all are highly specific for the charged phosphotyrosine residue. The chronic dosing required for treating diabetes requires the development of inhibitors with high selectivity for PTP1B over several highly related tyrosine phosphatases. So far, most of the reported PTP1B inhibitors are derived from synthetic compound library by HTS. The PTP1B inhibitors from plants especially from folklore medicinal plants are not reported yet. In this paper we report our effort in searching of PTP1B inhibitors from Traditional Chinese Medicine.

Discovery of PTP1B inhibitors from Traditional Chinese Medicine

National Center for Drug Screening has established an extract bank based on Traditional Chinese Medicine (TCM) since 2000. To date, about 2,000 medicinal plants around China have been collected and 15,000 fractions prepared. Each plant was extracted with 95% ethanol, and the residues partitioned between water and chloroform. The water phase was subjected to AB-8 macro porous resin, eluted with water-ethanol (0:100, 15:85, 30:70, 50:50, 75:25, 95:5) to yield six fractions. In our research work for natural products with anti-diabetes activity from our extract bank since 2001, we have screened about 10,000 fractions for inhibitors against PTP1B enzyme by High-Throughput

Japan Tobacco
Hydroxyphenyl azole derivatives
IC_{50} = 540 nM

Wyeth
Ertiprotafib (Phase II discontinued)
IC_{50} = 384 nM

Sugen
Trifluoromethyl sulfonyl derivatives
IC_{50} = 400 nM
ISIS
ISIS-113715
Antisense oligonucleotide (Phase II)

Ontogen
Cinnamic acid derivatives
IC_{50} = 72 nM

Takeda
Pyrrol phenoxy propionic
acid derivatives
IC50 = 90 nM

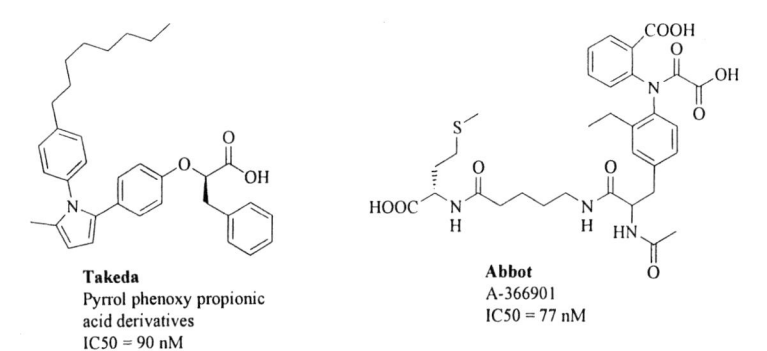

Abbot
A-366901
IC50 = 77 nM

Figure 1A. Companies involved in protein tyrosine phosphatases 1B (PTP1B) inhibitor discovery and representative compounds.

Merk-Frosst
Aryldifluoromethylphosphonic
acid derivatives
>90% inhibition at 10 uM

Albert Einstein College of Medicine
4'-phosphonyldifluoromethyl
phenylalanine derivatives
IC$_{50}$ = 14 μM

Pharmacia
3'-carboxy-4'-(O-carboxymethyl)
tyrosine derivatives
Ki = 870 nM

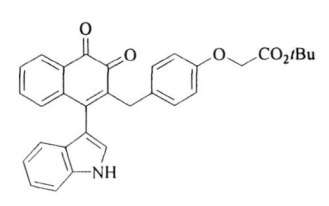

Korea Research Institute of Chemical Technology
1,2-naphtoquinone derivatives
IC$_{50}$ = 650 nM

Figure 1B. Companies involved in protein tyrosine phosphatases 1B (PTP1B) inhibitor discovery and representative compounds (continued).

Screening technique, and found that several fractions from *Ardisia japonia*, *Hypericum erectum*, *Broussonetia papyrifera*, *Ginkgo biloba*, *Eobotrya japonica* showed strong inhibitory activity.

p-Benzoquinonoid Inhibitors from *Ardisia japonica* and *Hypericum erectum*

The chloroform fraction from *A. japonica* showed 69% inhibition against PTP1B at the concentration of 5 µg/mL. Under bioassay-guided purification, four active benzoquinonoids, 5-ethoxy-2-hydroxy-3-[pentadec-10'(Z)-enyl]-1,4-benzoquinone (**1**), 5-ethoxy-2-hydroxy-3-[tridec-8'(Z)-enyl]-1,4-benzo-quinones (**2**), maesanin (**3**), and 2,5-dihydroxy-3-[(Z)-10'-pentadecenyl]-1,4-benzoquinone (**4**) were isolated (see Figure 2). We found compounds **1-4** have the same 2,5-dihydroxy-1,4-benzoquinones skeleton, and more nonpolar substituents of this skeleton increases their inhibitory activities [from **4** (a hydroxyl group at C-6) to **3** (a methoxyl group at C-6), to **2** (an ethoxyl group at C-6)] (see Table I). Two other 1,4-benzoquinone derivatives, erecquinone A (**5**) and B (**6**) were isolated from the active CHCl$_3$ fraction of *H. erectum* (see Figure 2) (*4*). Compound **5**, with two free hydroxyl groups, showed stronger activity than compound **6**, with two masked hydroxyl groups. Liu *et al.* reported that several 3,6-diaryl-2,5-dihydroxy benzoquinones showed significant reduction in hyperinsulinemia in *ob/ob* mice (*6*). In order to study the relationships between the structure of 2,5-dihydroxy-1,4-benzoquinone derivatives and the activity against PTP1B, we have synthesized several aromatic substituted 3,6-diaryl-2,5-dihydroxy benzoquinone derivatives using the same synthetic method reported by Liu *et al.* (*5*). Unfortunately, all of these aromatic substituted benzoquinones showed no inhibitory activity against PTP1B enzyme. These results indicated the long aliphatic chain on the 1,4-benzoquinone was helpful for the inhibitory activity. Further structural modification of 2,5-dihydroxy-1,4-benzoquinonoids will be carried out in our lab.

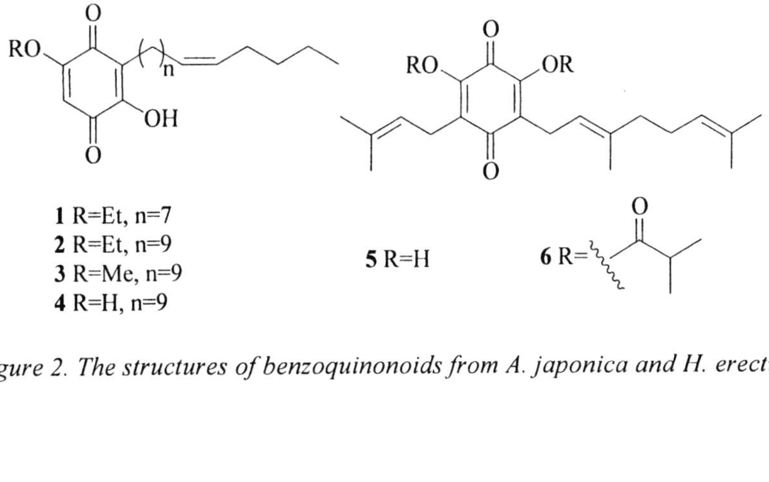

1 R=Et, n=7
2 R=Et, n=9
3 R=Me, n=9
4 R=H, n=9

5 R=H   **6** R=

Figure 2. The structures of benzoquinonoids from A. japonica and H. erectum.

Flavanoid Inhibitors from *Broussonetia papyrifera* and *Ginkgo biloba*

B. papyrifera (L.) (Moraceae) is a deciduous tree, and its fruits have been used for impotency and to treat ophthalmic disorders in China. Extracts of *B. papyrifera* have shown antifungal, antihepatotoxic, antioxidant and lens aldose reductase inhibitory activities. Also, several flavonoid constituents of this plant have been shown to inhibit lipid peroxidation, aromatase activity, and to exhibit antiplatelet effects (*6*). The chloroform fraction from the bark of *B. papyrifera* (L.) Vent. was found to significantly inhibit PTP1B enzyme (93% inhibition at 20 µg/mL). Bioassay-guided fractionation of the chloroform fraction led to the isolation of four flavonoids that were found to be active. They were characterized as quercetin (**7**), uralenol (**8**), 8-(1,1-dimethylallyl)-5'-(3-methylbut-2-enyl)-3',4',5,7-tetrahydroxyflanvonol (**9**) and 5',8-di(3-methylbut-2-enyl)-3',4',5,7-tetrahydroxyflanvonol (**10**) (see Figure 3 and Table I) (*7*). Analyzing their structures and activity, we found that with the introduction of unsaturated aliphatic substituents into the flavanol skeleton, the activity against PTP1B was improved.

G. biloba leaf extract is being widely used for memory improvement, mental alertness, vertigo and tinnitus, allergies-through ginkgo's ability to antagonize, or inhibit platelet activating factor (PAF), altitude sickness-through improved blood flow to the brain to compensate for low oxygen levels at high altitude, early stage Alzheimer's disease, asthma, impotence, intermittent claudication (lameness), macular degeneration, migraines, to deter aging overall-through anti-oxidant-nature of chemical compounds in ginkgo, as a mood enhancer, and more. It was also used in the prevention of complications of diabetes mellitus (*8*). The chloroform fraction of *G. biloba* leaf showed strong inhibitory activity with an IC$_{50}$ of 8.6 µg/mL. Its chemical study yielded two biflavonoid compounds: ginkgetin (**11**, IC$_{50}$ 4.0 µM) and sciadopitysin (**12**, IC$_{50}$ 6.0 µM) (Figure 3).

Triterpenoid Inhibitors from *Eriobotrya japonica*

The leaves of *E. japonica* has been documented for the treatment of various skin disease and diabetes mellitus in folk medicine (*9*). 3,6,19-trihydroxy-urs-12-en-28-oic acid and corosolic acid, isolated from this plant, were reported to reduce blood glucose levels in normoglycemic acid. We found the crude triterpene acids, prepared from this plant, showed high activity against PTP1B enzyme with an IC$_{50}$ of 2.5 µg/mL. Chemical analyses afforded four main active triterpenoids, oleanolic acid (**13**), ursolic acid (**14**), 2α-hydroxyoleanolic acid (**15**), and corosolic acid (**16**) (see Figure 4 and Table I). The quantitative analysis of the crude terpene acids from this plant by HPLC-ELSD indicated the

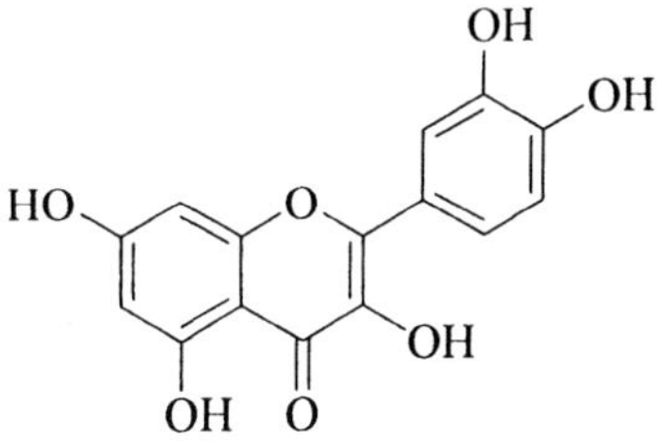

7

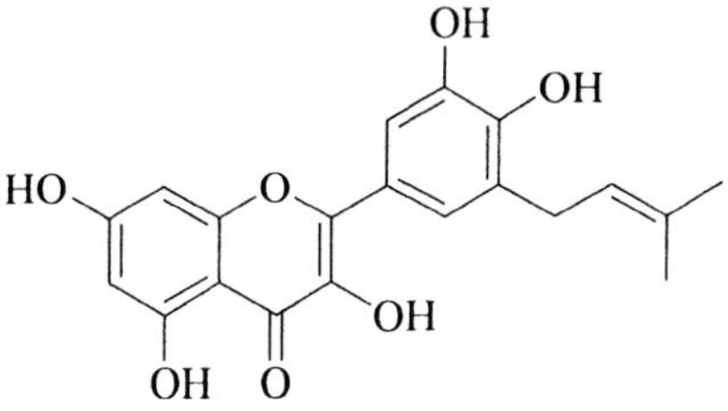

8

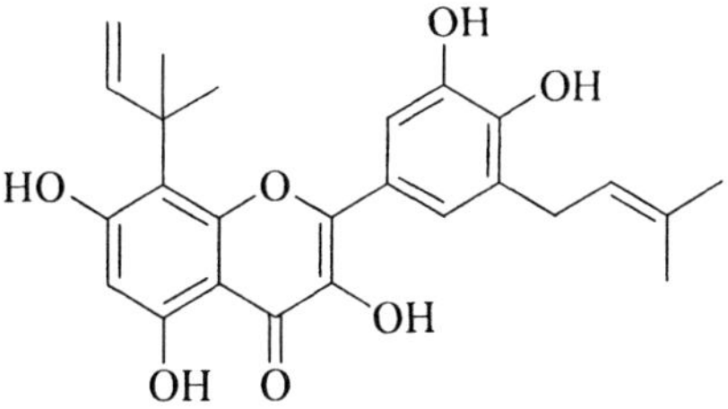

9

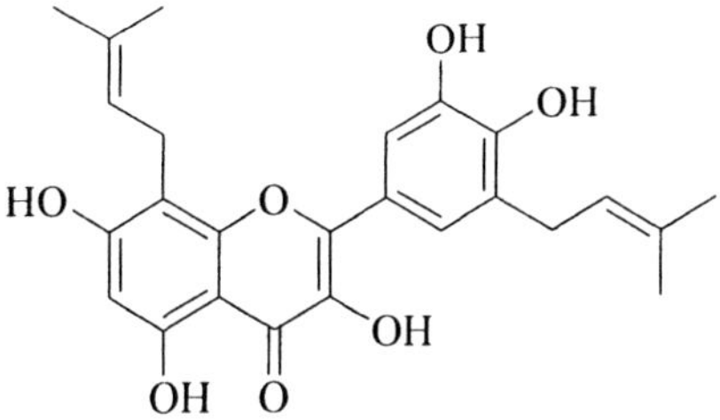

10

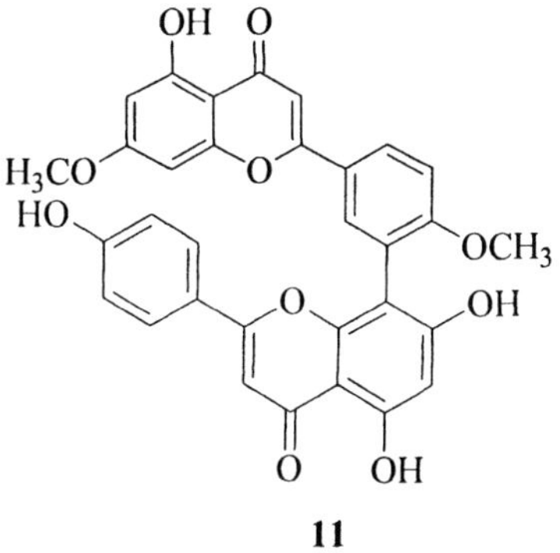

11

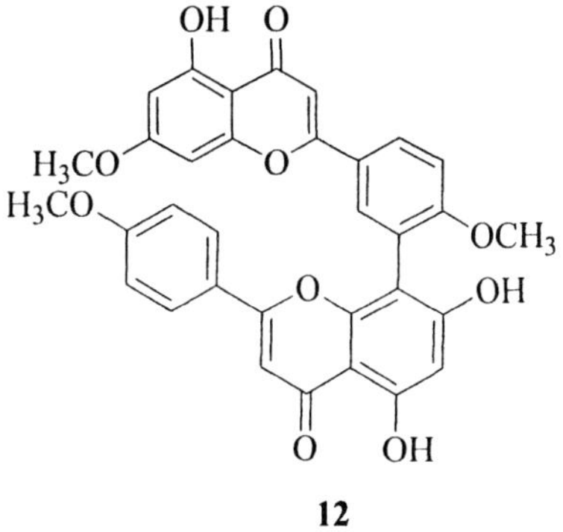

12

Figure 3. The structures of flavanoids from B. papyrifera and G. biloba.

content of oleanolic acid, 2α-hydroxyoleanolic acid, corosolic acid, and ursolic acid was about 1%, 5%, 10%, and 10% (w/w), respectively.

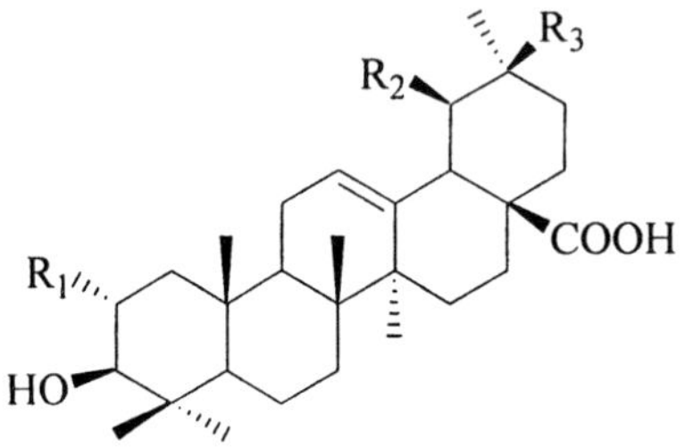

13 R_1=H, R_2=H, R_3=CH_3
14 R_1=H, R_2=CH_3, R_3=H
15 R_1=OH, R_2=H, R_3=CH_3
16 R_1=OH, R_2=CH_3, R_3=H

Figure 4. The structures of triterpenoids from E. japonica.

Table I. The Inhibitory Activity of Natural Compounds Against PTP1B Enzyme

No	IC$_{50}$ (μM)	No	IC$_{50}$ (μM)
1	3.01±1.25	9	4.38±1.05
2	3.63±0.97	10	5.20±2.02
3	4.62±2.41	11	4.03±0.62
4	19.18±3.77	12	6.08±0.39
5	4.40±1.32	13	21.96±1.47
6	12.74±5.37	14	8.11±3.64
7	25.78±2.53	15	4.57±0.56
8	21.55±6.48	16	13.43±4.77

Structural Modification of Oleanolic Acid

Compounds **13-16** are four very common triterpene acids existng in many higher plants. Corosolic acid, an anti-diabetes ingredient from *Lagerstroemia speciosa* leaf, showed significant glucose transport-stimulating activity in Ehrlich ascites tumor cells at a concentration of 1.0 μM (*10*). In a randomized clinical trial type II diabetics, the anti-diabetic activity of an *L. speciosa* leaf extract standardized to 1% corosolic acid (Glucosol[TM]) demonstrated a statistically significant reduction in blood glucose level at 48 mg per day dose for 2 weeks and clinical use suggests it is very safe (*11*). Some oleanolic acid glycosides were reported to have the activity of lowering blood glucose level in oral glucose-loaded rats (*12*). Oleanolic acid has been shown to protect against some hepatotoxicants and it has been used to treat hepatitis as a drug for decades in China (*13*). Oleanolic acid is very cheap and commercial available in China. In order to improve the activity and selectivity of oleanolic acid, structure-activity relationship studies were performed by our group.

Wrobel reported in the docking study of compound **17**, a strong inhibitor of PTP1B (IC_{50} = 61 nM), with the X-ray crystal structure of PTP1B that the carboxylic acid group of **17** binds to the side chain of active site Arg 221 via a charge-charge interaction (*14*). Comparing oleanolic acid with compound **17**, we found they had the similar stereo-structure. Their difference was that the distance between the carboxylic acid and five-ring skeleton in oleanolic acid is shorter than the distance between the carboxylic acid and tetracyclic ring in **17**. If we lengthen the distance between the carboxylic acid and the five-ring skeleton, perhaps its potency would increase. Therefore, we synthesiezed two series of oleanolic acid derivatives: the C-28 long-chain peptide derivatives and the C-28 long-chain acid derivatives. Their structures and activities are listed in **Tables II** and **III**, respectively.

17

When the carboxylic acid of oleanolic acid was masked by methyl group (**18**), it was inactive. We hypothesized that the carboxylic acid of oleanolic acid might bind to the side chain of active site Arg 221 via a charge-charge interaction. Then we linked some types of amino acid to find out which type of amino acid residue will improve the activity. In this experiment, we have found that the activities of oleanolic acid C-28 long-chain peptide derivatives were related to the length of amino acid chain. When the length of carbon chain were

4 or 6, such as compounds **19** and **20,** no activity was observed; and the length of the carbon chain was increased to 11 carbons, the active compound **21** (with the IC$_{50}$ of 2.5 µM) was obtained, whose activity was 10-fold greater than the parent compound **5**. According to the docking study of compound **17** with the X-ray crystal structure PTP1B (*14*), there is a large hydrophobic pocket beside the active site Arg 221. So we inserted a (*R*)-benzyl moiety on the α-carbon and obtained compound **22** (IC$_{50}$: 0.89 µM), which is 3-fold more potent than **21**.

Table II. Structures and Inhibitory Activity of Oleanolic Acid C-28 Long-chain Peptide Derivatives Against PTP1B Enzyme

No	R	IC$_{50}$ (µM)
5	OH	21.90±10.11
18	CH$_3$	>20
19	NH(CH$_2$)$_4$COOH	>20
20	NH(CH$_2$)$_6$COOH	>20
21	NH(CH$_2$)$_{10}$COOH	2.52±1.61
22	HN(H$_2$C)$_{10}$-C(O)-NH-CH(CH$_2$Ph)(R)-COOH	0.89±0.32

When the carbon chain of acid was prolonged by fatty acid instead of amino acid, a series of oleanolic acid C-28 long-chain aliphatic acid derivatives were synthesized. Their structure-activity relationship study results are listed in Table III. When the C-28 long-chain aliphatic acids are saturated acid, they are inactive (compounds **23–25**). Comparing to oleanolic acid C-28 long-chain peptide derivatives, we found the existence of amide bond very important for inhibitory activity against PTP1B enzyme. When the C-28 long-chain residues contain olefinic bonds, their activity was improved. For 3-OH and 3-OMe derivatives, compounds **27** and **32** were the most potent inhibitors, respectively. From above data, we hypothesized that different substituent at C-3 might bind to the different site of enzyme, and the distance between the carboxylic acid and the five-ring skeleton of oleanolic acid is also different.

Table III. Structures and Inhibitory Activity of Oleanolic Acid C-28 Long-chain Acid Derivatives Against PTP1B Enzyme

No	R_1	R_2	IC_{50} (μM)
23	H	~CH_2-COOH	13.45±7.23
24	H	$(CH_2)_8COOH$	>20
25	H	$(CH_2)_{12}COOH$	>20
26	H	~CH_2-CH=CH-COOH	7.80±1.33
27	H	~(CH_2)_2-CH=CH-COOH	0.98±0.34
28	H	~(CH=CH)_2-COOH	1.41±0.95
29	H	~(CH=CH)_3-COOH	>20
30	CH_3	~CH_2-CH=CH-COOH	7.80±2.45
31	CH_3	~(CH_2)_2-CH=CH-COOH	5.72±1.65
32	CH_3	~(CH=CH)_2-COOH	2.60±1.14
33	CH_3	~(CH=CH)_3-COOH	>20

The PTPase domains of receptor and nonreceptor are highly conserved with ~35% mean sequence identity among known phosphatases (*15*). Therefore it is critical that inhibitors of PTPases used for therapeutic purpose show requisite selectivity. Several potent oleanolic acid derivatives were tested on inhibitory activity against CDC25A, CDC25B and PTP-LAR (Table IV). They all have good selectivity over PTP-LAR, and **32** has good selectivity over CDC25A and CDC25B. So we surmised that the hydrophobic substituted at C-3 might interact at a special site on the PTP1B and this site is not existent in CDC25A, CDC25B and PTP-LAR. (See Table IV).

Table IV. The Inhibitory Activity (IC_{50} (µM)) of Compounds 21, 22, 27, 28, 31 and 32 Against PTP1B, CDC25 A, CDC25B, and PTP-LAR.

No	PTP1B	CDC25 A	CDC25 B	PTP-LAR
21	2.52±1.61	6.55	1.5±0.74	>100
22	0.89±0.32	5.50	1.1±0.11	>100
27	0.98±0.34	1.56	2.52±0.21	>100
28	1.41±0.95	>100	0.85±0.12	>100
31	5.72±1.65	>100	4.50±1.3	>100
32	2.60±1.14	>100	>100	>100

Biological Activity Evaluation

All PTPases are C-terminal truncated, soluble form of recombinant human PTPases. Their catalytic activities were routinely measured as rates of hydrolysis of *p*-nitrophenyl phosphate (pNPP) in a 96-well microtiter plate format (*16*). Na_3VO_4 acts as positive control (IC_{50} = 2 µM). Three independent measurements were performed for IC_{50} determinations. Similar results were obtained in multiple measurements. The reported values are the average of all experiments and the errors are standard deviations.

Conclusions

Several natural PTP1B inhibitors, *p*-benzoquinonoids, isoprenylflavanoids, biflavanois, and triterpenoic acids, were discovered from TCM. Modifying the structure of oleanolic acid, we have gotten more potent PTP1B inhibitors with high selectivity over CDC25A, CDC25B and PTP-LAR.

Acknowledgements

This work was supported by the National Natural Science Fund of China (30100229 and 30371679) and the Science and Technology Development Fund of Shanghai, China (01QB14051).

References

1. Elchebly, M.; Payette, P.; Michaliszyn, E.; Cromlish, W.; Collins, S.; Loy, A.L.; Normandin, D.; Cheng, A.; Himms-Hagen, J.; Chan, C.C.; Ramachandran, C.; Gresser, M.J.; Trmblay, M.L.; Kennedy, B.P. *Science* **1999**, *283*, 1544-1558.
2. Klaman, L.D.; Boss, O.; Peroni, O.D.; Kim, J.K.; Martino, J.L.; Zabotny, J.M.; Moghal, N.; Lubkin, M.; Kim, Y.-B.; Sharp, A.H.; Stricker-Krongrad, A.; Shulman, G.I.; Nell, B.G.; Kahn, B.B. *Mol. Cell Biol.* **2000**, *20*, 5479- 5489.
3. Huijsduijnen,, R.H.; Bombrun, A.; Swinnen, D. *Drug Discovery Today* **2002**, *7*, 1013-1019.
4. An, T.Y.; Shan, M.D.; Hu, L.H.; Liu, S.J.; Chen, Z.L *Phytochemistry* **2001**, *59*, 395-398.
5. Liu, K.; Xu, L.; Szalkowski, D.; Li, Z.; Ding, V.; Kwei, G.; Huskey, S.; Moller, D.E.; Heck, J.V.; Zhang, B.B.; Jones, A.B. *J. Med. Chem.* **2000**, *43*, 3487-3494.
6. Lee, D.H.; Bhat, K.P.L.; Fong, H.H.S.; Farnsworth, N.R.; Pezzuto, J.M.; Kinghorn, A.D. *J. Nat. Prod.* **2001**, *64*, 1286-1293.
7. Chen, R.M.; Hu, L.H.; An, T.Y.; Li, J.; Shen, Q. *Bio. Med. Chem. Lett.* **2002**, *12*, 3387-3390.
8. Savickiene, N.; Dagilyte, A.; Lukosius, A.; Zitkevicius, V. *Medicina* **2002**, *38*, 970-975.
9. Tommasi, N.D.; Simone F.D.; Cirino, G.; Cicala, C.; Pizza, C. *Planta. Med.* **1991**, *57*, 414-416.
10. Murakami, C.; Myoga, K.; Kasai, R.; Ohtani, K.; Kurokawa, T.; Ishibashi, S.; Dayrit, F.; Padolina, W.G.; Yamasaki, K. *Chem. Pharm. Bull.* **1993**, *41*, 2129-2131.
11. Judy, W.V.; Hari, S.P.; Stogsdill, W.W.; Judy, J.S.; Passwater, R. *J. Ethnopharmacol* **2003**, *87*, 115-117.
12. Yoshikawa, M.; Matsuda, H. *Biofactors* **2000**, *13*, 231-237.
13. Liu, J.; Liu, Y.; Mao, Q.; Klaassen, C.D. *Fundam. Appl. Toxicol.* **1994**, *22*, 34-40.

14. Wrobel, J.; Scredy, J.; Moxham, C.; Dietrich, A.; Li, Z.; Sawicki, D.R.; Seestaller, L.; Wu, L.; Katz, A.; Sullivan, D.; Tio, C.; Zhang, Z.Y. *J. Med. Chem.* **1999**, *42*, 3199-3202.
15. Barford, D.J.; Jia, A.; Tonks, N.K. *Nature Struct. Biol.* **1995**, *2*, 1043-1053.
16. Zhang, Z.Y.; Dixon, J.E. *Adv. Enzymol.* **1994**, *68*, 1-36.

Chapter 12

Thioglucosidase-Catalyzed Hydrolysis of the Major Glucosinolate of Maca (*Lepidium meyenii*) to Benzyl Isothiocyanate
Mini-Review and Simple Quantitative HPLC Method

Matthew W. Bernart[1,2]

[1]Analytical Laboratory, Herb Pharm, Williams, OR 97544
[2]Current address: Science Department, Rogue Community College, 3345 Redwood Highway, Grants Pass, OR 97527

This HPLC method is for the quantification of benzyl isothiocyanate (BITC, **1**) released by the action of the thioglucosidase enzyme on the substrate glucotropaeolin (**2**), the predominant glucosinolate of maca hypocotyls. Maca, a native Peruvian member of the cabbage family, is a popular herbal product. Details of the HPLC method for BITC, enzymatically produced from maca powder extracts, as well as highlights of BITC's relevance to human health, are provided.

Introduction

Maca root, or hypocotyl, (*Lepidium meyenii* Walpers), also known as *L. peruvianum*, is now receiving nearly worldwide popularity as a nutritional supplement, following centuries of obscurity (*1*). The glucosinolate content of maca, a potential indicator of herbal quality, has been recently shown to consist predominantly of glucotropaeolin (**2**, *2-4*). Best known in, but by no means restricted to the Brassicaceae, or cabbage family, glucosinolates constitute a structure class containing over 120 different molecules, all of which are

© 2006 American Chemical Society

substrates for the enzyme β-thioglucosidase (E.C.3.2.3.1), also called myrosinase (*3,5*). Depending on conditions but especially near pH 7, this water-dependent hydrolysis often yields volatile and bioactive isothiocyanates (ITC) in addition to other compounds (*5-8*). Considering the literature, we deduced that benzyl isothiocyanate (BITC, **1**) enzymatically produced from glucotropaeolin contained in maca (Figure 1), is a biologically relevant marker compound and so endeavored to devise a simple, rapid analytical high performance liquid chromatographic (HPLC) method that could be used to quantify BITC directly from hydroethanolic solutions containing thioglucosidase and maca extractives.

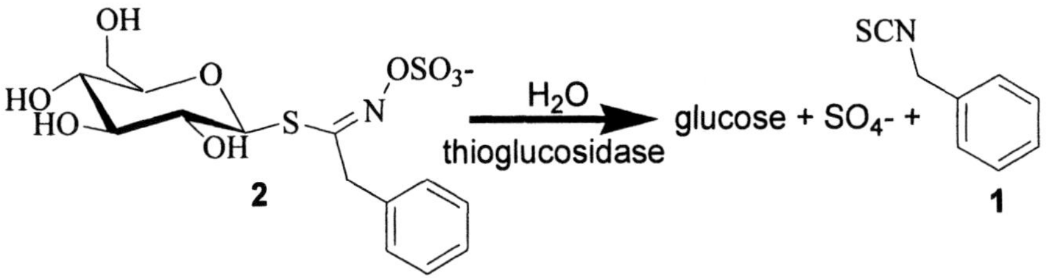

Figure 1. Thioglucosidase enzyme reaction using glucotropaeolin (2) substrate.

Rationale

The biological importance of BITC to human health hinges on the thioglucosidase activity resident in the intestinal flora (*5,7-9*), thus giving an organism consuming dietary glucotropaeolin at least the capability of releasing BITC into the digestive tract. Among the variety of ITCs released from glucosinolates by the enzyme, BITC possesses some unique biological activities. Over three decades ago, BITC from papaya fruit was reported as fungistatic to sporangia of *Phytophthora parasitica* (*10*). Papaya seed extract also contains BITC, which was shown to be anthelmintic to the nematode *Caenorhabditis elegans* (*11*). In vitro work has demonstrated that, among all ITCs and nitriles tested, BITC was the most potent antiproliferative against human erythroleukemic K562 cells (*12*). The mechanism of BITC's chemopreventive action may be exerted through the inactivation of a cytochrome P450 enzyme that is involved in the carcinogenesis process (*13*). Although it appears that most ITCs inhibit cancer when administered before or during carcinogen exposure, BITC inhibited mammary tumor formation in rats when administered subsequent to treatment with the carcinogen 7,12-dimethylbenz[*a*]anthracene (*14*). Taken together, these studies indicate significant bioactivity, and hence, a need for a simple analysis to quantify enzymatically produced BITC in foodstuffs and herbal products.

Methods of BITC and Glucosinolate Analysis

Historically, BITC was quantified by a gas chromatographic (GC) method that, in the case of papaya, required extensive sample cleanup (*10*). This methodology also required endogenous thioglucosidase activity in the biomass itself to produce the analyte, and so would not be applicable to dehydrated or otherwise processed samples, nor to hydroethanolic extracts or tinctures. An isocratic HPLC method, requiring a derivitization step, has also been reported (*15*). Isocratic elutions are often unsuitable for complex herbal matrices; hence, gradient elution on HPLC is preferred (*16,17*). HPLC methods for the nonvolatile glucosinolates have appeared in the literature (*2,18-20*); however, these methods may require extensive sample cleanup and ion-pairing conditions; moreover, the glucosinolate standards, if available, are quite expensive. As reported in the literature, major glucosinolates of maca are depicted in Figure 2:

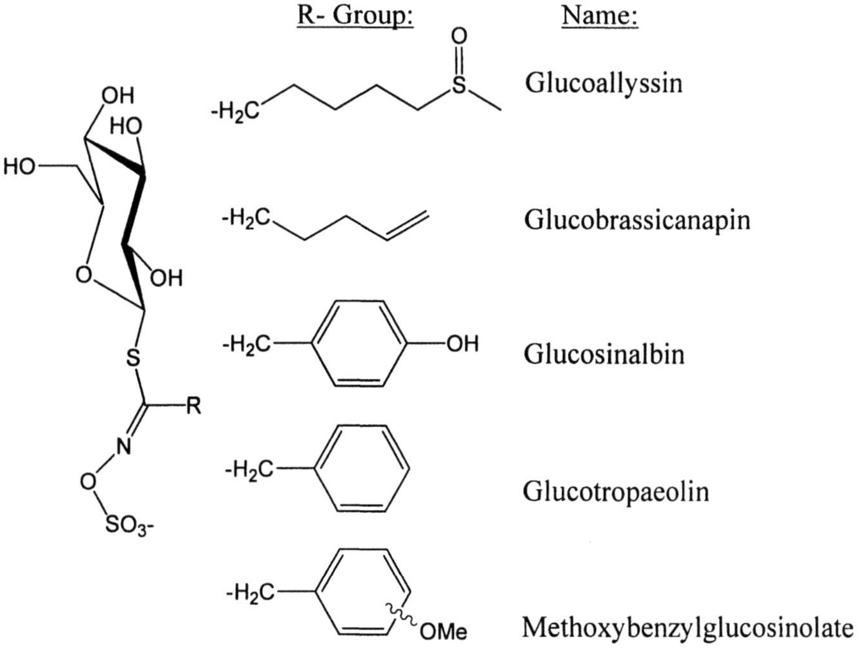

Figure 2: Major glucosinolates of maca: substrates for thioglucosidase hydrolysis to corresponding isothiocyanates.

HPLC-amenable glucosinolate derivatives from maca have been produced via a sulfatase enzyme reaction to produce desulfoglucosinolates (*4*); however, the availability of these compounds as standards is also in question. The same thioglucosidase enzyme used in the present work was recently applied to the quantification of maca glucosinolates by Dini *et al.* (*3*). In their work, this enzyme was demonstrated to exhibit hydrolytic activity in the presence an of organic solvent, dichloromethane. Furthermore, the glucotropaeolin content was estimated from the glucose liberated from the hydrolysis, as measured by GC-MS, not by direct detection of BITC. Dini's thioglucosidase-coupled technique requires extensive sample cleanup to ensure that other glucose pools in the plant material do not interfere with quantitative analysis.

A HPLC method for BITC and its metabolites formed by microsomal cytochrome P450 2B1 has been reported (*21*). A reversed-phase method on a C-18 column, it employs a linear gradient of acetonitrile in an aqueous buffer, monitored by liquid scintillation counting. Under those conditions, the retention time of BITC was 39.9 minutes. Other technologies that have seen utility in investigations of this structure class have included ^{1}H NMR (*22*), as well as micellar electrokinetic capillary chromatography of the parent glucosinolates (*23*). Because BITC is so reactive, especially with methanol or any substance capable of generating nucleophilic species in solution (*10,24*), we wanted to achieve reproducible chromatography using pure water with no modifiers in the aqueous phase, and to eliminate the use of methanol in any of the extraction and chromatographic procedures.

Experimental

Safety

A lab coat and safety glasses should be worn while performing all bench work, especially the heating of sealed EtOH-containing vials and the subsequent syringe filtering of the solutions contained therein. Good ventilation is necessary as the use of acetonitrile is required. General considerations for the use and storage of flammable solvents apply.

Materials

All organic solvents were hplc grade except for 95% EtOH, which was food grade. Distilled water (Sparkletts, Albertsons) was filtered through a 0.45 μm nylon membrane prior to use. Thioglucosidase, isolated from *Sinapis alba*, was purchased from Sigma. One enzyme unit is defined as producing 1.0 μmole glucose per min at pH 6, 25° C. The number of enzyme units per mg of enzyme preparation may vary from lot to lot. Glucotropaeolin (2) and BITC (1) were obtained from Chromadex. Maca dried-hypocotyl (maca powder) was obtained from commercial sources and from the Herb Pharm quality assurance department retain samples. Hydroethanolic maca extract was also provided from Herb Pharm stock.

Instrumentation and Equipment

A Dionex Summit HPLC system, consisting of a Gina 50T autosampler, a P580A LPG pump, and a UVD 170S UV-Vis detector, all controlled by Chromeleon™ software, was employed. An Eppendorf TC-30 column heater and TC-50 temperature controller was used to control the column temperature. The column was a YMC Pro™ C18, 5μ, 120 Å, 4.6 x 150 mm, protected by a guard column of the same stationary phase, 4 x 20 mm, and an Upchurch A-430 2μm PEEK inline filter. Samples were sonicated and heated in a Branson 2210 sonicator / heater. Volumetric flasks and pipets were class A. Fluid transfers of one ml or less were handled with a Gilson P-1000 Pipetman.

Chromatographic Conditions

The column temperature is maintained at 40° C, the mobile phase flow rate set to 1.0 ml / min, and the detector set to 246 nm (1 nm bandwidth). Five to 30 μl injections, typically made for expected BITC concentrations ranging from 0.02 – 0.2 mg / ml, fell within the linear range of the detector. The gradient elution profile is outlined in Table I below:

Table I. Linear Gradient Elution of BITC on YMC Pro™ C18 Column

Time (minutes)	% H_2O	% Acetonitrile
-10	97	3
0	97	3
5	92	8
9	52	48
23	38	62
25	1	99

Standard and Sample Preparation

BITC (**2**, oily liquid), kept refrigerated until use, was diluted with acetonitrile using volumetric glassware to 0.02 – 0.2 mg / ml, then kept in a freezer for as long as one week with no evidence of decomposition. Maca powder (100 mg) was sonicated for 1 hr at 50° C in 10 ml of 47.5% EtOH/H$_2$O (v/v). Separately, 10 enzyme units are dissolved in 10 ml H$_2$O. Using a volumetric pipet, transfer at least one enzyme unit, in its aqueous solution, into a volumetric flask, calculating that at least one enzyme unit is needed for every four ml of hydroethanolic maca extract. Bring the flask to volume with the maca extract, recording the dilution of the extract obtained, and mix well. Transfer an aliquot of extract-enzyme mixture to a scintillation vial or other suitable gas-tight container and incubate in a water bath at 45° C for 30 minutes, then filter through a 0.45 µm PTFE syringe-filter prior to injection on HPLC.

Results and Discussion

The chromatogram of an acetonitrile solution of the BITC standard is seen in Figure 3 below:

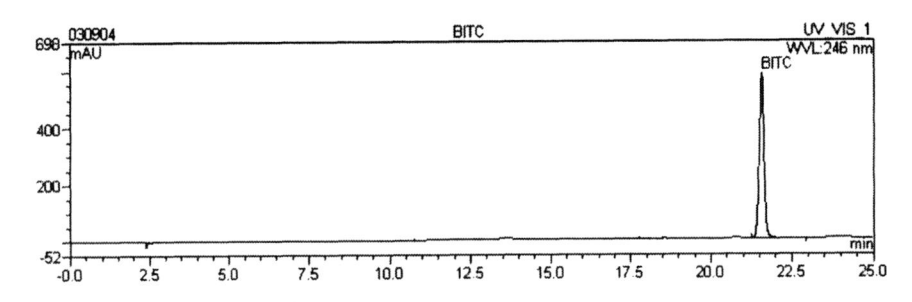

Figure 3. HPLC chromatogram at 246 nm of BITC standard in acetonitrile.

The typical chromatographic efficiency was calculated by the Chromeleon™ software to be 85,000 to 100,000 theoretical plates, according to the European Pharmacopoieal method. Using a standard least-squares linear regression including a y-intercept, a standard curve was generated for the quantification of each day's analyses. For example, a six-point standard curve, plotting peak area at 246 nm versus concentration, was generated for a set of low BITC concentration standards. The equation takes the form y = mx + b, where m is the slope and b is the y-intercept, giving the linear relationship depicted in Figure 4 below:

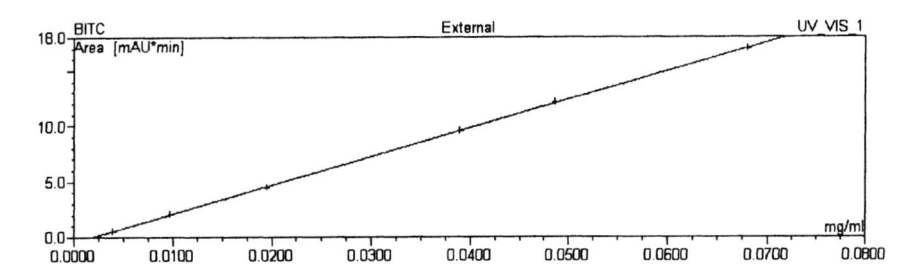

Figure 4. BITC standard curve with slope 257, y-intercept 0.4, and r^2 0.9999.

When a BITC solution in acetonitrile was kept in the freezer, it gave three standard curves with the same slope for assays conducted over three successive days.

Stability Indication

One of the important parameters to be addressed in a single laboratory method development project is that of stability indication. A valid analytical method should be able to distinguish between an analyte and its degradation products. Due to its inherent reactivity, it was simple to produce degradants of BITC with heat, degradants that were fortuitously separated using this hplc method. When BITC was heated at 80°C for one hour on a water bath in 50% EtOH/H$_2$O, the chromatogram (Figure 5) seen below resulted:

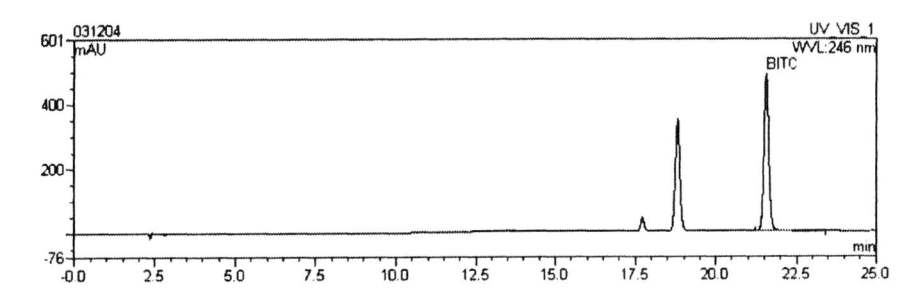

Figure 5. Chromatogram of hydroethanolic BITC solution following heating.

In addition to the prominent BITC peak eluting at 21.6 minutes, two peaks attributed to BITC degradants are observed in the 18-19 minute range. By this method, BITC was determined to be more heat-stable in acetonitrile than in the alcohols MeOH, EtOH, or iPrOH.

Development of the Method

Though thioglucosidase from *Sinapis alba* was active in organic solvent (*3*), we chose an aqueous environment for these reasons:

1. It was easier to obtain a homogeneous enzyme preparation using only distilled water for dissolution.
2. Maca is very sugary (*2*), with even more sugar produced in the hydrolysis.
3. A commercially available maca product was a hydroethanolic extract.

We found BITC levels in maca extracts made with 50/50 (vol/vol) aqueous mixtures of tetrahydrofuran, acetone, dioxane, or acetonitrile to be one half those seen in the hydroethanolic extract, all other variables held constant (data not shown). Given that the water / *n*-octanol partition coefficient of BITC has been determined at 0.006, the most lipophilic ITC tested (*25*), we needed a relatively high percentage of organic solvent in the sample prep. In a series of experiments where the ethanol /water ratio was varied, the optimum for production of BITC from maca extracts was determined to be 47.5% EtOH/water (vol/vol). Figure 6 below shows a chromatogram obtained from the thioglucosidase reaction with a maca hydroethanolic extract as per sample prep in the Experimental Section:

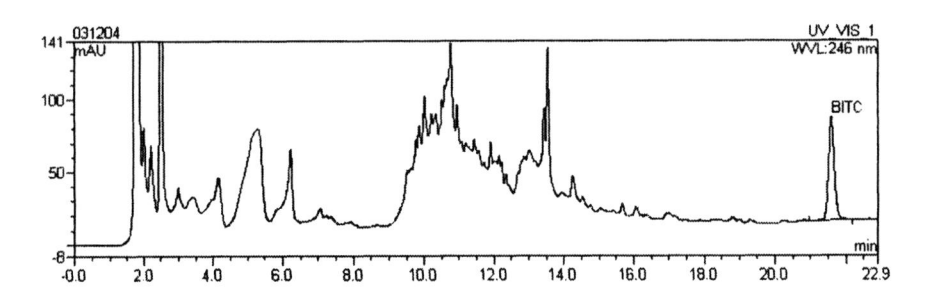

Figure 6. HPLC chromatogram obtained from a maca extract incubation with thioglucosidase enzyme preparation.

An earlier version of our hplc method did not incorporate the high-aqueous flush at the beginning of the gradient. We found flushing (Table 1, minus 10 to 5 minutes) to be a crucial component of the method. Without it, filters and precolumns became irreversibly clogged after a few samples. This presumably occurred when sugars in the samples precipitated in the HPLC system upon encountering a strong organic (40% acetonitrile) mobile phase. With the current modification to the gradient, which accommodates high sample throughput before any significant systemic pressure spikes, it appears that sugars are eluted from the column without precipitating. Further experiments showed that given the conditions in the experimental section, the biomass can be exhaustively extracted at 10 mg biomass per ml extraction solvent.

An Italian group (2) has reported the endogenous BITC content of dry maca tuber by GC/MS, without enzyme treatment, to be 32 ng/g. Such a small amount may be well below the limits of detection and quantification of the proposed method. In fact, when a maca extract was injected on HPLC without previous enzymatic treatment, we obtained the chromatogram seen in Figure 7 below:

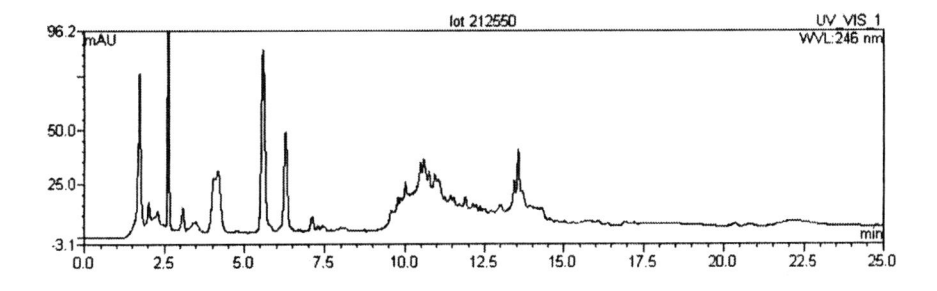

Figure 7. HPLC chromatogram of a maca extract obtained without pretreatment with thioglucosidase enzyme preparation.

Note that although the vertical scale of Figure 7 is magnified approx 150% relative to Figure 6, no peak corresponding to BITC was detected. Concentrated hexane extracts of maca powder were reduced *in vacuo* and redissolved in minimal acetonitrile for injection on hplc; however, no BITC peak was observed (data not shown). On the other hand, authentic glucotropaeolin standard was converted to BITC when subjected to the experimental enzyme treatment. The proposed HPLC system here is unsuitable for the separation of the glucotropaeolin substrate (data not shown). Not only has the change in chromophore, a result of the enzymatic transformation (Figure 1), rendered 246 nm an unsuitable wavelength for the detection of glucotropaeolin (2), but also the lack of ion pairing is not amenable to its separation. Early versions of the proposed HPLC method included the option of detecting at 211 nm, an alternative λ_{max} for BITC, but a sloping baseline eliminated that wavelength from consideration. Our evidence suggests that with this method, detectable amounts of BITC derived from maca result only from endogenous glucotropaeolin substrate reacting with the added thioglucosidase enzyme. A similar analysis of extract of fresh watercress, *Nasturtium officinale*, a related food plant, demonstrated thioglucosidase-dependent production of later-eluting compound than BITC. Presumably, this was the ITC resulting from gluconasturtiin hydrolysis, 2-phenylethylisothiocyanate, but being beyond the scope of this project, it was not pursued further.

Method Optimization

Although no method is validated before it has been run in another laboratory, there are a number of useful experiments that may be performed by a method initiator. A series of assays fixed the optimum temperature at 45°C, the same optimum reported for a thioglucosidase isolated from the cabbage aphid *Brevicoryne brassicae* (26). Our work showed incomplete hydrolysis below this temperature, and evidence of enzyme denaturation at 50°C and higher. Similar experiments showed a 30-minute incubation time to be optimum. No significant change in BITC level was detected following addition of ascorbic acid (26-30) to the maca extracts. It is not known whether enough Vitamin C is already present in maca to negate the effect of exogenous Vitamin C addition.

A C-16 amide column was at first employed in this research, but it exhibited lower retention and resolution of BITC than the C-18 column chosen. Spiked recovery experiments, in which BITC in acetonitrile was diluted with maca extract minus enzyme and injected on HPLC, showed recoveries as low as 103-106%. Apparently, a constant error in autosampler volumetric accuracy, resulting in significantly negative y-intercepts, introduced error at low BITC concentrations. More rigorous treatment and better technique may bring recoveries closer to 100%. The precision of the peak areas of quadruplicate enzyme incubations of maca extract gave a relative standard deviation (RSD) = 2.9%.

Quantification of Glucotropaeolin in Maca

The proposed method is different from most herbal analyses, as it is direct measurement of BITC, a metabolite that like most ITCs is generally not present in the herbal material, but is quickly produced by endogenous thioglucosidase in response to herbivory or other cellular disruption as a chemical defense (5,7,31). The characteristic pungent tastes of mustard oil, wasabi, horseradish, and other cabbage family plants arise from isothiocyanates derived from glucosinolate storage molecules. The proposed method may be used to quantify the amount of glucotropaeolin (**2**). This is an ionic species, and like all glucosinolates occurs mainly as its potassium salt (7), MW 447.5. The method assumes its quantitative conversion to BITC (**1**), MW 149.2. Therefore, the BITC result for a sample (mg/ml, % wt/wt, etc.) is multiplied by three to give the content of glucotropaeolin potassium salt. By this measure, the commercial hydroethanolic extract assayed at 0.6 mg/ml glucotropaeolin potassium, whereas the three different dry root powders, run in duplicate, assayed at 0.5, 0.6, and 0.75% wt/wt glucotropaeolin potassium, uncorrected for moisture content or loss-on-drying.

Conclusion

Values for the glucosinolate content of maca root have only been published recently. Li et al.'s assay (4) of 17 μmol/g of compound **2** in maca fresh hypocotyl is in the range of what we observed in dry hypocotyl samples. The admittedly semi-quantitative figures of Piacente et al. (2) are slightly lower but may have been compromised by the high ethanol content, 70%, used to perform the extractions. In their comprehensive review on the diversity of glucosinolates in the plant kingdom, Fahey el al. (5) found excellent correlation between assays based on enzymatic conversion to ITCs and those based on standard methods for glucosinolates. The proposed method is relatively simple, inexpensive, and precise, and shows promise for the measurement of many different bioactive ITCs produced in other plant materials. Advances in HPLC detector technology such as evaporative light-scattering, diode-array, and mass spectrometric detectors may already make it possible to evaluate the complex glucosinolate profiles found in food and crop plants around the world. BITC, although it can be produced by a number of species (5), is important as a marker compound because of its biological activity.

The importance of drying maca root material following harvest cannot be overemphasized for the preservation of the glucosinolates. The time-honored drying method of Peruvian Altiplano inhabitants is to clean the roots and spread them on tarps to dry in the sun and the cold, dry atmosphere in that extremely high altitude, protecting them from rain when necessary (32-35). Evidently, this practice is sufficient to remove enough water from the biomass and prevent premature hydrolysis of the glucosinolates. In fact, in the two months required for fresh roots to dry, they lose 75% of their original weight (1), prior to their being milled into the typical powder of commerce. Whether there is a direct connection between the reported aphrodisiac and fertility enhancing properties of maca (*35-38*) and its glucosinolate content deserves further study.

Acknowledgement

We are grateful to Professor Steven C. Petrovic, Department of Chemistry, Southern Oregon University, for recording the UV spectrum of benzyl isothiocyanate.

References

1. Smith, E. *J. Am. Herbalists Guild* **2003**, *4*, 15.
2. Piacente, S.; Carbone, V.; Plaza, A.; Zampelli, A.; Pizza, C. *J. Agric. Food. Chem.* **2002**, *50*, 5621.
3. Dini, I.; Tenore, G.C.; Dini, A. *Biochem. Syst. Ecol.* **2002**, *30*, 1087.
4. Li, G-y.; Ammermann, U.; Quirós, C.F. *Econ. Bot.* **2001**, *55*, 255.
5. Fahey, J.W.; Zalcmann, A.T.; Talalay, P. *Phytochemistry* **2001**, *56*, 5.
6. Fahey, J.W.; Zhang, Y-s.; Talalay, P. *Proc. Nat. Acad. Sci. U.S.A.* **1997**, *97*, 10367.
7. Van Etten, C.H.; Tookey, H.L. In *Herbivores: Their Interaction with Secondary Plant Metabolites;* Rosenthal, G.A.; Janzen, D.H., Eds.; Academic Press, New York, 1979; pp 471-500.
8. Keck, A-S.; Finley, J.W. *Integ. Cancer Ther.* **2004**, *3*, 5.
9. Kassie, F.; Rabot, S.; Kundi, M. Chabicovsky, M.; Qin, H-M.; Knasmüller, S. *Carcinogenesis* **2001**, *22*, 1721.
10. Tang, C-S. *Phytochemistry* **1971**, *10*, 117.
11. Kermanshai, R; McCarry, B.E.; Rosenfeld, J.; Summers, P.S.; Weretilnyk, E.A.; Sorger, G.J. *Phytochemistry* **2001**, *57*, 427.
12. Nastruzzi, C; Cortesi, R.; Esposito, E.; Menegatti, E.; Leoni, O.; Iori, R.; Palmieri, S. *J. Agric. Food Chem.* **2000**, *48*, 3572.
13. Goosen, T.C.; Kent, U.M.; Brand, L.; Hollenberg, P.F. *Chem. Res. Toxicol.* **2000**, *13*, 1349.
14. Hecht, S.S. *J. Nutr.* **1999**, *129*, 768S.
15. Zhang, Y.; Wade, K.L.; Prestera, T.; Talalay, P. *Anal. Biochem.* **1996**, *239*, 160.
16. He, X-g.;Lin, L.; Bernart, M.W.; Lian, L. *J. Chromatogr. A* **1998**, *796*, 327.
17. He, X-g.; Bernart, M.W.; Lin, L.; Lian, L. *J. Chromatogr. A* **1998**, *815*, 205.
18. Lykkesfeldt, J.; Møller, B.L. *Plant Physiol.* **1993**, *102*, 609.
19. Du, L-c.; Halkier, B.A. *Plant Physiol.* **1996**, *111*, 831.

20. Betz, J.M.; Fox, W.D. In *Food Phytochemicals for Cancer Prevention I: Fruits and Vegetables*; Huang, M.T.; Osawa, T.; Ho, C.-T.; Rosen, R.T., Eds.; ACS Symp. Ser. 546; American Chemical Society: Washington, D.C., 1994; pp. 181-196.
21. Goosen, T.C.; Mills, D.E.; Hollenberg, P.F. *J. Pharmacol. Exp. Ther.* **2001**, *296*, 198.
22. Combourieu, B.; Elfoul, L.; Delort, A-M.; Rabot, S. *Drug Metab. Disp.* **2001**, *29*, 1440.
23. Paugam, L.;Menard, R.; Larue, J.P.; Thouvenot, D. *J. Chromatogr. A* **1998**, *864*, 155.
24. Tang, C-S.; Bhothipaksa, K.; Frank, H.A. *Appl. Microbiol.* **1972**, 23, 1145.
25. Nastruzzi, C; Cortesi, R.; Esposito, E.; Menegatti, E.; Leoni, O.; Iori, R.; Palmieri, S. *J. Agric. Food Chem.* **1996**, *44*, 1014.
26. Pontoppidan, B.; Ekbom, B.; Erikkson, S.; Meijer, J. *Eur. J. Biochem.* **2001**, *268*, 1041.
27. Botti, M.G.; Taylor, M.G.; Botting, N.P. *J. Biol. Chem.* **1995**, *270*, 20530.
28. Shikita, M.; Fahey, J.W.; Golden, T.R.; Holtzclaw, W.D.; Talalay, P. *Biochem. J.* **1999**, *341*, 725.
29. Burmeister, W.P.; Cottaz, S.; Rollin, P.; Vasella, A.; Henrissat, B. *J. Biol. Chem.* **2000**, *275*, 39385.
30. Björkman, R.; Lönnerdal, B. *Biochem. Biophys. Acta* **1973**, *327*, 121.
31. Bones, A.M.; Rossiter, J.T. *Physiol. Plant.* **1996**, *97*, 194.
32. León, J. *Econ. Bot.* **1964**, *18*, 122.
33. Ochoa, C. *Econ. Bot.* **2001**, *55*, 344.
34. Quirós, C.F.; Cárdenas, R.A. In *Andean Roots and Tubers: Ahipa, arracacha, maca and yacon*; Hermann, M.; Heller, J., Eds.; International Plant Genetic Resources Institute: Rome, Italy, 1997; pp. 173-197.
35. Zheng, B.L.; He, K.; Kim, C.H.; Rogers, L.; Shao, Y.; Huang, Z.Y.; Lu, Y.; Yan, S.J.; Qien, L.C.; Zheng, Q.Y. *Urology*, **2000**, *55*, 598.
36. Cicero, A.F.G.; Bandieri, E.; Arletti, R. *J. Ethnopharmacology* **2001**, *75*, 225.
37. Gonzales, G.F.; Córdova, A.; Vega, K.; Chung, A.; Villena, A.; Góñez, C.; Castillo, S. *Andrologia* **2002**, *34*, 367.
38. Gonzales, G.F.; Córdova, A.; Vega, K.; Chung, A.; Villena, A.; Góñez, C. *J. Endocrocrinology* **2003**, *176*, 163.

Chapter 13

Studies on Chemical Constituents of Jiaogulan (*Gynostemma pentaphyllum*)

Feng Yin and Lihong Hu

National Center for Drug Screening, Shanghai Institute of Materia Medica, Chinese Academy of Sciences, Shanghai 201203, People's Republic of China

Thirty-three new dammarane glycosides, together with nine known compounds, gypenosides IV, VIII, XLVIII, XLIX, LXIX, LXXI, gylongiposide I, allantion and vitexin were isolated from the methanol extract of the aerial parts of *Gynostemma pentaphyllum*. Their structures were elucidated by NMR spectroscopy and mass spectrometry, as well as by chemical degradation.

Introduction

Jiaogulan (*Gynostemma pentaphyllum* (Thunb.) Makino) (Cucurbitaceae) is praised in China as xiancao, the herb of immortality. It is a perennial creeping herb which grows in Japan, Korea, China and Southeast Asia. Jiaogulan was once used as a sweetener in Japan (*1*). In 1976, while looking for a sugar alternative, Japanese researchers studied Jiaogulan for this reason. They discovered that Jiaogulan has many properties similar to ginseng, even though they are unrelated as plants. These findings sparked years of research and provided evidence that Jiaogulan is a powerful adaptogenic and antioxidant herb with many health-promoting properties (*1-17*). *G. Pentaphyllum* was also reported to prevent cardiac myocytes from ischemic damage by inhibiting "calcium overload" (*18*), and to have weak activity in preventing Ha-ras cancer gene mutation in rats (*19*). Recently, certain gypenosides isolated from Jiaogulan were reported to inhibit the proliferation of Hep-3B and HA22T cells, by affecting calcium and sodium currents in a dose-dependent manner (20).

Since the early 1990s, many brands of drugs, functional foods and tea based on G. pentaphyllum have been available in Chinese markets. However, there is lack of quality control for these products. It is know from the literature that the chemical constituents of G. Pentaphyllum planted in different areas are very different in composition. In our previous studies, we used HPLC-UV-ESIMS to analyze three commercial products of G. pentaphyllum manufactured in Hunan, Zhejiang and Guangxi Provinces (16-17) and found that the HPLC profile of G. pentaphyllum, product from Henan was very different from those of the other two. We then investigated the chemical components G. pentaphyllum from Hunan and the study led to the isolation of forty-two compounds comprising thirty-three new compounds (**1-33**) and nine known compounds (18-21). The structure elucidation was accomplished mainly on the basis of the interrelation of 2D NMR spectral data, including ^{1}H-^{1}H and ^{1}H-^{13}C chemical shift correlation spectroscopy.

Results and Discussion

The known compounds, gypenosides (Gyp) IV (2), VIII (2), XLVIII (9), XLIX (9), LXIX (12), LXXI (12), gylongiposide I (22), allantion (23) and vitexin (24) were identified by comparison of their spectral data with those described in the literature. New compounds (**1-33**) were identified as shown in Figure 1 on the basis of the interrelation of 1D NMR, 2D NMR and MS spectral data (^{13}C NMR see Tables I-V).

1. R_1 = Rha, R_2 = Xyl, R_3 = CH$_3$CO, R_4 = Glc', R_5 = CH$_3$
2. R_1 = Rha, R_2 = Glc', R_3 = H, R_4 = Glc", R_5 = CH$_3$
3. R_1 = Rha, R_2 = Xyl, R_3 = H, R_4 = Glc', R_5 = CH$_3$

4. R_1 = Rha, R_2 = Xyl, R_3 = Glc, R_4 = CH$_2$OH
5. R_1 = Rha, R_2 = Xyl, R_3 = H, R_4 = Glc', R_5 = CH$_2$OH

6. R_1 = Rha, R_2 = Xyl, R_3 = Glc
26. R_1 = Rha, R_2 = Xyl, R_3 = H

8. R_1 = Glc(2→1)Glc', R_2 = Glc''(6→1)Xyl

7. R_1 = Rha, R_2 = Xyl, R_3 = Glc

9. R_1 = Xyl, R_2 = Xyl', R_3 = OH
10. R_1 = Glc', R_2 = Xyl, R_3 = OH
15. R_1 = Xyl, R_2 = Xyl', R_3 = OAc

11. R_1 = Xyl(2→1)Glc, R_2 = OH
12. R_1 = Glc(2→1)Xyl, R_2 = OH
13. R_1 = Xyl(2→1)Xyl', R_2 = OAc
14. R_1 = Glc(2→1)Xyl, R_2 = OAc

16. R$_1$ = Rha, R$_2$ = Xyl, R$_3$ = CH$_3$CO

17. R$_1$ = Rha, R$_2$ = Xyl

18. R$_1$ = Rha, R$_2$ = Xyl, R$_3$ = CH$_3$CO

19. R$_1$ = Rha, R$_2$ = Xyl

20. R$_1$ = Rha, R$_2$ = Xyl

21. R$_1$ = Rha, R$_2$ = Xyl, R$_3$ = H
22. R$_1$ = Rha, R$_2$ = Xyl, R$_3$ = CH$_3$CO

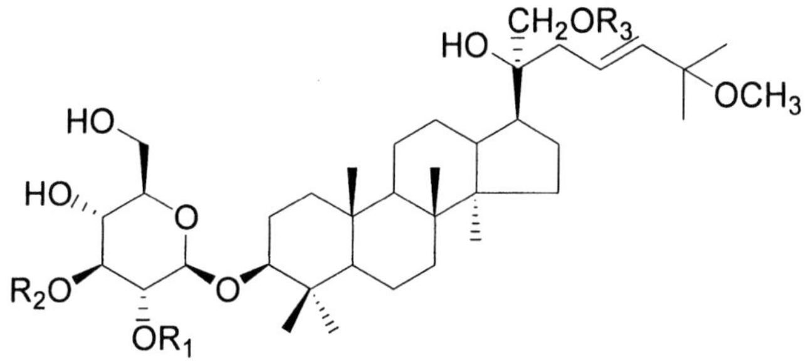

23. R₁ = Rha, R₂ = Xyl, R₃ = Xyl'

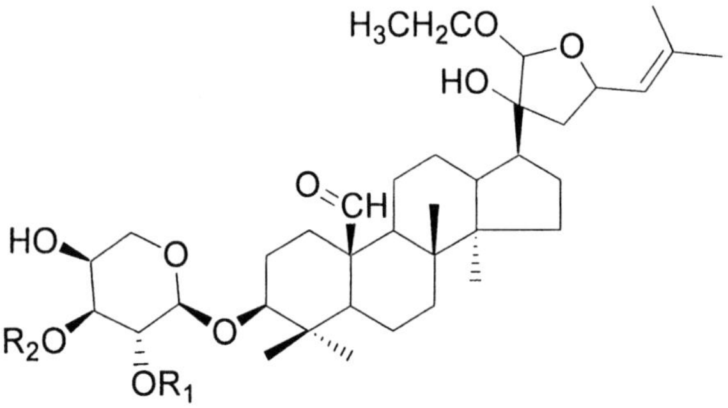

24. R₁ = Rha, R₂ = Xyl

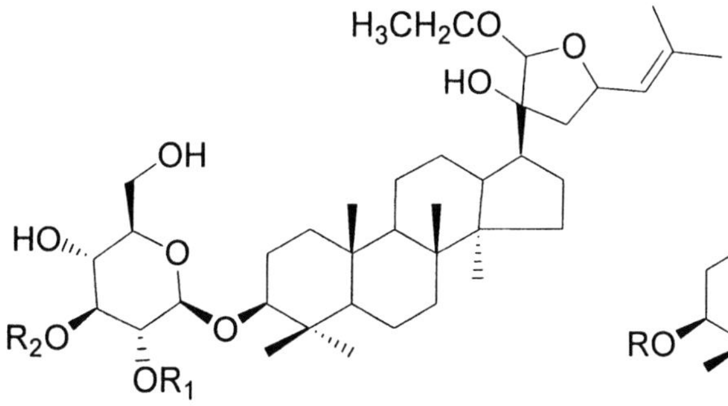

25. R₁ = Rha, R₂ = Xyl

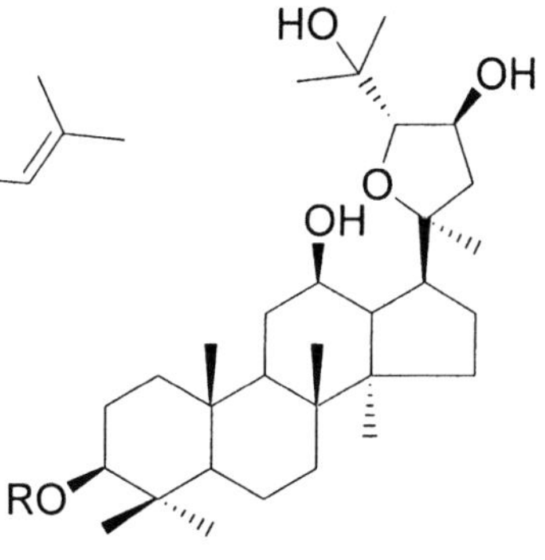

27. R = Glc(2→1)Xyl

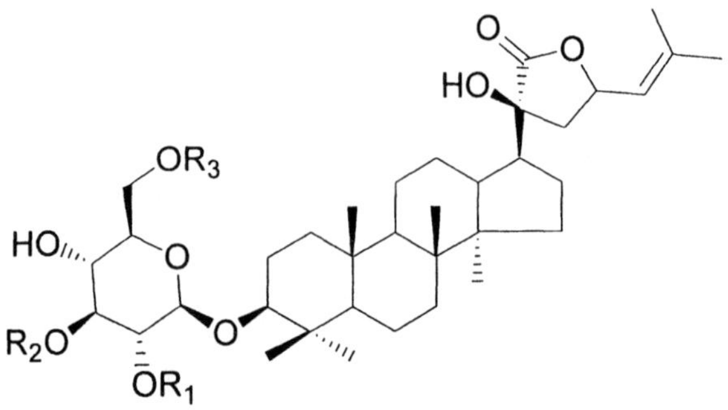

28. R_1 = Rha, R_2 = Xyl, R_3 = CH$_3$CO
30. R_1 = Rha, R_2 = Xyl, R_3 = H
33. R_1 = Rha, R_2 = Glc, R_3 = H

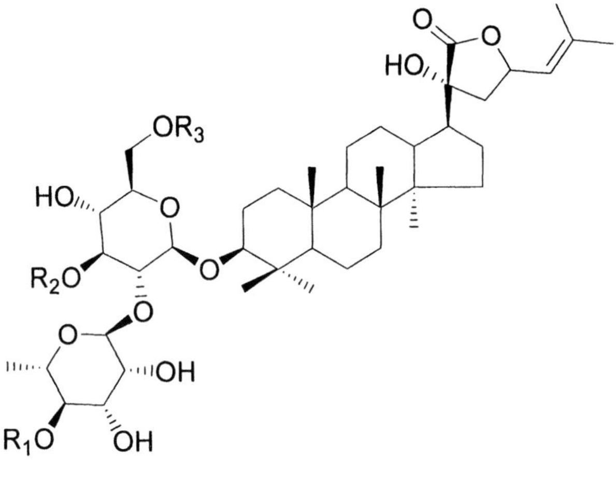

29. R_1 = H, R_2 = Xyl, R_3 = CH$_3$CO
31. R_1 = H, R_2 = Xyl, R_3 = H
32. R_1 = CH$_3$CO, R_2 = Xyl, R_3 = CH$_3$CO

Figure 1. Structures of new compounds 1- 33

Table I. ^{13}C NMR Data of Compounds 1-8 in C_5D_5N

Carbon	1	2	3	4	5	6	7	8
1	39.9	39.9	39.9	35.1	35.3	33.7	33.7	39.4
2	27.0	27.0	27.2	27.7	28.0	27.8	27.8	26.9
3	89.0	88.9	89.0	88.9	89.5	87.3	87.3	89.2
4	39.9	39.8	40.0	39.9	39.9	40.7	40.5	39.9
5	57.0	56.8	56.9	57.5	57.6	55.0	55.0	56.6
6	18.7	18.7	18.7	18.5	18.5	17.8	17.8	18.7
7	35.8	36.2	35.8	36.4	36.4	34.8	34.8	35.4
8	40.9	40.8	41.0	41.2	41.2	40.2	40.2	40.3
9	51.3	51.1	51.3	53.3	53.3	53.0	53.0	50.4
10	37.0	37.0	37.3	42.3	42.3	52.9	52.9	37.1
11	22.0	21.9	22.0	24.9	25.0	22.4	22.4	31.1
12	24.8	24.8	24.8	27.7	28.0	24.5	24.8	70.7
13	41.9	41.9	42.0	42.4	42.5	41.8	41.7	49.8
14	50.6	50.6	50.7	51.0	51.0	50.3	50.4	51.7
15	31.7	31.7	31.9	32.2	32.2	32.0	32.1	30.8
16	27.0	27.0	27.2	29.0	29.0	27.7	27.7	26.6
17	46.3	46.3	46.4	46.3	46.3	46.3	46.4	52.1
18	15.9	15.9	16.0	16.3	16.3	16.1	16.2	16.2
19	16.7	16.7	16.7	61.8	61.8	205.6	205.6	16.5
20	76.5	76.5	76.4	76.7	76.7	76.8	76.3	83.5
21	76.5	76.5	76.6	76.6	76.7	76.3	76.3	23.5
22	36.8	36.7	36.7	36.8	36.8	39.9	32.8	40.3
23	23.5	23.4	23.6	23.6	23.6	126.8	30.5	126.9
24	126.1	126.1	126.0	126.2	126.3	138.3	76.1	138.3
25	131.1	131.0	131.0	131.0	131.0	81.4	150.1	81.6
26	26.0	25.9	25.9	26.0	26.0	25.3	110.3	25.6
27	17.9	17.9	17.9	18.0	17.9	25.2	18.4	25.3
28	28.0	28.0	28.1	28.9	28.8	26.5	26.5	28.3
29	16.9	16.8	17.0	17.2	17.2	16.5	16.5	16.8
30	16.8	17.0	16.8	17.4	17.4	17.4	17.4	17.4
	C-3-Glc	C-3-Glc	C-3-Glc	C-3-Ara	C-3-Glc	C-3-Ara	C-3-Ara	C-3-Glc
1	105.1	105.1	105.1	104.9	105.2	104.8	104.9	105.3
2	76.8	77.0	77.0	74.8	77.1	74.7	74.7	83.5
3	87.9	89.6	88.0	81.6	88.4	81.7	81.8	78.3
4	70.0	70.0	69.9	68.3	69.9	68.4	68.5	71.9
5	74.5	78.3	78.1	64.9	78.1	65.1	65.2	78.1
6	64.3	62.4	62.0		62.2			63.0
OCOCH$_3$	170.9							
OCOCH$_3$	20.9							
	Rha	Rha	Rha	Rha	Rha	Rha	Rha	Glc'
1	101.9	101.2	101.0	102.1	101.9	102.1	102.1	106.2
2	72.6	72.5	72.5	72.7	72.7	72.7	72.6	77.3
3	72.5	72.5	72.6	72.5	72.5	72.5	72.5	78.4
4	74.0	73.9	74.0	74.0	74.0	74.0	73.9	71.9
5	70.0	70.0	69.0	70.2	69.9	70.2	70.1	78.2
6	18.8	18.7	18.7	18.7	18.7	18.7	18.7	62.9
	Xyl	Glc'	Xyl	Xyl	Xyl	Xyl	Xyl	C-20-Glc''
1	105.1	104.0	105.0	105.2	105.0	105.3	105.3	98.4
2	74.9	75.2	74.9	74.6	75.0	74.5	74.5	75.0
3	78.4	77.9	78.4	77.7	78.4	77.8	77.8	79.1
4	70.7	71.5	70.7	71.0	70.8	71.0	71.2	71.7
5	67.4	78.5	67.0	67.1	67.4	67.1	67.0	76.9
6		62.8						70.1
	C-21-Glc'	C-21-Glc''	C-21-Glc'	C-21-Glc'	C-21-Glc	C-21-Glc'	C-21-Glc	Xyl
1	106.3	106.1	106.0	106.2	106.3	106.2	106.1	105.7
2	75.5	75.5	75.6	75.6	75.6	75.6	75.5	75.2
3	78.7	78.6	78.7	78.6	78.8	78.6	78.6	78.1
4	71.9	71.8	71.9	71.8	71.8	71.8	71.8	71.3
5	78.6	78.6	78.6	78.7	78.7	78.7	78.6	67.2
6	63.0	62.9	63.0	63.0	62.9	62.9	62.9	

^{a} 125 MHz; referenced to δ 135.9 (C_5D_5N).

Source: Reproduced with permission from reference 5. Copyright 2003 Wiley.

Table II. ^{13}C NMR Data of Compounds 9-15 in C$_5$D$_5$N^a

Carbon	9	10	11	12	13	14	15
1	39.5	39.3	39.5	39.6	39.5	39.4	39.4
2	27.1	26.9	27.1	27.1	27.1	26.9	27.0
3	89.3	89.3	89.0	89.2	89.1	89.3	89.3
4	39.9	39.8	40.1	40.1	40.1	39.9	39.9
5	56.7	56.5	56.8	56.8	56.8	56.6	56.6
6	18.7	18.6	18.8	18.7	18.7	18.6	18.6
7	35.3	35.2	35.4	35.4	35.3	35.2	35.2
8	40.2	40.2	40.3	40.3	40.3	40.2	40.2
9	50.6	50.5	50.6	50.7	50.5	50.4	50.4
10	37.2	37.1	37.3	37.2	37.3	37.0	37.1
11	32.2	31.9	32.4	32.1	32.1	32.0	31.8
12	70.8	70.6	70.8	70.8	70.8	70.7	70.7
13	49.5	49.5	49.7	49.3	49.8	49.7	49.7
14	52.5	52.4	52.5	52.2	52.5	52.3	52.3
15	32.0	32.2	32.0	32.4	31.8	31.7	31.7
16	27.1	27.9	28.0	27.1	27.8	27.6	27.7
17	52.7	52.6	52.7	52.7	52.8	52.6	52.7
18	15.9	15.8	15.9	15.9	16.0	15.8	15.9
19	16.4	16.6	16.8	16.5	16.8	16.6	16.6
20	79.6	79.5	79.7	79.7	79.5	79.4	79.4
21	27.7	27.7	27.8	28.0	27.3	27.9	27.0
22	37.5	37.4	37.6	37.6	34.6	34.5	34.5
23	67.3	67.3	67.4	67.4	71.8	71.8	71.7
24	81.0	81.1	81.0	81.0	77.1	77.1	77.0
25	79.4	79.3	79.5	79.5	79.7	79.5	79.6
26	24.8	24.8	24.7	24.9	24.9	24.7	24.8
27	30.6	30.5	30.7	30.7	30.6	30.4	30.4
28	28.0	28.2	28.3	27.8	28.0	27.9	27.9
29	16.7	16.7	16.8	16.8	16.4	16.3	16.3
30	18.1	18.0	18.1	18.2	18.0	17.9	17.9
-OCOCH$_3$					171.1	171.1	171.1
-OCOCH$_3$					21.4	21.3	21.3
	C-3-Glc	C-3-Glc	C-3-Xyl	C-3-Glc	C-3-Xyl	C-3-Glc	C-3-Glc
1	105.1	105.1	106.0	105.1	105.9	105.3	105.2
2	83.9	83.4	83.4	84.1	84.2	84.3	84.0
3	78.3	78.3	78.2	78.7	78.4	78.5	78.4
4	71.5	71.2	71.2	71.8	71.3	71.7	71.5
5	78.3	78.3	66.9	78.5	67.0	78.3	78.4
6	70.1	70.1		63.1		63.1	70.1
	Xyl	Glc'	Glc	Xyl	Xyl'	Xyl	Xyl
1	107.1	106.1	106.3	107.1	107.3	107.2	107.2
2	76.6	76.7	78.2	76.8	76.8	76.7	76.7
3	78.3	78.1	77.3	78.4	78.5	78.2	78.4
4	71.3	71.9	71.8	71.4	71.2	71.2	71.2
5	67.7	78.2	78.5	67.8	67.8	67.7	67.6
6		62.9	62.9				
	Xyl'	Xyl					Xyl'
X$_1$	106.0	106.0					106.1
X$_2$	74.9	74.9					75.0
X$_3$	76.7	77.1					76.7
X$_4$	71.3	71.5					71.2
X$_5$	67.2	67.2					67.2

^{a} 125 MHz; referenced to δ 135.9 (C$_5$D$_5$N).

Source: Reproduced with permission from reference 5. Copyright 2003 Wiley.

Table III. ^{13}C NMR Data of Compounds 16-20 in C$_5$D$_5$N

Carbon	16	17	18	19	20
1	39.8	33.7	39.9	33.9	33.8
2	26.9	27.7	27.9	27.9	27.8
3	89.6	87.3	89.7	87.4	87.4
4	39.8	40.6	39.8	40.7	40.7
5	56.9	55.0	57.0	55.1	55.1
6	18.6	17.8	18.6	17.9	17.9
7	35.8	34.9	35.9	35.0	35.0
8	40.9	40.2	40.8	40.3	40.3
9	51.2	53.0	51.4	53.3	53.1
10	37.2	53.0	37.3	53.1	53.1
11	21.9	22.4	22.0	22.8	22.4
12	25.6	25.6	25.9	25.9	25.2
13	42.8	42.6	44.1	43.9	41.5
14	50.5	50.3	50.4	50.3	50.3
15	31.6	32.1	31.9	32.4	32.3
16	27.6	27.4	26.9	27.8	27.8
17	50.4	50.3	47.8	47.8	45.3
18	15.8	16.2	15.8	16.3	16.2
19	16.8	205.8	16.7	205.8	205.7
20	80.7	80.7	83.3	83.3	84.9
21	76.2	76.2	78.0	78.1	103.0
22	47.2	47.2	32.3	32.4	45.3
23	71.7	71.7	22.7	22.6	73.6
24	60.4	60.3	54.7	54.8	130.3
25	144.6	144.6	71.8	71.9	132.5
26	112.8	113.0	29.9	30.1	25.9
27	23.6	23.7	27.5	27.7	18.2
28	27.9	26.6	28.0	26.7	26.6
29	16.9	16.7	16.9	16.8	16.8
30	16.7	17.4	16.6	17.4	17.7
	C-3-Glc	C-3-Ara	C-3-Glc	C-3-Ara	C-3-Ara
1	105.0	104.9	105.0	104.8	105.0
2	76.8	74.7	76.8	74.8	74.8
3	87.9	81.9	87.9	81.6	81.9
4	70.0	68.6	70.0	68.4	68.6
5	74.5	65.3	74.5	65.0	65.3
6	64.2		64.2		
-OCOCH$_3$	170.7		170.8		
-OCOCH$_3$	20.8		20.9		
Rha					
1	101.9	102.1	101.9	102.1	102.2
2	72.6	72.6	72.6	72.6	72.8
3	72.5	72.5	72.5	72.5	72.6
4	74.0	74.0	74.0	74.0	74.1
5	70.0	70.2	70.0	70.2	70.3
6	18.7	18.7	18.7	18.7	18.8
Xyl					
1	105.0	105.4	105.0	105.2	105.5
2	74.9	74.5	74.9	74.5	74.7
3	78.4	77.8	78.4	77.7	77.9
4	70.7	71.0	70.7	71.0	71.1
5	67.4	67.1	67.4	67.0	67.2

^{a} 125 MHz; referenced to δ 135.9 (C$_5$D$_5$N).

Table IV. ^{13}C NMR Data of Compounds 21-27 in C_5D_5N

Carbon	21	22	23	24	25	26	27
1	40.0	40.0	39.9	33.8	39.4	33.9	39.4
2	27.3	27.5	27.1	27.7	27.2	27.9	27.1
3	89.1	89.8	89.1	87.3	89.2	87.5	89.3
4	39.9	40.0	39.9	40.2	40.0	40.8	39.9
5	56.8	57.1	56.8	55.0	56.9	55.2	56.8
6	18.7	18.8	18.7	17.8	18.8	18.2	18.6
7	36.0	36.1	35.9	34.9	35.9	35.0	35.3
8	40.9	41.1	40.9	40.6	41.0	40.4	40.1
9	51.2	51.4	51.2	53.0	51.2	53.3	50.7
10	37.2	37.4	37.2	53.0	37.3	53.1	37.1
11	21.9	22.0	22.0	22.4	22.0	22.6	32.7
12	25.2	25.3	24.8	25.0	26.0	24.8	70.9
13	41.6	41.7	42.0	41.3	42.2	41.7	49.8
14	50.4	50.6	50.7	50.2	51.1	50.5	52.6
15	31.9	32.0	31.7	32.2	31.7	32.3	32.6
16	27.2	27.1	27.2	26.8	28.0	28.2	28.9
17	45.3	45.4	46.4	44.9	46.4	46.5	50.1
18	16.1	16.2	16.0	17.5	17.1	16.3	15.6
19	16.7	16.8	16.8	205.6	16.8	205.5	16.7
20	84.9	85.0	77.0	84.2	83.3	72.8	85.5
21	103.0	103.2	76.7	108.2	104.6	67.2	29.8
22	45.3	45.4	40.0	45.4	43.8	40.3	42.2
23	73.6	73.7	127.3	73.7	74.0	127.3	70.6
24	130.3	130.4	138.7	129.5	129.6	138.1	91.8
25	132.4	132.5	75.3	132.9	134.1	81.5	70.3
C-25-OCH$_3$			50.7				
26	25.8	25.9	26.5	25.7	26.0	25.5	27.7
27	18.1	18.2	26.2	18.1	18.2	25.4	26.7
28	28.1	28.0	28.0	26.5	28.2	26.7	27.9
29	16.9	17.0	16.9	16.1	17.2	16.8	16.3
30	17.1	17.1	16.8	16.6	16.1	17.6	18.2
C-21-CH$_2$CH$_3$				62.4	63.7		
C-21-CH$_2$CH$_3$				15.7	15.9		
	C-3-Glc	C-3-Glc	C-3-Glc	C-3-Ara	C-3-Glc	C-3-Ara	C-3-Glc
1	105.1	105.2	105.2	104.8	105.3	105.0	105.3
2	77.1	76.9	77.3	74.7	77.3	74.9	84.3
3	88.4	88.1	88.5	81.6	88.5	81.8	78.5
4	69.9	70.2	70.1	68.4	70.1	68.7	71.7
5	78.1	74.7	78.8	65.0	78.3	65.3	78.3
6	62.7	64.4	63.1		62.9		63.1
-OCOCH$_3$		170.9					
-OCOCH$_3$		21.1					
	Rha	Rha	Rha	Rha	Rha	Rha	Xyl
1	101.9	102.1	102.1	102.1	102.1	102.3	107.2
2	72.7	72.8	72.8	72.7	72.8	72.8	76.7
3	72.5	72.7	72.7	72.5	72.7	72.7	78.2
4	74.0	74.1	74.2	74.0	74.2	74.2	71.2
5	69.9	70.2	70.1	70.2	70.2	70.3	67.8
6	18.8	18.9	18.9	18.7	19.0	18.9	
Xyl							
1	105.0	105.2	105.1	105.3	105.2	105.5	
2	75.0	75.1	75.1	74.5	75.2	74.7	
3	78.4	78.6	78.5	77.8	78.6	78.0	
4	70.8	70.9	70.9	71.0	70.4	71.2	
5	67.4	67.6	67.5	67.0	67.6	67.3	
C-21-Xyl							
1			106.4				
2			75.7				
3			78.9				
4			72.0				
5			67.2				

[a] 125 MHz; referenced to δ 135.9 (C_5D_5N).

Table V. ^{13}C NMR Data of Compounds 28-33 in C$_5$D$_5$N

Carbon	28	29	30	31	32	33
1	39.7	39.9	39.8	39.8	40.0	40.0
2	26.9	26.9	27.0	27.0	27.2	27.2
3	89.6	89.6	89.0	89.0	89.8	89.0
4	39.8	39.8	39.8	39.8	40.0	40.0
5	56.8	56.9	56.8	56.8	57.1	56.9
6	18.5	18.6	18.5	18.5	18.8	18.7
7	35.7	35.8	35.8	35.8	35.8	36.0
8	40.9	40.9	40.8	40.8	41.1	41.0
9	51.3	51.3	51.2	51.2	51.5	51.4
10	37.1	37.3	37.1	37.2	37.5	37.3
11	21.8	21.9	21.8	21.9	22.1	22.0
12	25.8	26.4	25.9	26.3	26.6	26.1
13	43.4	45.1	43.4	45.1	45.3	43.6
14	50.7	50.3	50.7	50.3	50.5	50.9
15	31.7	31.9	31.8	31.8	32.0	32.0
16	27.4	28.1	27.4	28.1	28.2	27.6
17	46.0	45.5	45.9	45.4	45.6	46.1
18	15.7	15.8	15.7	15.7	16.0	15.9
19	16.6	16.7	16.7	16.7	17.0	16.8
20	79.1	81.2	79.1	81.2	81.5	79.3
21	179.4	178.4	179.5	178.4	178.6	179.7
22	40.7	39.2	40.8	39.2	39.4	41.0
23	74.2	75.3	74.2	75.3	75.3	74.2
24	125.5	124.1	125.5	124.1	124.2	125.7
25	138.5	139.6	138.5	139.5	139.8	138.7
26	25.6	25.7	25.7	25.7	25.7	25.9
27	18.2	18.3	18.2	18.3	18.3	18.4
28	27.8	28.0	28.0	28.0	28.2	28.2
29	16.8	16.9	16.9	17.0	16.9	17.1
30	16.6	16.4	16.6	16.4	16.6	16.8
C-3-Glc						
1	105.0	105.01	105.1	105.1	105.0	105.2
2	76.8	76.8	77.1	77.1	75.8	77.1
3	87.9	87.9	88.3	88.3	88.3	89.8
4	69.8	70.0	69.9	69.9	70.0	70.0
5	74.5	74.5	78.1	78.1	74.8	78.1
6	64.2	64.2	62.7	62.7	64.3	62.6
-OCOCH$_3$	170.7	170.8			171.0	
-OCOCH$_3$	20.8	20.9			21.2	
Rha						
1	101.9	101.9	101.9	101.9	101.4	101.8
2	72.6	72.6	72.7	72.7	72.2	72.6
3	72.4	72.4	72.5	72.5	70.1	72.6
4	74.0	74.0	74.0	74.0	76.1	74.1
5	69.9	70.0	69.9	69.9	67.1	70.1
6	18.7	18.7	18.7	18.7	18.2	18.8
-OCOCH$_3$					171.0	
-OCOCH$_3$					20.9	
	Xyl	Xyl	Xyl	Xyl	Xyl	Glc
1	105.0	105.0	105.0	105.0	105.3	104.1
2	74.9	74.9	75.0	75.0	74.6	75.4
3	78.4	78.4	78.4	78.4	78.8	78.6
4	70.7	70.7	70.7	70.7	70.7	71.7
5	67.3	67.4	67.4	67.4	67.4	78.8
6						62.9

[a] 125 MHz; referenced to δ 135.9 (C$_5$D$_5$N).

As a result of this investigation we isolated and characterized 33 new saponins from *G. Pentaphyllum*. Among these, the aglycons, 21,24-cyclopentyldammar-25-ene of **16** and **17**, and the 21,24-cyclopentyldammarane of **18** and **19**, were novel. The aglycon, 20*S*,25-epoxydammarane of **9-15** were isolated from this plant for the first time. Furthermore, we proposed a plausible biogenetical pathway (Scheme 1) for new saponins isolated from this plant. We found that the chemical constituents of *G. pentaphyllum* planted in different areas were very different. Therefore, it is not acceptable for the drugs, functional foods and teas prepared from *G. pentaphyllum* grew in different area to be labeled the same. A standard of quality control with chromatographic fingerprint analysis techniques for each *G. pentaphyllum* product should be established soon.

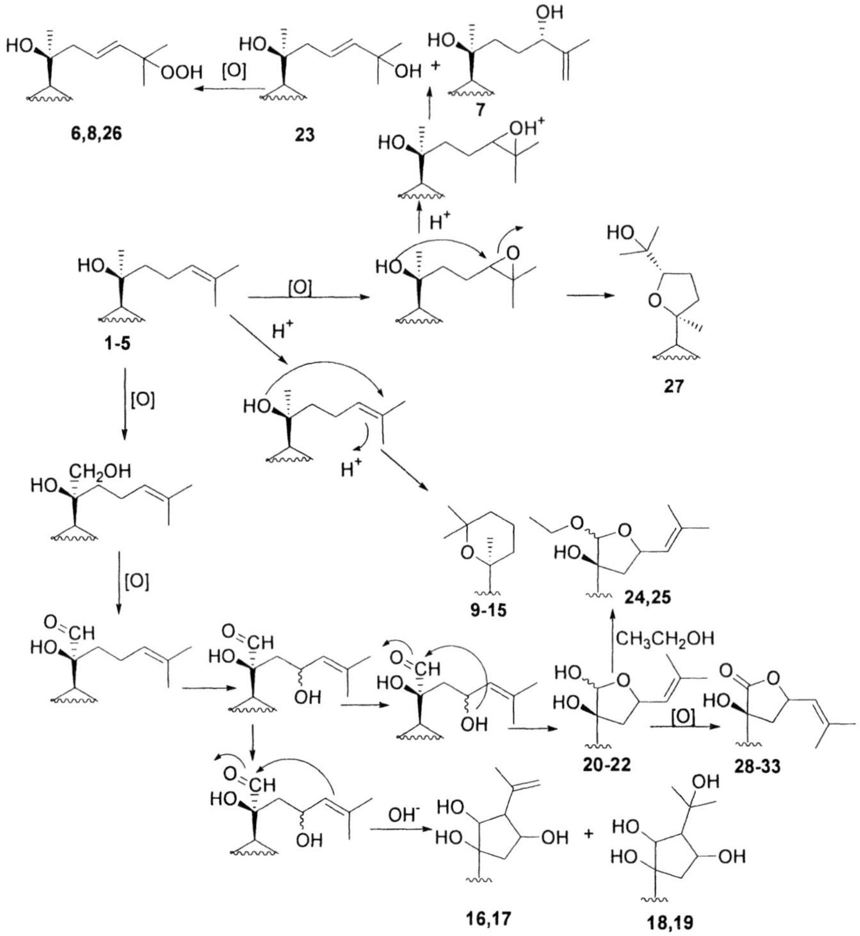

Scheme 1. Plausible biogenetical pathway proposed for saponins isolated from G. pentaphyllum

References

1 Nagai, M.; Izawa, K.; Nagumo, S.; Sakurai, N.; Inoue, T. *Chem. Pharm. Bull.* **1981**, *29*, 779-783.
2 Takemoto, T.; Arihara, S.; Nakajima, T.; Okuhira. M. *Yakugaku Zasshi* **1983**, *103*, 173-185.
3 Takemoto, T.; Arihara, S.; Nakajima, T.; Okuhira. M. *Yakugaku Zasshi* **1983**, *103*, 1015-1023.
4 Takemoto, T.; Arihara, S.; Yoshikawa, K.; Nakajima, T.; Okuhira, M. *Yakugaku Zasshi* **1984**, *104*, 325-331.
5 Takemoto, T.; Arihara, S.; Yoshikawa, K.; Nakajima, T.; Okuhira, M. *Yakugaku Zasshi* **1984**, *104*, 332-339.
6 Takemoto, T.; Arihara, S.; Yoshikawa, K.; Nakajima, T.; Okuhira, M. *Yakugaku Zasshi* **1984**, *104*, 724-730.
7 Takemoto, T.; Arihara, S.; Yoshikawa, K.; Nakajima, T.; Okuhira, M. *Yakugaku Zasshi* **1984**, *104*, 939-945.
8 Takemoto, T.; Arihara, S.; Yoshikawa, K.; Kawasaki, J.; Nakajima, T.; Okuhira, M. *Yakugaku Zasshi* **1984**, *104*, 1043-1049.
9 Takemoto, T.; Arihara, S.; Yoshikawa, K.; Hino, K.; Nakajima, T.; Okuhira, M. *Yakugaku Zasshi* **1984**, *104*, 1155-1162.
10 Takemoto, T.; Arihara, S.; Yoshikawa, K *Yakugaku Zasshi* **1986**, *106*, 664-670.
11 Yoshikawa, K.; Takemoto, T.; Arihara, S. *Yakugaku Zasshi* **1986**, *106*, 758-763.
12 Yoshikawa, K.; Takemoto, T.; Arihara, S. *Yakugaku Zasshi* **1987**, *107*, 262-267.
13 Yoshikawa, K.; Arimitsu, M.; Kishi, K.; Takemoto, T.; Arihara, S. *Yakugaku Zasshi* **1987**, *107*, 361-366.
14 Kuwahara, M.; Kawanishi, F.; Komiya, T.; Oshio, H. *Chem. Pharm. Bull.* **1989**, *37*, 135-139.
15 Piacente, S.; Pizza, C.; De Tommasi, N.; De Simone, F. *J. Nat. Prod.* **1995**, *58*, 512-519.
16 Hu, L. H.; Chen, Z. L.; Xie, Y. Y. *J. Nat. Prod.* **1996**, *59*, 1143-1145.
17 Hu, L. H.; Chen, Z. L.; Xie, Y. Y. *Phytochemistry* **1997**, *44*, 667-670.
18 Zhao, Y.; Hu, D. *Zhongguo Yaolixue Tongbao* **1998**, *14*, 60-62.
19 Zhou, Z. T.; Tang, G. Y.; Zhong, W. J.; Ding, D. Y.; Mao, D. H.; Li, W. G. *Zhonghua Kouqiang Yixue Zazhi* **2000**, *35*, 91-94.
20 Chou, C. J.; Gung, C. J.; Der, C. L. *Cytobios* **1999**, *100*, 37-48.

21 Yin, F.; Hu, L. H.; Lou, F. C.; Pan, R. X. *J. Nat. Prod.* **2004**, *67*, 942-952.
22 Guo, X. L.; Wang, T. J.; Bian, B. L. *Acta Pharm. Sinica* **1997**, *32*, 524-529.
23 Coxon, B.; Fatiadi, A. J.; Sniegoski, L. T.; Hertz, H. S.; Schaffer, R. *J. O. C.* **1977**, *42*, 3132-3140.
24 Osterdahl, B. G. *Acta Chemica Scandinavica Series B* **1978**, *B 32*, 93-97.

Chapter 14

Chemical Components of Noni (*Morinda citrifolia* L.) Root

Shengmin Sang[1,2] and Chi-Tang Ho[1]

[1]Department of Food Science, Rutgers, the State University of New Jersey, 65 Dudley Road, New Brunswick, NJ 08901–8520
[2]Department of Chemical Biology, Ernest Mario School of Pharmacy, Rutgers, The State University of New Jersey, 164 Frelinghyusen Road, Piscataway, NJ 08854–8020

Morinda citrifolia (Rubiaceae), commonly known as noni, is a plant typically found in the Hawaiian and Tahitian islands. It is believed to be one of the most important plants brought to Hawaii by the first Polynesians. The bark, stem, root, leaf and fruit have been used traditionally as a folk remedy for many diseases including diabetes, hypertension, and cancer. In this research, we reported the constituents of the roots of this plant, which included 3 new naphthoquinone derivatives and one new anhydride. Their structures were identified by the combination of APCI-MS, 1D and 2D-NMR (COSY, TOCSY, ROSEY, HMQC and HMBC).

Introduction

Morinda citrifolia (Rubiaceae), commonly known as noni, is a plant typically found in the Hawaiian and Tahitian islands. It is believed to be one of the most important plants brought to Hawaii by the first Polynesians (*1*). Native from the Indian Ocean, this species is not present in continental Africa where it is replaced by a related species, *M. lucida*. The plant is a small evergreen tree growing in the open coastal regions and in forest areas up to about 1300 feet above the sea level. This plant is identifiable by its straight trunk, large green leaves and its distinvtive, ovid, "grenade-like" yellow fruit. The fruit can grow to a size of 12 cm and results from coalescence of the inferior ovaries of many closely packed flowers. It has a foul taste and a soapy smell when mature. The bark, stem, root, leaf, and fruit have been used traditionally as a folk remedy for many diseases including diabetes, hypertension, and cancer (*2,3*). The fruits of this plant were also used as foods in time of famine, whereas the roots were used to produce a yellow or red dye for cloth. In earlier studies, ricinoleic acid was found in the seeds (*4*), whereas anthraquinones, including morenone 1, morenone 2, damnacanthal, and 7-hydroxy-8-methoxy-2-methylanthraquinone, have been identified in the root of noni (*5,6*). From the heartwood, two known anthraquinones (morindone and physcion) and one new anthraquinone glycoside have been isolated from the heartwood of *Morinda citrifolia* (*7*). Studies on the chemical components of the flowers of noni have resulted in the identification of one anthraquinone glycoside and two flavone glycosides from the flowers of noni (*8,9*), whereas β-sitosterol and ursolic acid have been isolated from the leaves (*10*). Although the fruits of noni have been used as a food, very few reports on the chemical components of the fruits are available (*1*). Several nonvolatile compounds including acetyl derivatives of asperuloside, glucose, caproic acid, and caprylic acid have been identified in fruits (*1*). More than 50 volatile components were identified by GC-MS from the ripe fruit (*11,12*). Recently, our research group reported 4 new glycosides, one known iridoid, asperulosidic acid, and one known flavonoid, rutin, from the hawaiin noni fruits (*13,14*). In addition, 4 new iridoids, as well as 2 known iridoids and 5 known flavonoids were identified from the Indian noni leaves (*15-18*). Among them, two new iridoids showed significant inhibition of UVB-induced Activator Protein-1 (AP-1) activity. To continue study the bioactive constituents from noni, the present report discusses the chemical components of the Tahitian noni root (*19*).

Materials and Methods

Chemicals

Silica gel (130-270 mesh), Sephadex LH-20 (Sigma Chemical Co., St. Louis, MO) and Lichroprep RP-18 column were used for column chromatography. All solvents used for chromatographic isolation were analytical grade and purchased from Fisher Scientific (Springfield, NJ).

General Procedures

^{1}H NMR and ^{13}C NMR spectra were obtained on a VXR-600 (Varian Inc.), operating at 600 and 150 MHz, respectively. Compounds were analyzed in CDCl$_3$, CD$_3$OD and DMSO-d_6 with tetramethylsilane (TMS) as an internal standard. ^{1}H-^{1}H COSY, NOESY, HMQC and HMBC were performed on a VXR-600 (Varian Inc.). APCI-MS was obtained on a Fisons/VG Platform II mass spectrometer. Preparative thin-layer chromatography was performed on Sigma-Aldrich TLC plates (1000 µm thickness, 2-25 µm particle size), Thin-layer chromatography was performed on Sigma-Aldrich TLC plates (250 µm thickness, 2-25 µm particle size), with compounds visualized by spraying with 5% (v/v) H$_2$SO$_4$ in ethanol solution.

Results

By the combination of solvent extraction, partition and modern chromatograph methods, including normal phase and reverse phase silica gel column chromatography, prepared thin layer chromatography, and sephadex LH-20 column chromatograph, a total of 18 compounds were isolated from noni root. Four of them are new compounds. The structures of the isolated compounds were elucidated by spectral methods which include APCI-MS, ^{1}H-NMR, ^{13}C-NMR, ^{1}H-^{1}H COSY, TOCSY, ROESY, HMQC and HMBC.

Extraction and Isolation Procedure for Noni Roots

The extraction procedure for noni root is shown in Figure 1. The dried noni roots (1.5 Kg) were extracted with 95% ethanol (4 L) at 50°C for 1 day. The extract was concentrated to dryness under reduced pressure, and the residue was suspended in water (500 mL) and partitioned successively with hexane (3 × 500 mL), ethyl acetate (3 × 500 mL), and *n*-butanol (3 × 500 mL). The ethyl acetate fraction was subjected to a silica gel column, eluted with Hexane-Chloroform-Methanol (5:1:0, 2:1:0, 1:1:0, 1:2:0, 0:1:0, 0:3:1, and 0:0:1) solven system. Twenty nine fractions were obtained. Fraction 5 was rechromatographed on a silica gel column using hexane-chloroform (5:1) to obtain pure compounds **5** (91 mg) and **6** (46 mg). Fraction 8 was filtered to get pure compound **9** (800 mg). Fraction 9 was first subjected to a normal phase silica gel column, eluted by hexane-chloroform (10:1) to get three subfractions, then subfraction 2 was subjected to a Sephadex LH-20 column eluted with 95% ethanol to give 30 mg pure **7**. Fractions 10-12 were combined and then rechromatographed on a silica gel column eluted with hexane-chloroform (10:1 and 5:1) to give 13

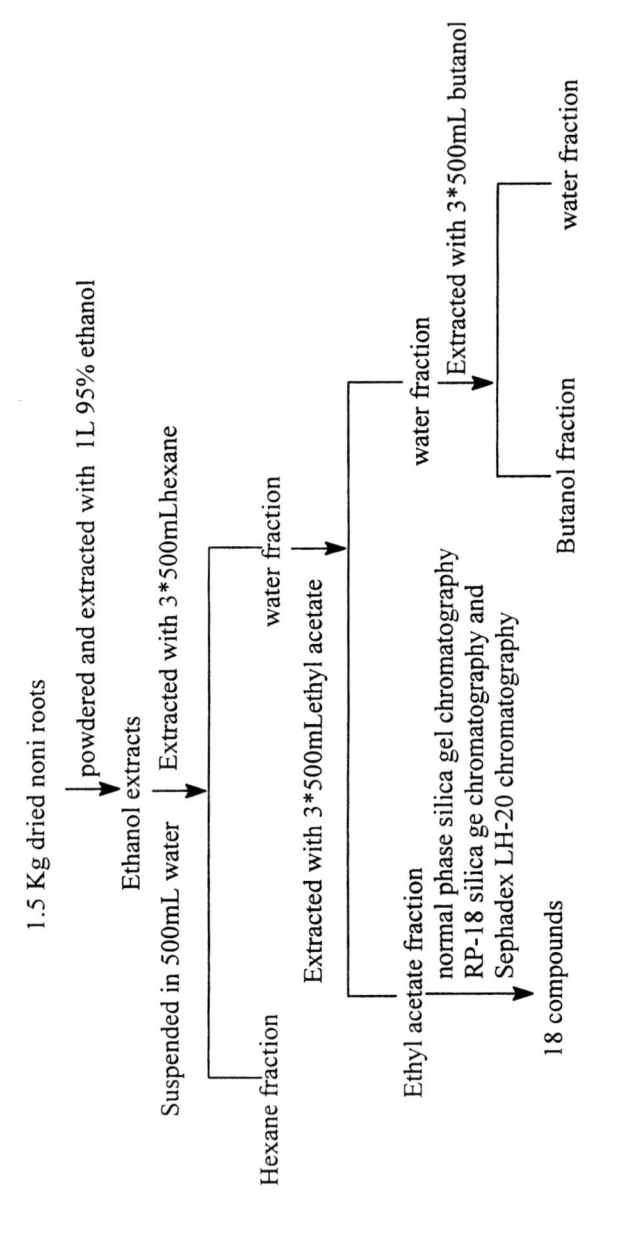

1.5 Kg dried noni roots
↓ powdered and extracted with 1L 95% ethanol
Ethanol extracts
↓ Extracted with 3*500mL hexane
Suspended in 500mL water

Hexane fraction

water fraction
Extracted with 3*500mL ethyl acetate

Ethyl acetate fraction
normal phase silica gel chromatography
RP-18 silica ge chromatography and
Sephadex LH-20 chromatography

18 compounds

water fraction
Extracted with 3*500mL butanol

Butanol fraction   water fraction

Figure 1. Extraction and isolation procedure of noni roots.

subfractions. Compound **10** (1.5 g) was crystalized from subfraction 12. Subfraction 4 was subjected to a Sephadex LH-20 column eluted with 95% ethanol to get 35 mg pure **8**. Subfraction 6 was also subjected to a Sephadex LH-20 column to get a crystal, **18** (100 mg), and a pure compound **1** (18 mg). Compound **2** (22 mg) was abtained from subfraction 9 by subjected to a Sephadex LH-20 column. Fraction 14 was was first subjected to a Sephadex LH-20 column eluted with 95% ethanol to give three subfractions, subfraction 2 was subjected to a RP C-18 column eluted with 75% methanol/water solvent system first, to give 40 mg compound **17**, then eluted with 80% methanol/water to give compounds **3** (24 mg) and **13** (50 mg); subfraction 3 was subjected to a normal phase silica gel column, eluted with hexane-chloroform (1:1.5) to get 80 mg compound **11**. Fraction 16 was packaged on the RP C-18 column eluted with 75% methanol/water, two pure compounds **14** (30 mg) and **15** (25 mg). Fraction 20 was purified by RP C-18 column eluted with methanol/water (65%-80%) system to get 8 subfractions. Fraction 21 was subjected to RP C-18 column eluted with methanol/water (65%-80%) system to get 11 subfractions. 20 mg compound **4** was obtained after subjected subfraction 2 to Sephadex LH-20 column eluted with 95% ethanol. Subfraction 3 was subjected to sephadex LH-20 column eluted with 95% ethanol first to get 2 fractions (21-3-1 and 21-3-2). Fraction 21-3-1was packaged on the preparative TLC eluted with hexane-ethyl acetate (1.5:1) solvent system to give 20 mg compound **16**. Fraction 21 was subjected to RP C-18 column eluted with methanol/water (65%-80%) system to get 10 subfractions. Subfraction 4 was subjected to Sephadex LH-20 column eluted with 95% ethanol to afford 20 mg compound **12**.

Three new naphthoquinones derivatives, one new anhydride and thirteen known anthraquinones and one sterol have been identified from noni roots (Figures 2-6).

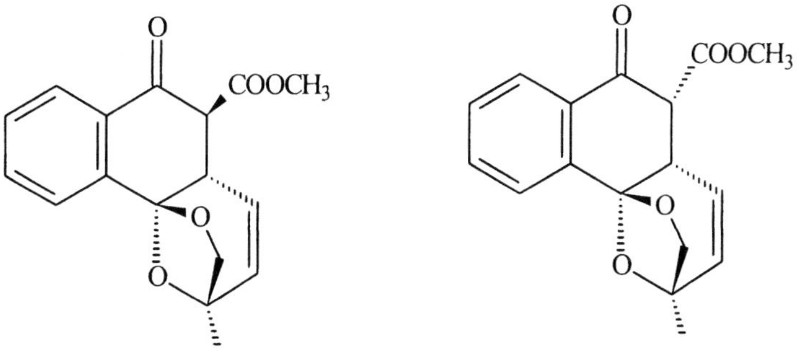

Figure 2. Structures of new compounds (1) and (2) isolated from noni roots.

Figure 3. Structures of new compounds (3) and (4) isolated from noni roots.

1-hydroxy-2-methylanthraquinone (5)

tectoquinone (6)

1-hydroxy-2-formylanthraquinone (7)

2-formylanthraquinone (8)

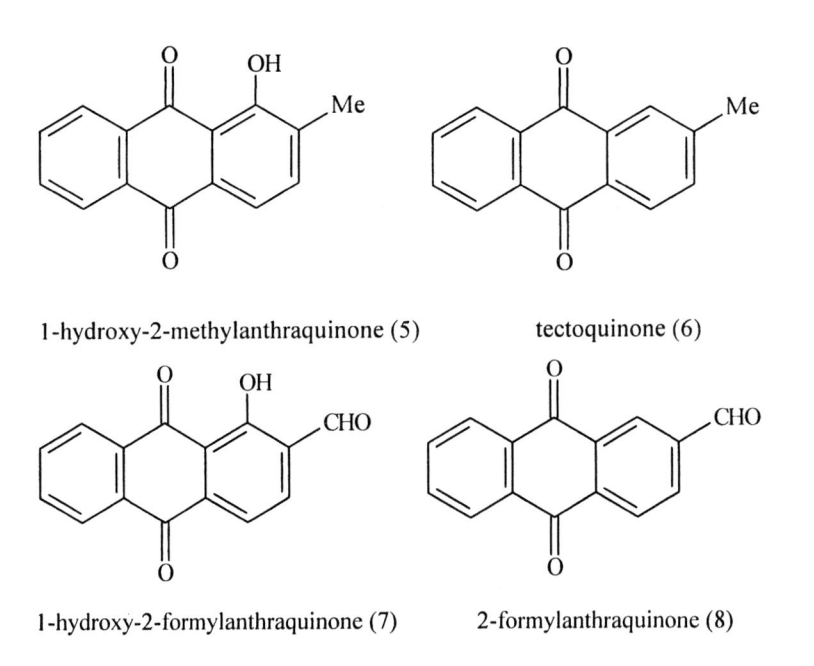

Figure 4. Structures of known compounds (5)-(8) isolated from noni roots.

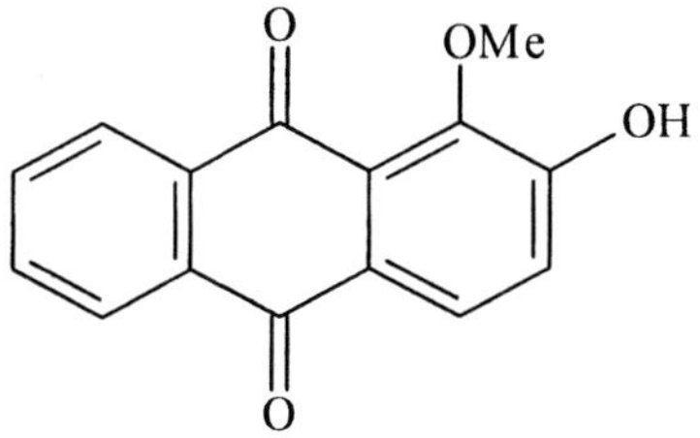

alizarin-1-methyl ether (11) 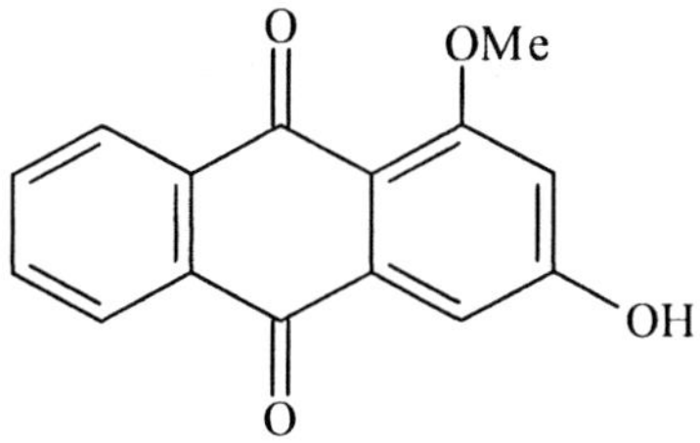   1-methoxy-3-hydroxyanthraquinone (12)

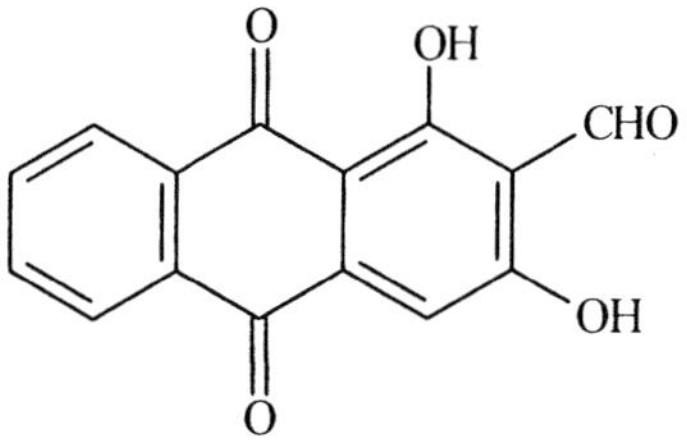

nor-damnacanthal (9) 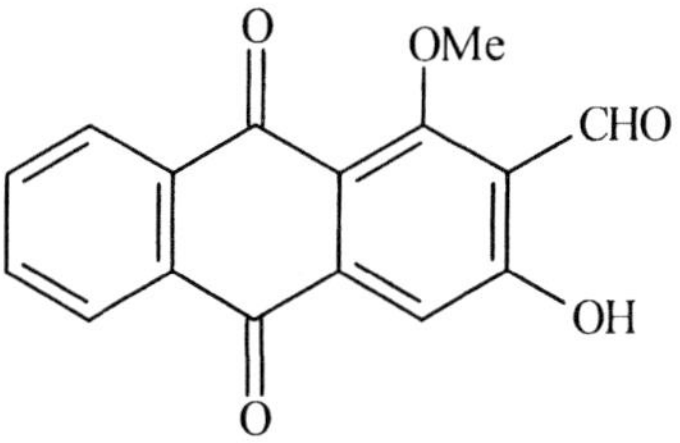   damnacanthal (10)

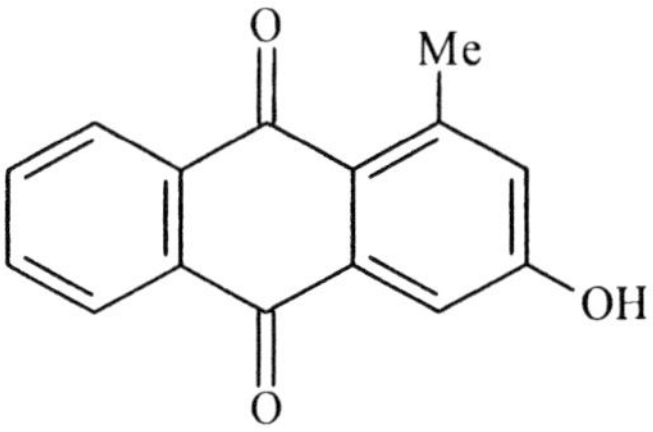

1-methyl-3-hydroxyanthraquinone (13) 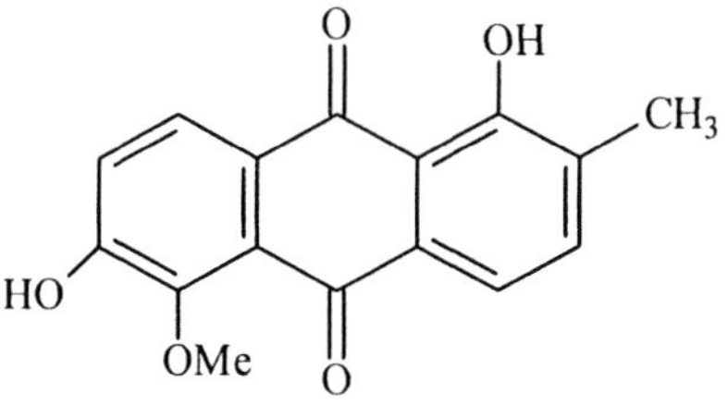   morindone-5-methylether (14)

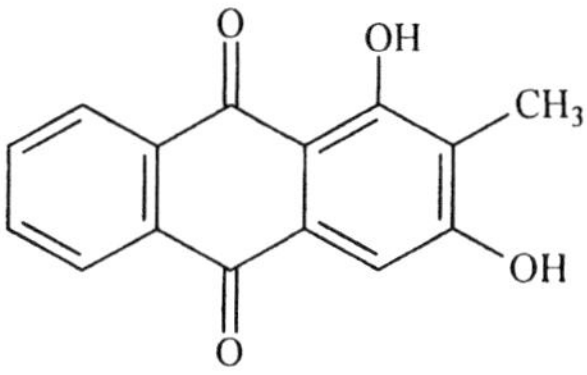

rubiadin (15) 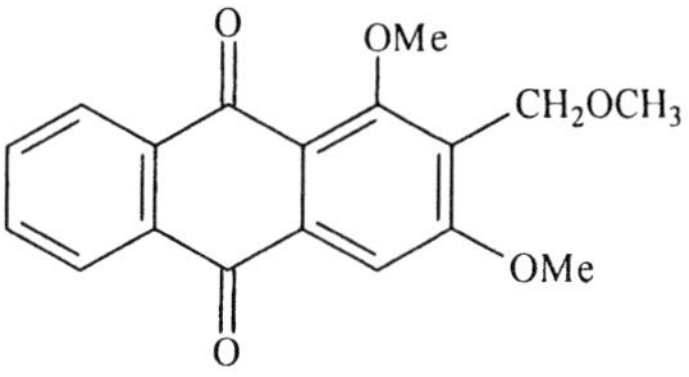   1,3-dimethoxy-2-methoxymethylanthraquinone (16)

Figure 5. Structures of known compounds (9)-(16) isolated from noni roots.

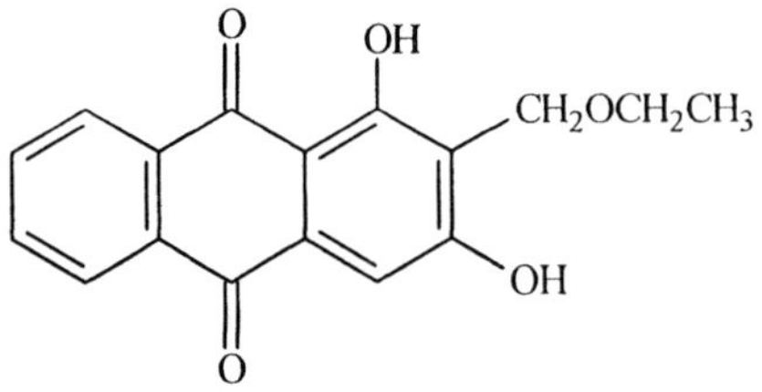

ibericin (17)

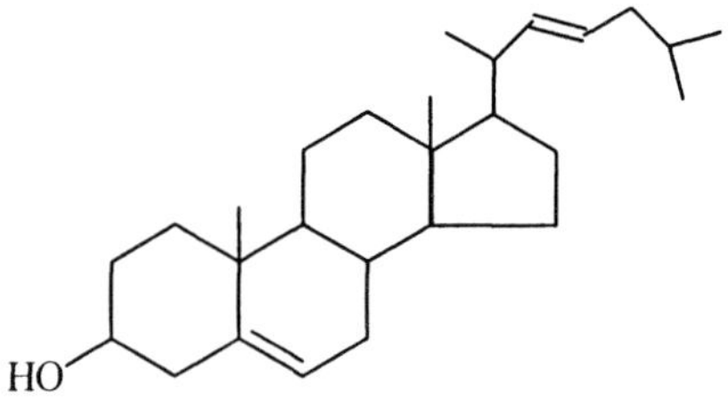

cholesta-5,22-dien-3-ol (18)

Figure 6. Structures of known compounds (17) and (18) isolated from noni roots.

Conclusion

In this research, we systematically studied the chemical components of Tahitian noni roots. A total of 18 compounds were isolated. 4 of them are new compounds. The structures of the isolated compounds were eludidated by spectral methods which include APCI-MS, ^{1}H-NMR, ^{13}C-NMR, ^{1}H-^{1}H COSY, TOCSY, ROESY, HMQC and HMBC. Numerous biological activities have been reported for anthraquinons, which are the major constituents of plants of this family (20). For example, damnacantha (**10**), the major constituent of noni root, was reported to be a poten and selective inhibitor of p56(lck) tyrosine kinase in a variety of tissues (21); to be a inhibitor of Plasmodium falciparum (22); to be a new inhibitor of ras function (23); to show an intensive inhibitory effect against topoisomerase II (24); Alizarin-1-methyl ether (**11**) demonstrated significant activity in the P-388 lymphocytic leukemia screen (25); Rubiadin (**15**), 1-hydroxyl-2-methylanthraquinone (**14**) and 1,3-dimethoxy-2-methoxymethyl-anthraquinone (**16**) were reported to have mutagenic activity (26). Thus, noni roots are a rich source of bioactive anthraquinons, further study on their bioactivities will be necessary.

References

1. Levand, O.; Larson, H. *Planta Med.* **1979**, *36*, 186-187.
2. Hirazumi, A.; Furusawa, E.; Chou, S.C.; Hokama, Y. *Proc. Western Pharm. Soc.* **1994**, 37, 145-146.
3. Hirazumi, A.; Furusawa, E.; Chou, S.C.; Hokama, Y. *Proc. Western Pharm. Soc.* **1996**, 39, 7-9.
4. Daulatabad, C.D.; Mulla, G.M.; Mirajkar, A.M.. *J. Oil Technol. Assoc. (India)* **1989**, *21*, 26-27.
5. Rusia, K.; Srivastava, S.K. *Curr. Sci.* **1989**, *58*, 249-251.
6. Jain, R. K.; Ravindra, K.; Srivastava, S. D. *Proc. Natl. Sci. India, Sect. A* **1992**, *62*, 11-13.
7. Srivastava, M.; Singh, J. *Int. J. Pharmacogn.* **1993**, 182-184.
8. Tiwari, R.D.; Singh, J. *J. India Chem. Soc.* **1977**, *54*, 429-430.
9. Singh, J.; Tiwari, R.D. *J. India Chem. Soc.* **1976**, *52*, 424-425.
10. Ahmad, V.U.; Bano, M. *J. Chem. Soc. Pak.* **1980**, *2*, 71-72.
11. Farine, J.P.; Legal, L. *Phytochemistry* **1996**, *41*, 433-438.
12. Wei, G.J.; Huang, T.C.; Huang, A.S.; Ho, C.-T. In *Nutraceutical Beverages: Chemistry, Nutrition, and Health Effects.* Shahidi, F.; Deepthi K. Weerasinghe, D.K., Eds., ACS Symp. Ser. No. 871, American Chemical Society: Washington, D.C., 2004, pp. 52−61.
13. Wang, M.F.; Kikuzaki, H.; Csiszar, K.; Boyd, C.D.; Maunakea, A.; Fong, S. F.T.; Ghai, G.; Rosen, R.T.; Nakatani, N.; Ho, C.-T. *J. Agric. Food Chem.* **1999**, *47*, 4880-4882.
14. Wang, M.F.; Kikuzaki, H.; Jin, Y.; Nakatani, N.; Zhu, N.Q.; Csiszar, K.; Boyd, C.; Rosen, R.T.; Ghai, G.; Ho, C.-T. *J. Nat. Prod.* **2000**, *63*, 1182 - 1183.
15. Sang, S.M.; He, K.; Liu, G.M. Zhu, N.Q.; Cheng, X.F.; Wang, M.F.; Zheng, Q.Y.; Dong, Z.G.; Ghai, G.; Rosen, R.T.; Ho, C.-T. *Tetrahedron Lett.* **2001**, *42*, 1823-1825.
16. Sang, S.M.; He, K.; Liu, G.M.; Zhu, N.Q.; Cheng, X.F.; Wang, M.F.; Zheng, Q.Y.; Dong, Z.G.; Ghai, G.; Rosen, R.T.; Ho, C.-T. *Org. Lett.* **2001**, *3*, 1307-1309.
17. Sang, S.M.; Cheng, X.F.; Zhu, N.Q.; Wang, M.F.; Zheng, Jhoo, J.W.; Stark, R.E.; Vladimir, B.; Ghai, G.; Rosen, R.T.; Ho, C.-T. *J. Nat. Prod.* **2001**, *64*, 799-800.
18. Sang, S.M.; Cheng, X.F.; Zhu, N.Q.; Vladimir B.; Stark, R.E.; Ghai, G.; Rosen, R.T.; Ho, C.-T. *J. Agric. Food Chem.* **2001**, *49*, 4780-4783.
19. Liu, G.; Bode, A.; Ma, W.Y.; Sang, S.; Ho, C.-T.; Dong, Z. *Cancer Research* **2001** *61*, 5749−5756.
20. Thomson, R.H. *Naturally occurring quinines.* Academic Press, London & New York, **1971**, pp. 365-535.

21. Aoki, K; Parent, A.; Zhang, J. *Eur. J. Pharm.* **2000**, *387*, 19-124.
22. Koumaglo, K.; Gbeassor, M.; Nikabu, O.; De Souza, C.; Werner, W. *Planta Med.* **1992**, *58*. 533-534.
23. Hiramatsu, T.; Imoto, M.; Koyano, T.; Umezawa, K. *Cancer Letters*, **1993**, *73*, 161-166.
24. Tosa, H.; Iinuma, M.; Asai, F.; Tanaka, T.; Ikeda, S.; Tsutsui, K.; Yamada, M.; Fujimori, S. *Bio. Pharm. Bull.* **1998**, *21*, 641-642.
25. Chang, P.; Lee, K.H. *Phytochemistry* **1984**, *23*, 1733-1736.
26. Kawasak, Y.; Goda, Y.; Yoshihira, K. *Chem. Pharm. Bull.* **1992**, *40*, 1504-1509.

Chapter 15

Characterization of Chemical Components of *Ixeris denticulata*

Hang Chen[1], Shiming Li[1], Zhu Zhou[2], Naisheng Bai[1], and Chi-Tang Ho[1]

[1]Department of Food Science, Rutgers, The State University of New Jersey, 65 Dudley Road, New Brunswick, NJ 08901–8520
[2]College of Pharmacy, Fudan University, Shanghai 200032, People's Republic of China

The extract of *Ixeris denticulata* was subjected to successive column chromatography to obtain pure compounds. Five compounds with potential bioactivities were isolated. They were identified as the flavonoids apigenin and luteolin, the sterols β-sitosterol and stigmasterol, and the chlorinated iridoid rehmaglutin D. With the exception of luteolin, the other compounds were reported from *Ixeris denticulate* for the first time. A systematic investigation on the structure-activity relationships among six flavonoids, namely, quercetin, kaempferol, catechin, epicatechin, apigenin and luteolin was also undertaken, with respect to antioxidant activity. The combined results of DPPH radical scavenging assay and ORAC antioxidant capacity assay indicated the importance of at least four structural elements to the antioxidant activity of these compounds.

There are about 50 species of *Ixeris*, a genus belonging to the family Compositae, across the world. Among these, 20 species are mainly distributed in Eastern and Western China. These plants are well known as wild vegetables in certain areas of China, and thus have been incorporated in the diet of the populations indigenous to these regions. *Ixeris* plants have also been widely used in Chinese and Korean as traditional folk medicine to treat many diseases. *Ixeris denticulata* is one of the most important species of *Ixeris* and is of particular interest to researchers (*1*). This plant was recently found to possess a wide spectrum of therapeutic properties, including antioxidant, cytotoxic, analgesic, antipyretic, anti-inflammatory, anti-tumor and anticancer activities (*2,3*).

These findings validate the long and widespread use of *Ixeris denticulata* as a popular folk medicine in China. Thus, there is increasing interest in investigating the chemical components of *Ixeris denticulata*. Ma and co-workers isolated three new sesquiterpene lactone glucoside, Ixerin X, Y and Z, as well as two known flavonoids, luteolin and luteolin-7-*O*-glucoside from *Ixeris denticulata* (*4-6*).

Many of the biological properties of phytochemicals can be attributed, in large part, to their antioxidant activity and their ability to scavenge free radicals (*7*). Widespread awareness of the role of reactive oxygen species (ROS) in both the aging process and many chronic diseases has given rise to a high degree of public concern. Flavonoids have been reported to possess the ability to quench and scavenge ROS and various radicals (*8*).

In this paper, we report the isolation and structure identification of bioactive compounds from *Ixeris denticulata*. In addition, the structure-activity relationships among several structure-related flavonoids are elucidated.

Materials and Methods

Reagents

Chloroform, ethyl acetate, methanol, hexane and acetone were purchased from Fisher Scientific (Pittsburgh, PA). 4-Hydroxy-3-methoxybenzyl-aldehyde (vanillin), 1,1-diphenyl-2-picrylhydrazyl radical (DPPH), fluorescein, 2,2′-azobis-(2-amidinopropane) dihydrochloride (AAPH), quercetin, kaempferol, epicatechin and catechin standards were obtained from Sigma Chemical Co. (St. Louis, MO). 6-Hydroxy-2,5,7,8-tetramethylchroman-2-carboxylic acid (Trolox) was purchased from Aldrich Chemical Co. (Milwaukee, WI). All chemicals were used without further purification.

Extraction

The fresh, whole *Ixeris denticulata* plants were collected by Dr. Zhu Zhou of the Pharmaceutical Institute in Shanghai, China. This material was dried, ground into powder and extracted three times with 95% ethanol for 3 weeks at room temperature. The ethanol extracts were pooled, filtered, concentrated and further extracted with aqueous hexane and ethyl acetate. The final extract was obtained upon drying of the ethyl acetate layer.

Column Chromatography

The ethyl acetate extract was subjected to normal phase silica gel column chromatography. The solvent system employed to elute the compounds of interest consisted of mixtures of chloroform and methanol wherein the content of methanol increased with each successive volume (specifically, 1%, 5%, 10%, 20%, and 50%), and finally, 100% methanol. Fractions obtained from normal phase silica gel chromatography were screened for compounds of interest by thin layer chromatography (TLC) using a solvent system comprised of chloroform: methanol: water: acetic acid (500:100:1:1 (v/v)) and vanillin spray as the visualization agent. Fractions containing compounds of interest that were determined by TLC, to be impure were subjected to further purification by Sephadex LH-20 column chromatography with 95% ethanol as the mobile phase or normal phase silica gel chromatography. The above chromatographic procedures were repeatedly applied to the fractions until pure compounds were obtained.

Mass Spectrometry (MS)

MS analysis was used to elucidate structure information of compounds of interest. The parameters of MS are the following. (1) Instruments: MicroMass AutoSpec HF (EI-MS) and MicroMass Platform II (ES-MS) (MicroMass Co., MA); (2) Mass scan Range: 100-900 amu; (3) Scan rate: 0.4 sec (ES-MS); (4) Cone voltage: 36 Volts (ES-MS); (5) Corona voltage: 3.59 K Volts (ES-MS); (6) Source temperature: 250°C (EI-MS) and 150°C (ES-MS); (6) Mobile phase: 50/50 acetonitrile/water.

Nuclear Magnetic Resonance (NMR)

^{1}H NMR and ^{13}C NMR were used to identify the structures of the compounds of interest. The spectra were obtained on Varian (300 MHz, 400 MHz and 500 MHz) instruments.

DPPH Assay

The 1,1-diphenyl-2-picrylhydrazyl radical (DPPH) is a stable radical which exhibits a dark purple color in alcoholic solution and has a maximum absorbance at 517 nm. Interaction of this radical with antioxidants results in a decrease in its absorbance intensity, thus providing a basis for the measurement of the antioxidant activity of investigational compounds. A Model 301 spectrophotometer from Milton Roy (Ivyland, PA) was used to obtain absorbance measurements at 517 nm in the DPPH assay. Each test compound was added to a solution of DPPH radical in 3 mL of 95% ethanol to make a sample solution having 0.1 mM of DPPH and 1 mM of the test compounds. Samples were then shaken vigorously with a Vortex and left in the dark for 30 minutes at room temperature. The absorbance of the test samples at 517 nm was compared to that of a control lacking antioxidants (*9*). Each test compound (luteolin, apigenin, quercetin, kaempferol, epicatechin, catechin) was assayed in triplicate and the results were averaged. The % inhibition of the DPPH radical by each sample was calculated according to the formula of Yen and Duh (*10*).

$$\% \text{ Inhibition} = [(A_0 - A_t) / A_0] \times 100 \qquad (1)$$

ORAC Assay

The oxygen radical absorbance capacity (ORAC) assay is a recently developed method that measures antioxidant scavenging activity against peroxyl radicals induced by 2,2'-azobis-(2-amidinopropane) dihydrochloride (AAPH) at 37°C. The underlying basis for this assay is the measurement of oxidative damage to a fluorescent probe caused by a peroxyl radical. The ability of a compound to inhibit this loss of fluorescence intensity is used as a measure of its antioxidant capacity against free radicals (*11*). The ORAC analysis was performed on a HITACHI F-3010 Fluorescence Spectrophotometer (Hitachi, Japan) with excitation and emission wavelengths of 493 nm and 515 nm, respectively. The fluorescence intensities of all standards and samples were monitored over the course of 60 minutes. Fluorescein working solution (8.16 x 10^{-5} mM) was prepared daily with 75 mM phosphate buffer (pH=7.4). The 153 mM solution of AAPH was kept in an ice bath and discarded after eight hours. Trolox was used as a standard and diluted with 75 mM phosphate buffer (pH=7.4) to give a 10 μM Trolox working solution. Samples of pure flavonoids were prepared by dissolving in a 1:1 (v/v) mixture of acetone and water and then diluting with 75 mM phosphate buffer (pH 7.4) to a final concentration of 5 μM which was used for analysis. A mixture consisting of 3000 μL of working fluorescein solution, 500 μL of AAPH solution, 100 μL of 75 mM phosphate buffer (pH 7.4), and 400 μL of sample/standard/blank solution was assayed by the Fluorescence Spectrophotometer. The fluorescence decay curve for each sample was generated by plotting the intensity data obtained by the instrument at each time point, and the final ORAC value of each standard and sample was calculated for comparison according to the formulas presented below:

Antioxidant Capacity = $AUC_{antioxidant} - AUC_{blank}$ (2)

Relative ORAC Value = $[(AUC_{antioxidant} - AUC_{blank})/(AUC_{trolox} - AUC_{blank})]$ x
Molarity of Trolox/Molarity of Sample (3)

Results and Discussion

Isolation of Compounds from *Ixeris denticulate*

Upon fractionation of 120 g ethyl acetate extract of *Ixeris denticulata* by normal phase silica gel column chromatography, we identified three fractions of interest, via TLC, for further analysis. The first fraction of interest contained 277.6 mg of material. This fraction was subjected to further purification by successive column chromatographies to yield 4 mg of compound (**1**) and 13.9 mg of compound (**5**). Upon spraying with vanillin, compound (**1**) became yellow while compound (**5**) became red, indicating that they may be a flavonoid and a terpene, respectively. The second fraction of interest was dried and weighed 235.6 mg. It was determined by TLC to be primarily composed of a compound that became yellow upon spraying with vanillin, indicating that this compound may also be a flavonoid. 40.8 mg of compound (**2**) was obtained after purification by Sephadex LH-20 column chromatography. The final fraction of interest contained 1.59 g of material. Compound (**3**) (19.4 mg) and compound (**4**) (13.4 mg) were isolated from this fraction. Given a purple color was formed upon spraying with vanillin, these two compounds were identified as steroids.

Identification of Compounds Isolated from *Ixeris denticulata*

A total of 5 compounds were isolated from *Ixeris denticulata*. The chemical structures of these compounds are shown in Figure 1, and the spectral data for each of the isolated compounds are listed and discussed below.

Compound (1) was isolated as a white powder. We estimated the molecular formula for this compound from the results of ES-MS (*m/z* 271 [M+H]$^+$) to be $C_{15}H_{10}O_5$. The NMR data corresponding to compound (1) are as follows: ^{1}H NMR (300 MHz in methanol-d_4): δ_H 6.21 (1H, s, H-6), 6.46 (1H, s, H-8), 6.60 (1H, s, H-3), 6.93 (2H, d, *J* =9 Hz, H-3',5'), 7.85 (2H, d, *J* =9 Hz, H-2',6'). ^{13}C NMR (300 MHz in DMSO-d_6): δc 94.47 (C-8), 99.38 (C-6), 103.51 (C-3), 104.35 (C-10), 116.64 (C-3,5), 121.83 (C-1'), 129.15 (C-2',6'), 157.98 (C-9), 161.97 (C-4'), 164.39 (C-2), 164.86 (C-7). This spectral data is identical to that published for apigenin (*5*). Thus, we have confirmed the identity of compound (1) as apigenin. Our research constitutes the first report of the isolation of apigenin from *Ixeris denticulata*.

Compound (2) was isolated as a yellow powder. We determined the molecular formula of this compound from the results of EI-MS (*m/z* 286 [M$^+$]) to be $C_{15}H_{10}O_6$. The NMR data corresponding to compound (2) are as follows: ^{1}H NMR (500 MHz in DMSO-d_6): δ_H 6.19 (1H, d, *J* =2.0 Hz, H-6), 6.43 (1H, d, *J* =2.0 Hz, H-8), 6.64 (1H, s, H-3), 6.89 (1H, d, *J* =8.0 Hz, H-5'), 7.39 (1H, dd, *J*= 4.0, 8.0 Hz, H-6'), 7.40 (1H, d, *J* =2.0 Hz, H-2'). ^{13}C NMR (300 MHz in DMSO-d_6): δc 93.84 (C-8), 98.84 (C-6), 102.85 (C-3), 103.65 (C-10), 113.34 (C-2'), 116.00 (C-5'), 118.98 (C-6'), 121.48 (C-1'), 145.75 (C-3'), 149.73 (C-4'), 157.29 (C-9), 161.47 (C-5), 163.87 (C-2), 164.22 (C-7), 181.64 (C-4). This spectral data is identical to that published for luteolin (*5*). Thus, we have confirmed the identity of compound (2) as luteolin.

Compound (3) was isolated as a white crystalline powder. The molecular formula of this compound was determined from the results of EI-MS (*m/z* 414 [M$^+$]) to be $C_{29}H_{50}O$. The NMR data corresponding to compound (3) are as follows: ^{1}H NMR (300 MHz in chloroform-d_1): δ_H 0.68 (3H, s, H-18), 0.81(6H, d, *J* = 4.1 Hz, H-26 and 27), 0.84 (3H, t, *J* =3.9 Hz, H-29), 0,92 (3H, d, J= 3.9 Hz, H-21), 0.99 (3H, s, H-19), 3.52 (1H, m, H-3), 5.35 (1H, distorted triplet, H-6). ^{13}C NMR (300 MHz in chloroform- d_1): δc 37.2 (C-1), 31.6 (C-2), 71.7 (C-3), 38.8 (C-4), 140.7 (C-5), 121.6 (C-6), 31.9 (C-7), 31.8 (C- 8), 50.1 (C-9), 36.5 (C-10), 21.1 (C-11), 39.7 (C-12), 42.3 (C-13), 56.7 (C-14), 24.2 (C-15), 28.2 (C-16), 56.0 (C-17), 11.9 (C-18), 19.3(C-19), 36.1 (C-20), 18.8 (C-21), 33.9 (C-22), 26.0 (C-23), 45.8 (C-24), 29.1 (C-25), 19.8 (C-26), 19.0 (C-27), 23.0 (C-28), 11.9 (C-29). This spectral data is identical to that published for β-sitosterol (*12*). Thus, the identity of compound (3) was confirmed as β-sitosterol. Although β-sitosterol is widely distributed in plants, our research constitutes the first report of its isolation from *Ixeris denticulata*.

Compound (4) was isolated as a white crystalline powder. The combined MS and NMR data indicate that, despite showing only a single spot on TLC analysis, our sample of compound (4) contains this compound in admixture with β-sitosterol. Furthermore, we have confirmed the identity of compound (4) as stigmasterol by comparison of our data with published spectra. The EI/MS of the sample shows two similar molecular ion peaks having *m/z* of 414.5 and 412.5. This is in accord with the fact that the sample is a mixture of β-sitosterol, which has a molecular weight of 414.72, and stigmasterol, which has a molecular weight of 412.7. In addition, the ^{1}H NMR and ^{13}C NMR spectra show a combination of these two compounds. The NMR data corresponding to the components are as follows: ^{1}H NMR of stigmasterol (300 MHz in chloroform-d_1): δ_H 0.67 (3H, s, H-18), 0.82(6H, d, *J* = 6 Hz, H-26 and 27), 0.82 (3H, t, *J* =6 Hz, H-29), 0,91(3H, d, J= 6.3 Hz, H-21), 1.00 (3H, s, H-19), 3.51 (1H, m, H-3), 5.00 (1H, dd, J=14.1 , 8.1 Hz, H-23), 5.14 (1H, dd, J= 15.4 , 8.5 Hz, H-22), 5.35 (1H, distorted triplet, H-6). ^{13}C NMR of stigmasterol (300 MHz in chloroform-d_1): δ_c 37.3 (C-1), 32.4 (C-2), 72.5 (C-3), 43.0 (C-4), 141.5 (C-5), 122.4 (C-6), 32.7 (C-7), 32.6 (C-8), 52.0 (C-9), 38.0 (C-10), 21.9 (C-11), 40.5 (C-12), 41.4 (C-13), 57.6 (C-14), 25.1 (C-15), 29.9 (C-16), 56.8 (C-17), 12.8 (C-18), 19.8 (C-19), 36.9 (C-20), 21.9 (C-21), 139.0 (C-22), 130.0 (C-23), 46.6 (C-24), 32.4 (C-25), 19.6 (C-26), 22.0 (C-27), 26.2 (C-28), 13.0 (C-29). ^{1}H NMR of β-sitosterol (300 MHz in chloroform-d_1): δ_H 0.67 (3H, s, H-18), 0.82(6H, d , *J* = 6 Hz, H-26 and 27), 0.82 (3H, t, *J* =6 Hz, H-29), 0,91(3H, d, J= 6.3 Hz, H-21), 1.00 (3H, s, H-19), 3.51 (1H, m, H-3), 5.35 (1H, distorted triplet, H-6). ^{13}C NMR of β-sitosterol (300 MHz in chloroform-d_1): δ_c 38.0 (C-1), 32.4 (C-2), 72.5 (C-3), 43.0 (C-4), 141.5 (C-5), 122.4 (C-6), 32.7 (C-7), 32.6 (C-8), 50.9 (C-9), 37.3 (C-10), 21.9 (C-11), 40.5 (C-12), 43.8 (C-13), 57.6 (C-14), 25.1 (C-15), 29.0 (C-16), 56.8 (C-17), 12.6 (C-18), 20.2 (C-19), 36.9 (C-20), 19.6 (C-21), 34.7 (C-22), 26.9 (C-23), 46.6 (C-24), 29.9 (C-25), 19.8 (C-26), 20.6 (C-27), 23.8 (C-28), 12.8 (C-29). There are only two extra peaks in the ^{1}H NMR spectrum of the mixture compared to that of β-sitosterol, specifically, at δ_H 5.00 and 5.14 ppm. These peaks are due to minor structural variations in stigmasterol and β-sitosterol. Stigmasterol has one more double bond than β-sitosterol at C-22-C-23, which likely gives rise to the additional peaks evident in the mixture. As was the case with β-sitosterol, although stigmasterol is widely distributed in plants, our research constitutes the first report of its isolation from *Ixeris denticulata*.

Compound (5) was isolated as a colorless oil. The molecular formula of this compound was determined from the results of ES(+)-MS (*m/z* 185 (100%) [M-Cl]$^+$) and ES(-)-MS (m/z 255(100%), 257 (67%), 259 (11%) [M+Cl]$^-$) to be $C_9H_{13}O_4Cl$. The peak of 255 is the molecular ion plus one chlorine, which maybe presented in the sample while the peak of 265 is the molecular ion plus formate,

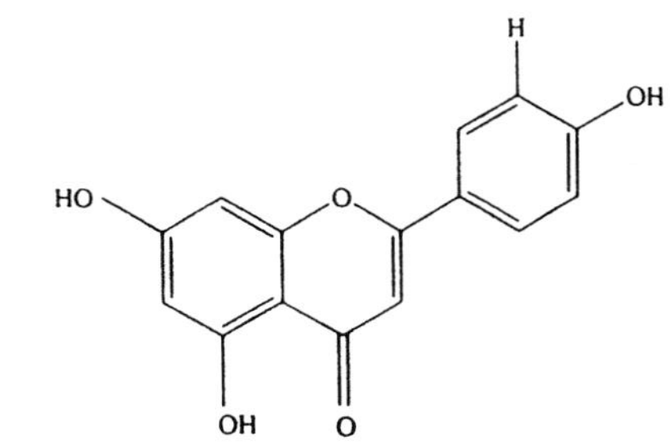

202

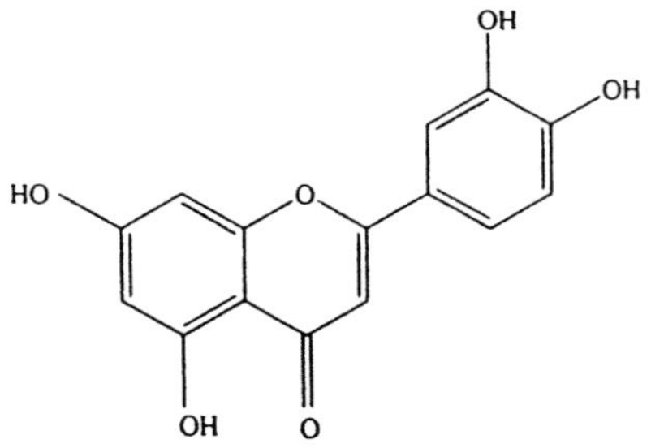

Compound 2: Luteolin

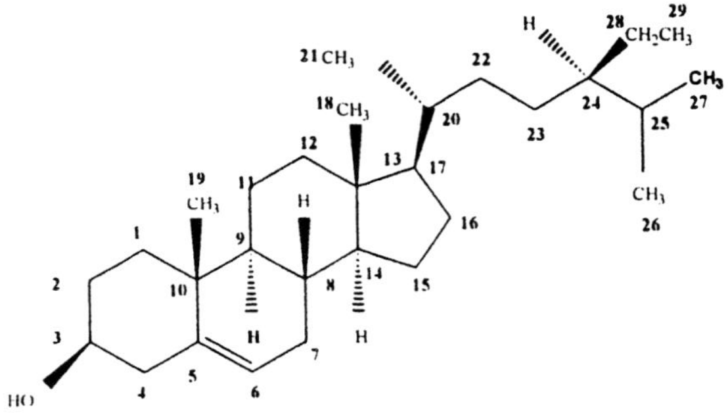

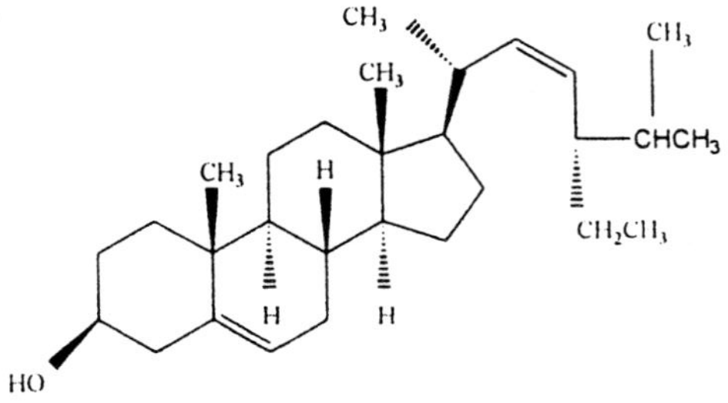

Compound 4: Stigmasterol

Compound 1: Apigenin

Compound 3: β-sitosterol

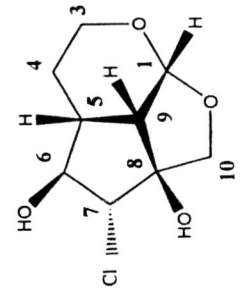

Compound 5: Rehmaglutin D

Figure 1: Chemical structures of compounds 1-5

which presented in the mobile phase. The NMR data corresponding to compound (5) are as follows: ^{1}H NMR (500 MHz in methanol-d_4): δ_H 5.30 (1H, d, J= 5.5 Hz, H-1), 4.39 (1H, d, J= 10.5 Hz, H-10α), 4.06 (1H, dd, J= 1.5, 10 Hz, H-7), 3.85 (1H, dd, J= 9, 10 Hz, H-6), 3.54 (2H, m, H-3), 3.43 (1H, dd, J= 1.5, 10 Hz, H-10β), 2.29 (1H, dd, J= 5.5, 10.5 Hz, H-9), 2.15 (1H, m, H-5), 1.78 (2H, m, H-4). 13C NMR (500 MHz in methanol-d_4): δ_C 101.4 (C-1), 56.5 (C-3), 22.0 (C-4), 36.7 (C-5), 72.7 (C-6), 74.1 (C-7), 85.2 (C-8), 45.9 (C-9), 76.2 (C-10). This spectral data is identical to that published for rehmaglutin D, which was previous isolated from Chinese Rehmanniae Radix (13,14). Thus, we have confirmed the identity of compound (5) as rehmaglutin D. This chlorine-containing iridoid has mainly been found in scrophulariaceae plants, such as *Verbascum wiedemannianum*, Rehmanniae Radix and *Cymbaria mongolica* (13-16). Most of these plants are traditional Chinese medicines of great importance for their antipyretic and antianemic properties, and have long been used as tonics as well as for the treatment of rheumatism (13,17). Our research constitutes the first report of the isolation of a chlorinated iridoid from *Ixeris* plants. Thus, our determination of the presence of rehmaglutin D in *Ixeris denticulata* may be of great pharmacological importance at both the genus and species levels.

Antioxidant Activity of Flavonoids from *Ixeris denticulata*

Two flavonoid compounds, apigenin and luteolin, isolated from *Ixeris denticulate* differ only in the presence of the catechol moiety on the B ring in luteolin. As relationships between flavonoid structure and antioxidant activity have yet to be elucidated, we decided to investigate these structure-activity relationships among six structurally related flavonoids, including quercetin, epicatechin, catechin, and kaempferol, in addition to apigenin and luteolin. The structures of these compounds are shown in Figure 2, along with their antioxidant activities as determined by the DPPH and ORAC assays discussed more fully below.

The free radical scavenging activities of six structurally related flavonoids determined by the DPPH assay are shown in Figure 3. All results are triplicate and averaged. Trolox was used as a standard.

The AAPH induced fluorescence decay curves of the blank, Trolox standard, and six structurally related flavonoid samples are presented in Figure 4. The relative ORAC values of these samples were calculated from the area under each of the decay curves using the formula presented previously. The relative ORAC values for these compounds are presented in Table I.

Table I. Relative ORAC Values of Six Strcturally-related Flavonoids

Compound	Relative ORAC Value
Trolox	1
Quercetin	7.33
Epicatechin	5.24
Catechin	5.22
Kaempferol	5.07
Luteolin	4.74
Apigenin	4.58

ORAC values are expressed as Trolox equivalens

Structure-Activity Relationship

The results we obtained from the analysis of six flavonoids in the DPPH and ORAC assays enabled us to identify a number of structural elements in these compounds that impart improved antioxidant activity. The order of the free radical scavenging activity of these compounds as determined by the DPPH assay is: quercetin > kaempferol > luteolin > apigenin > catechin ~ epicatechin. The order of the antioxidant capacity of these compounds as determined by the ORAC assay is: quercetin > epicatechin ~ catechin > kaempferol > luteolin > apigenin. Although the rank order of the antioxidant activities of these compounds differs for the DPPH and ORAC assays, the cumulative results indicate the importance of the presence of at least three separate structural elements: (1) an extended conjugated system comprising a 4-oxo (carbonyl) group on the C ring in conjugation with a C2-C3 double bond; (2) a 3-OH group on the C ring; and (3) a catechol (dihydroxy) group on the B ring. The results of the DPPH and ORAC assays are presented, with reference to the presence of these structural elements, in Table II.

As is evident from Table II, the different rank order of compounds in each assay reflects the apparent degree of importance of each structural element in that assay. In the DPPH assay, the conjugated 4-oxo group on the C ring appears to be of primary importance, while the 3-OH group on the C ring appears of secondary importance. Moreover, compounds possessing the catechol group on the B ring are more active than their respective analogues lacking that moiety. In the ORAC assay, the 3-OH group on the C ring appears to be of primary importance, while the catechol group on the B ring seems of secondary importance. Within this framework, compounds possessing the conjugated 4-oxo group on the C ring are more active than their respective analogues lacking that moiety. It is also noted that the increasing antioxidant capacity of molecules in the ORAC assay parallels an increasing overall number of hydroxyl groups. As these two assays utilize different free radicals, our findings are in accord with the

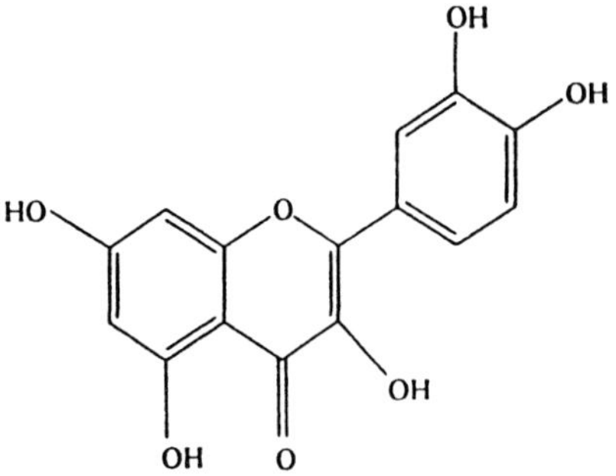

QUERCETIN

DPPH: 95.7%
ORAC: 7.33

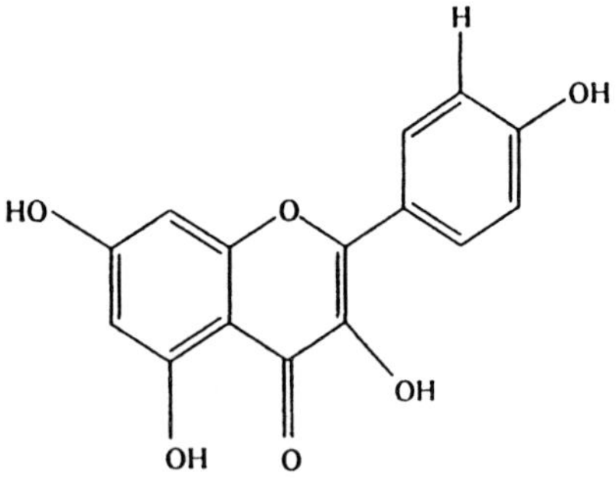

KAEMPFEROL

DPPH: 95.2%
ORAC: 5.07

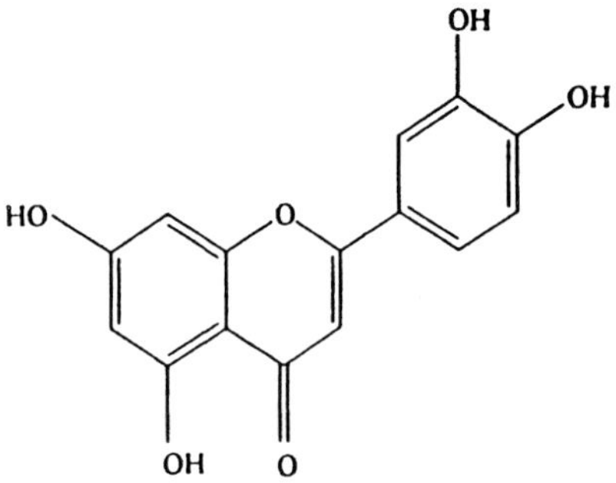

LUTEOLIN

DPPH: 94.5%
ORAC: 4.74

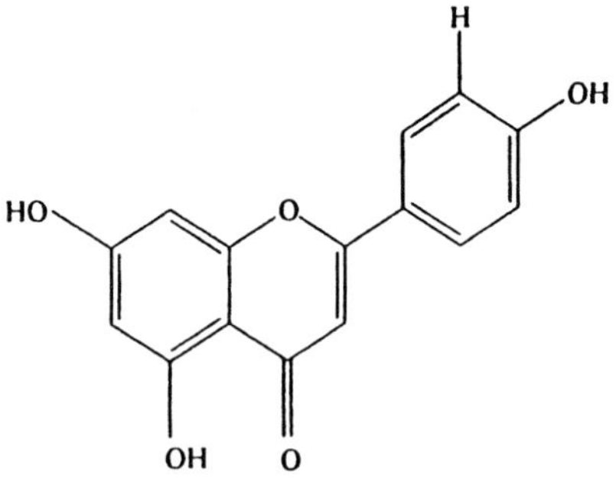

APIGENIN

DPPH: 93.9%
ORAC: 4.58

Figure 2: Chemical structures of six flavonoids and their DPPH and ORAC values.

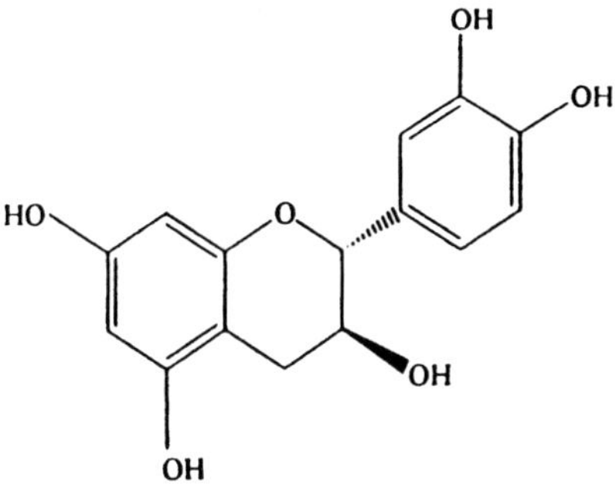

CATECHIN

DPPH: 92.9%
ORAC: 5.22

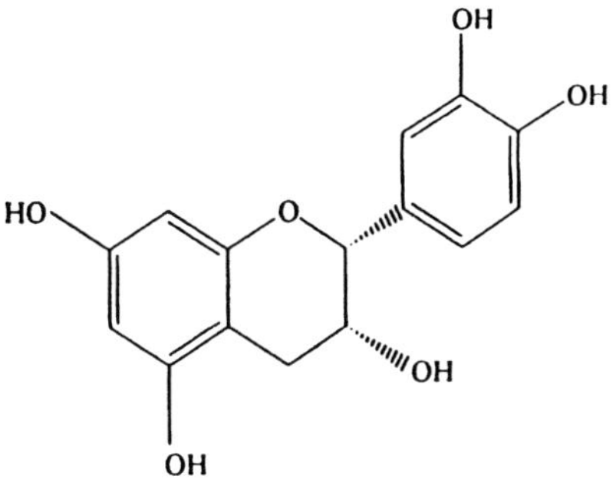

EPICATECHIN

DPPH: 92.9%
ORAC: 5.24

Figure 2. *Continued.*

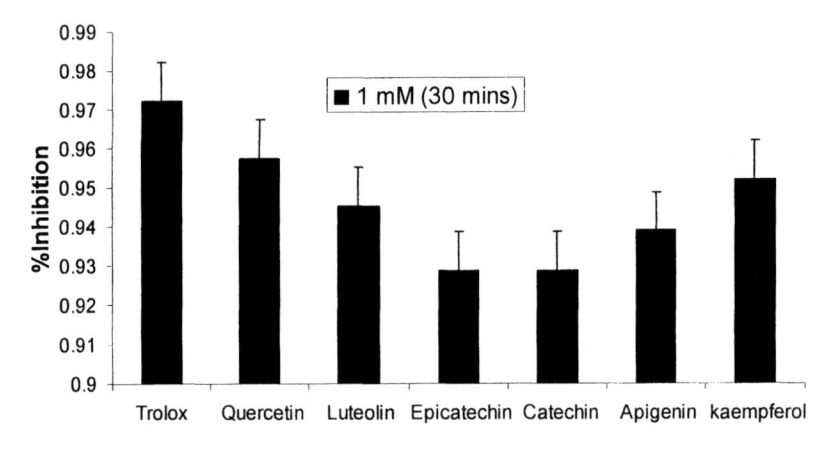

Figure 3: Free radical scavenging activity of flavonoids in the DPPH assay.

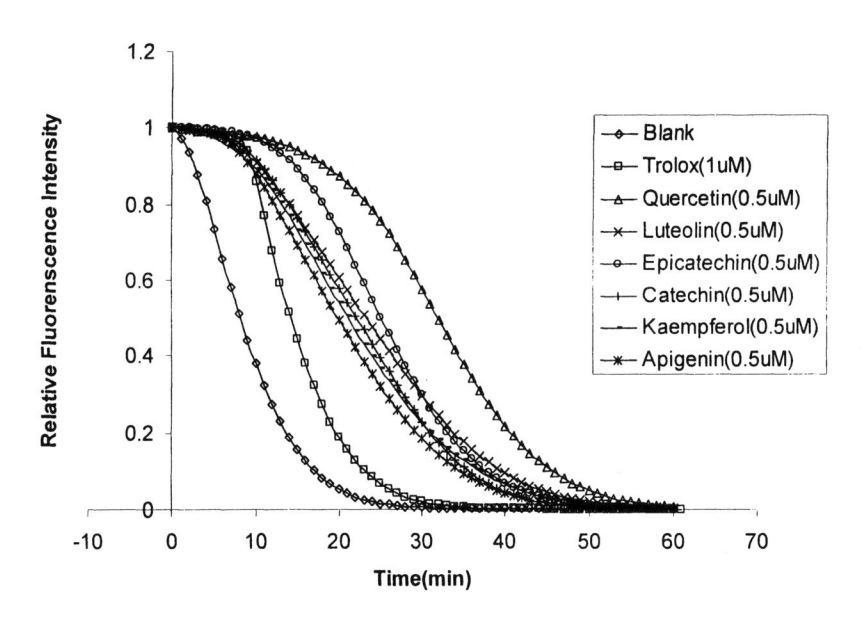

Figure 4: AAPH induced fluorescence decay curves from the ORAC assay.

conclusion drawn by Halliwell and Gutteridge (*18*) that the antioxidant activity of a compound depends on the type of free radical used in the assay.

Table II: Importance of Structural Eelements of Flavonoids to Antioxidant Activity

Compound	DPPH	C ring 4-oxo	C ring 3-OH	B ring Catechol
Quercetin	95.7	+	+	+
Kaempferol	95.2	+	+	
Luteolin	94.5	+		+
Apigenin	93.9	+		
Epicatechin	92.9		+	+
Catechin	92.9		+	+
	ORAC			
Quercetin	7.33	+	+	+
Epicatechin	5.24		+	+
Catechin	5.22		+	+
Kaempferol	5.07	+	+	
Luteolin	4.74	+		+
Apigenin	4.58	+		

References

1. Li, C.X.; Liu, H.B.; Guan, H.S. *Nat. Prod. Res. Develop.* **1999**, *12*, 95-100.
2. Shiojima, K.; Suzuki, H.; Kodeara, N.; Kubota, K.I.; Tsushima, S.; Ageta, H.; Chang, H.C.; Chen, Y.P. *Chem. Pharm. Bull.* **1994**, *42*, 2193-2195.
3. Ma, J.Y.; Wang, Z.T.; Qi, S.H.; Xu, L.S.; Xu, G.J. *J. China Pharm. Univ.* **1998**, *29*, 167-169.
4. Ma, J.Y.; Wang, Z.T.; Qi, S.H.; Xu, L.S.; Xu, G.J. *Phytochem.* **1999**, *50*, 113-115.
5. Ma, J.Y.; Wang, Z.T.; Xu, L.S.; Xu, G.J.; Wang, Y.X. *Zhongguo Yaoke Daxue Xuebao* **1998**, *29*, 94-96.
6. Ma, J.Y.; Wang, Z.T.; Qi, S.H.; Xu, L.S.; Xu, G.J. *Stud. Plant Sci.* **1999**, *6*, 394-398.

7. Saija, A.; Scalese, M.; Lanza, M.; Marzullo, D.; Bonina, F.; Castelli, F. *Free Rad. Biol. Med.* **1995**, *19*, 481-486.
8. Middleton, E. Jr.; Kandaswami, C.; Theoharides, T.C. *Pharmacol Rev.* **2000**, *52*, 673-751.
9. Chen, C.W.; Ho, C.-T. *J. Food Lipids* **1995**, *2*, 35-46.
10. Yen, G.C; Duh, P.D. *J. Agric. Food Chem.* **1994**, *42*, 629-632.
11. Huang, D.J.; Ou, B.X.; Woodill, M.H.; Flanagan, J.A.; Prior, R.L. *J. Agric. Food Chem.* **2002**, *50*, 4437-4444.
12. Ali, M.S.; Saleem, M.; Ahmad, W.; Parvez, M.; Yamdagni, R. *Phytochemistry* **2002**, 59, 889-895.
13. Kitagawa, I; Fukuda, Y; Taniyama, T; Yoshikawa, M. *Chem. Pharm. Bull.* **1986**, *34*, 1399-1402.
14. Kitagawa, I; Fukuda, Y; Taniyama, T; Yoshikawa, M. *Chem. Pharm. Bull.* **1991**, *39*, 1171-1176.
15. Dai, J.Q.; Liu, Z.L.; Yang, L. *Phytochem.* **2002**, *59*, 537-542.
16. Gazar, H.A; Tasdemir, D.; Ireland, C.M.; Calis, I. *Biochem. Systematics Ecol.* **2003**, *31*, 433-436.
17. Ma, Y.C.; Fu, H.C.; Chen, S. *Flora Intramongolia.* Intramongolia People's Press: Huhhot. **1982**, pp. 305-306.
18. Halliwell, B.; Gutteridge, J.M. *Methods Enzymol.* **1990**, *186*, 1–85.

Chapter 16

Bioavailability, Metabolism, and Pharmacokinetics of Glycosides in Chinese Herbs

Pei-Dawn Lee Chao[1], Su-Lan Hsiu[1], and Yu-Chi Hou[2]

Schools of [1]Pharmacy and [2]Chinese Medicine, China Medical University, Taichung, Taiwan 404, Republic of China

Glycosides are a group of natural products in Chinese herbs. In preparing decoctions of traditional Chinese medicine, glycosides are soluble in hot water. However, due to their hydrophilic nature, the biological fate of glycosides is of great interest. After oral administration of glycosides to rats, rabbits or humans, serum or urine samples collected at specific time points were analyzed by HPLC prior to and after hydrolysis by sulfatase and glucuronidase. The results indicated that the glycosides of flavonoids and monoterpenes were not present as their parent forms in the circulation. Moreover, the aglycones were negligibly present in the circulation except paeoniflorgenin, a monoterpene. The predominant metabolites of flavonoid glycosides circulating in the blood are the sulfate and glucuronide conjugates of their aglycones, which often reside in the body for prolonged time due to enteric recycling. In contrast, a triterpene glycoside glycyrrhizin and its aglycone glycyrrhetic acid were both present in the blood stream. In conclusion, glycosides are often subject to deglycosylation in the intestine prior to absorption and phenolic aglycones are further sulfated/glucuronidated by gut and/or liver. Therefore, the *in vitro* bioactivities of alcoholic aglycones and conjugated metabolites of phenolic aglycones should be more predictive for the *in vivo* efficacy of glycosides in Chinese herbs.

© 2006 American Chemical Society

Introduction

Glycosides are a group of natural products widely distributed in Chinese herbs, fruits and vegetables. The structure of a glycoside consists of a sugar moiety and an aglycone that is mainly a phenol or an alcohol. In preparing decoctions of traditional Chinese medicine, glycosides are quite soluble in the hot water. However, due to their hydrophilic nature, the biological fates and bioactivities of glycosides are of great interest.

There has recently been a proliferation of *in vitro* studies on herbal components, concerning mainly effects on cell signaling pathways. These studies are largely focused on commercially available compounds including various glycosides and aglycones. However, in recent years, there is an increasing recognition that flavonoid glycosides are generally not absorbed *per se* and most flavonoid aglycones are not present in the circulation, whereas their conjugated metabolites predominate (*1*). Therefore, whether the *in vitro* effects produced by glycosides and aglycones can be extrapolated to the *in vivo* bioactivities still remains unclear and needs to be clarified. This review would deal with the bioavailability, metabolism and pharmacokinetics of naturally occurring glycosides of flavonoids and terpenes in Chinese herbs.

Bioavailabilities of Glycosides

The definition of bioavailability for a drug is the fraction of an oral dose that reaches the systemic circulation in its parent form. In terms of glycosides, the presence of sugar moiety confers the hydrophilic property that makes them soluble in hot water. However, based on drug absorption theory, glycosides are generally too lipophobic to be absorbed through enterocytes *via* transcellular diffusion (*2*).

Most of flavonoid glycosides (e.g. rutin, naringin, hesperidin, baicalin, daidazin and phellamurin whose structures are shown in Figure 1) are hydrophilic and thus may not be transported across the phospholipid bilayers of cell membranes by passive diffusion. Naringin and hesperidin are flavanone glycoside constituents in many *Citrus* herbs like the fruits of *C. aurantium, C. grandis* and *C. reticulata*. The bioavailabilities of naringin and hesperidin after oral administration were investigated in rabbits. The results showed that naringin and hesperidin were not absorbed *per se* (*3,4*). Another flavanone glycoside phellamurin, a major constituent of *Phellodendron wilsoni*, was orally

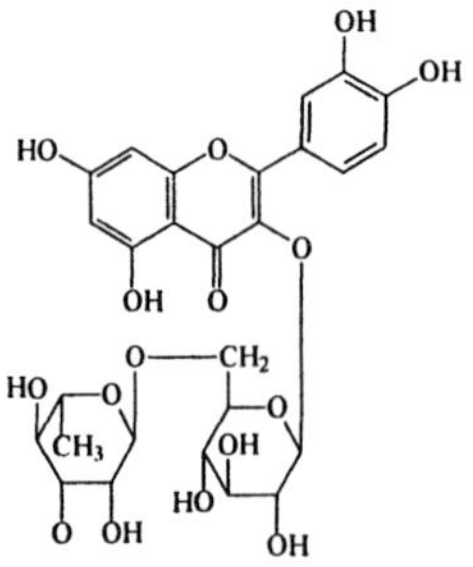

Rutin

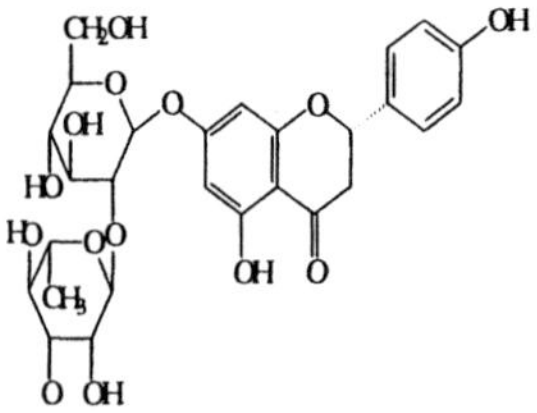

Naringin

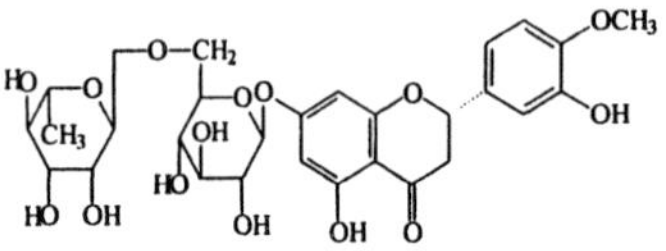

Hesperidin

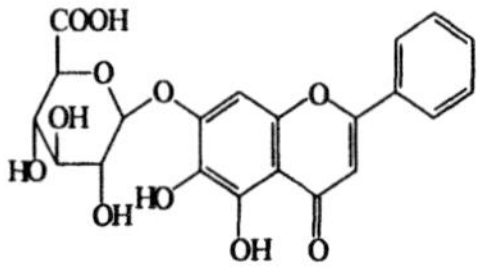

Baicalin

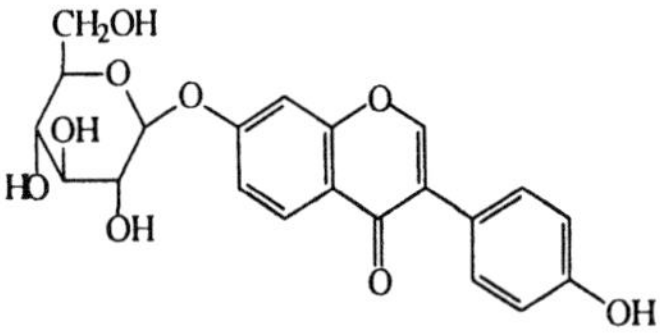

Daidzin

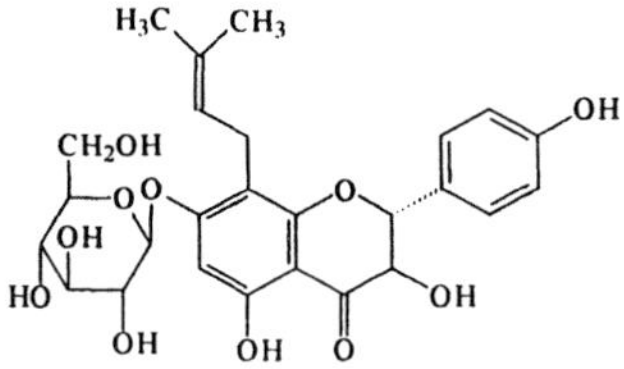

Phellamurin

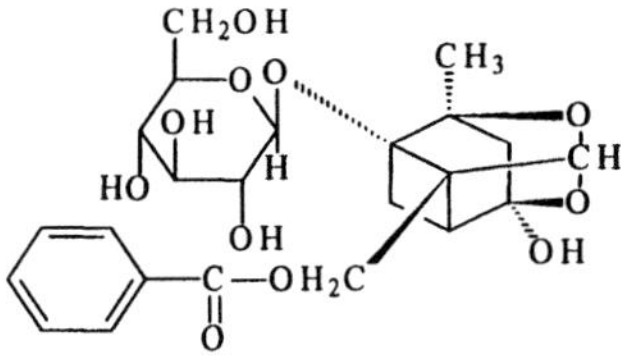

Paeoniflorin

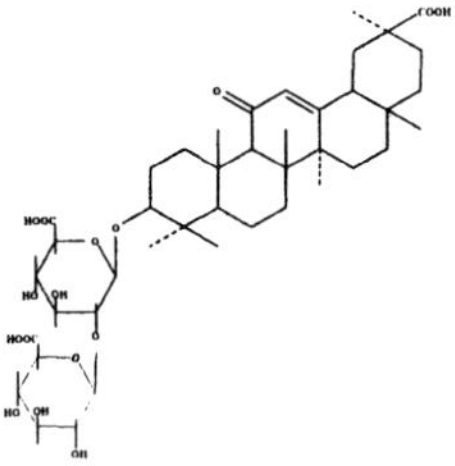

Glycyrrhyzin

Figure 1. Chemical structures of glycosides.

administered to rats. The parent form phellamurin was not absorbed (5). Likewise, rutin, quercetin 3-glucoside and 4'-glucoside as well as soy isoflavone glycosides i.e. genestin, daidzin were reportedly not present in human plasma (6-8). In contrast, the absorption of quercetin glucosides in their parent forms was reported and led to a speculation that they were transported across gut wall by the intestinal sodium- dependent glucose transporter (SGLT-1) (9). However, Walle et al. (10) later reported that quercetin glucosides are completely hydrolyzed in ileostomy patients before intestinal absorption.

Paeoniflorin and glycyrrhizin are terpene glycosides. The biological fates of paeoniflorin was investigated in rats and indicated that it was not present in the blood, however, glycyrrhizin was found to be absorbed *per se* in rabbits and rats (11-13).

Presystemic Metabolism of Glycosides in Gastrointestinal Tract

The gastrointestinal metabolism of glycosides has been reported to be dependent on the intestinal microflora (14-17). The microflora residing in the intestine can release enzymes to gradually hydrolyze the glycosides into aglycones that were absorbed by the intestine. Besides, in recent years, two β-glucosidases with activity toward flavonoid glycosides were isolated from human small intestinal mucosa: lactase phlorizin hydrolase (LPH) and cytosolic β-glucosidase (CBG), indicating a role of human LPH and CBG in glycoside metabolism and absorption (18-24).

The flavonoid aglycones that are not absorbed in the small intestine can thereafter be degraded by colonic microflora into phenolic acids (25-28). From human feces two phenotypically different type of bacteria utilizing quercetin-3-glucoside as carbon and energy source were isolated (17). Isolates of one type were identified as strains of *Enterococcus casseliflavus*. They utilized the sugar moiety of the glycoside, but did not degrade the aglycone further. The second type of isolate was identified as *Eubacterium ramulus*. This organism was capable of degrading the aromatic ring system by detachment of the A ring from the residual flavonoid molecules and the opening of the heterocyclic C ring (26, 29-32). 3,4-Dihydroxybenzaldehyde, phloroglucinol and ethanol were detected in small amount as breakdown products of quercetin-3-glucoside (15, 26, 33). Another *in vitro* fermentation study using human faecal flora indicated that rutin, naringin and naringenin were completely metabolized within the 72-h fermentation period (15). Our recent study reported that twelve tested flavonoid aglycones were all significantly degraded upon incubation with feces of rat, rabbit and human (34), whereas paeoniflorgeninin, a monoterpene being hydrolyzed from paeoniflorin by the fecal flora, was not further degraded (11).

This suggested that terpene aglycones are more resistant to bacterial degradation than flavonoid aglycones.

Determination of Aglycones and Conjugated Metabolites in Biological Fluids

After oral administrations of glycosides or glycoside-containing herbs to rabbits, rats or humans, the parent forms of glycosides and their aglycones as well as the conjugated metabolites of various aglycones were measured in serum or urine. The glycosides in serum were determined after deproteinization with four-fold volumes of methanol. Regarding the assay of conjugated metabolites, because authentic standards of conjugated metabolites of various aglycones are not commercially available, their concentrations were determined through hydrolysis with β-glucuronidase and sulfatase. Because many polyphenol aglycones (e.g. quercetin, naringenin, hesperetin, baicalein, daidzein and neophellamuretin) are prone to oxidize at 37 °C, the optimum quantitation methods for their conjugated metabolites in serum and urine were investigated in our laboratory (*35-37*). In literature, a mixed enzyme containing predominately β-glucuronidase and little sulfatase was commonly used for the hydrolysis of glucuronides/sulfates of flavonoids in serum. Recently, it was increasingly realized that significant amounts of flavonoid sulfates were present in biological fluids (*37,38*). Therefore, separate hydrolysis of biological fluids using glucuronidase and sulfatase, respectively, would be more appropriate in order to measure the sulfates and glucuronides accurately. Regarding the optimum condition for enzymolysis, air was removed and ascorbic acid was added to protect the polyphenol aglycones from oxidation, and the optimum time for hydrolysis was investigated by time course studies on each subject. The time needed for enzymolysis of various polyphenol conjugates was found rather different. In general, sulfate conjugates need less time to be completely hydrolyzed than the correspondent glucuronides. For example, baicalein glucuronides needed 7 h, whereas the sulfates needed only 2 h; daidzein glucuronides needed 14 h, whereas the sulfates needed only 2 h (*36,37*).

In regard to paeoniflorin, a monoterpene glucoside in the roots of *Paeonia lactiflora*, its aglycone paeoniflorgenin was prepared through hydrolysis of paeoniflorin with β-glucosidase. After oral administration of paeonoflorin-containing decoction to rats, serum was assayed for paeoniflorgenin prior to and after hydrolysis with sulfatase and glucuronidase, respectively. The result indicated that conjugated metabolites of paeoniflorgenin were not present in serum (*11*). This fact showed that the biological fate of monoterpene glycoside was rather different from those of flavonoid glycosides.

The HPLC/UV systems used for determination of aglycones were established for various polyphenols and alcohols in our laboratory. In general, the mobile phase needs to be acidified with acetic acid or phosphoric acid in order to obtain peaks in better shape for polyphenols, whereas acidification was not necessary for the analysis of alcohol aglycones (*35-37*).

Further Metabolism of Phenolic Aglycones by Gut/Liver

In order to investigate the metabolism of flavonoid aglycones, naringenin was administered to rabbits by oral and intravenous routes (*3*). After intravenous bolus, naringenin glucuronides/sulfates formed instantaneously and serum concentration decreased with time. After oral administration, naringenin glucuronides/sulfates formed rapidly as well and was present predominately in the blood stream. However, the patterns of serum profiles of naringenin glucuronides/sulfates were rather different between naringin and naringenin, indicating that the aglycone naringenin was absorbed more rapidly to result in much earlier T_{max} and higher C_{max} of its conjugated metabolites. The biological fates of hesperidin and hesperetin were investigated in rabbits. The major molecules circulating in the bloodstream were hesperetin glucuronides/sulfates (*4*). When phellamurin was orally administered to rats, the major metabolites were glucuronides/sulfates of its aglycone neophellamuretin, which emerged rapidly after dosing (*5*). These metabolites were immediately distributed into the brain, indicating that they may cross the blood brain barrier like morphine glucuronide, probably mediated by an active transport (*39,40*).

Another study administered baicalein to rats orally and intravenously (*36*). The results indicated that after intravenous bolus, baicalein glucuronides/sulfates formed instantaneously and the concentration decreased with time. The serum concentrations of the conjugated metabolites were much higher than the parent compound since the first sampling time. After oral dosing, only transient presence of baicalein was observed and the sulfates/glucuronides of baicalein were found predominantly in the circulation. Moreover, the biological fates were compared between baicalein and baicalin, a glucuronide of baicalein. The results indicated that the patterns of the serum profiles of baicalein glucuronides/sulfates were rather different between baicalin and baicalein, indicating that the aglycone baicalein was absorbed more rapidly to result in much earlier T_{max} and higher C_{max} of its conjugated metabolites and eliminated faster than baicalin (*36*).

When other flavonoid aglycones like quercetin, morin, naringenin and hesperetin were intravenously administered to rabbits or rats, they were all metabolized by conjugation very rapidly and intensively. The major metabolites were sulfates and glucuronides (*4,5,35-37,41*). In particular, the conjugated metabolites of quercetin showed much higher serum levels than the parent forms since the first blood sampling time after intravenous bolus (*41*). It is a clear indication that conjugation reactions with sulfate and/or glucuronic acid seem to be the most common pathways for flavonoid metabolism. Recently, sulfation metabolism was found more predominant than glucuronidation for quercetin, morin, baicalein, hesperetin and daidzein in our laboratory. Besides, flavonoid aglycones may also undergo reactions such as hydroxylation, methylation and reduction. Circulating glucuronides, sulfates and *O*-methylated forms are believed to be most likely to exert bioactivity *in vivo* (*42, 43*).

When terpene glycosides e.g. glycyrrhizin and paeoniflorin were given orally to rabbits and rats, respectively, no conjugated metabolites of their aglycones were detected, indicating that conjugation metabolism toward alcoholic aglycones was negligible. It is apparent that terpene aglycones exhibit different biological fates from phenolic aglycones (*11,12*).

Excretion of Flavonoid Conjugates into Bile and Urine

The sulfates and glucuronides of flavonoids are ionized under physiological pH and are very soluble in water, therefore they are readily excreted into bile and urine. When excreted into bile, the conjugated metabolites pass into the duodenum and are hydrolyzed by enterobacteria to transform into aglycones which may be reabsorbed and enter an enterohepatic circulation to result in a second peak of serum profile. The structures of flavonoid conjugates determine the extent of biliary excretion and enterohepatic circulation. The half-life of elimination can thus be prolonged and the plasma levels of quercetin metabolites have been detected up to 24 h after flavonoid consumption, indicating a possible build-up of quercetin metabolites in plasma after repeated intake of onion (*44*).

Urinary excretion of the conjugated metabolites of flavonoid glycosides in the naringin-containing and baicalin-containing decoctions were investigated in humans (*37,45,46*). The total urinary recovery of conjugated metabolites of naringenin was 22.4% and the elimination half-life of naringenin conjugates was 3.8 h. No free form of naringenin was detected in urine prior to hydrolysis. The urinary recoveries of sulfates/glucuronides of baicalein and wogonin were 7.2% and 11.6%, respectively. The half-lives of sulfates/glucuronides of baicalein and wogonin were about 8 and 10 h, respectively. No free form of baicalein and wogonin was detected in urine prior to hydrolysis (*37*). The amount of baicalein sulfates was much higher than glucuronides, whereas the amounts of sulfates and glucuronides of wogonin were comparable (*46*). The great interindividual variations of urinary excretion of flavonoid metabolites may be explained by the complex biological metabolism of glycosides. Our previous studies have shown that the urinary excretion of flavonoid conjugated metabolites can vary from trace to 24.6 % depending on the sources and structures of flavonoids (*45-47*).

Effect of Sugar Structure on Glycoside Absorption

The absorptions of quercetin and various forms of glycosides were compared in human volunteers. It was reported that both quercetin and its glycosides could be absorbed and the absorption from onion (52%) was greater than that from quercetin aglycone (24%), whereas the absorption from rutin, a quercetin rutinoside, was the poorest (17%) in healthy ileostomy subjects (9). In contrast, Walle et al. did not detect any quercetin glucosides in the ileostomy fluid, whereas substantial amount of quercetin aglycone was identified, suggesting that quercetin glucosides were hydrolyzed to quercetin in the small intestine and then absorbed (10).

A study reported that quercetin glucosides from onions were absorbed more efficiently than those from apples or quercetin glycoside supplements, which could be accounted for by that the attached sugar moiety on the flavonoid glycoside affected the rate of hydrolysis of glycosides and thus the absorption of their aglycones (48). Morand et al. reported that the nature of the glycosylation markedly influences the efficiency of quercetin absorption in rats (49). Quercetin 3-glucoside can be absorbed in the small intestine and the plasma level of quercetin glucuronides/sulfates was three times higher than quercetin itself. This fact can be explained by the higher water solubility of quercetin 3-glucoside than quercetin. Olthof et al. suggested that the quercetin glucosides were rapidly absorbed in humans irrespective of the position of the glucose moiety (50). However, the absorption of rutin, a quercetin glycoside containing a 3-glucose-rhamnose moiety, was even lower than its aglycone quercetin. This might be ascribed to the steric hindrance of 3-glucose-rhamnose moiety to enzymolysis. In contrast, 3-rhamnose moiety could not be hydrolyzed in the small intestine, indicating the lack of rhamnosidase in rats (49).

Metabolic Pharmacokinetics of Glycosides

High hydrophilicity of glycosides makes them satisfactorily soluble in herbal decoction and in the gastrointestinal juice. Then, glycosides are hydrolyzed in the intestine by the intestinal enzymes like LPH and CBG as well as glucosidase released from enterobacteria and transformed into absorbable aglycones. After entry to gut and liver, the phenolic aglycones are further metabolized into sulfates/glucuronides which then enter enterohepatic circulation and often reside in body for a pretty long duration. Moreover, the conjugated metabolites might cross the blood brain barrier to exert bioactivities in the central nervous system. *In vitro* studies on putative bioactive metabolites of glycosides at appropriate concentration should be able to provide valuable information for predicting the *in vivo* efficacy of glycosides in Chinese herbs.

Based on previous reports on metabolic pharmacokinetics of glycosides, the use of monoterpene aglycone and sulfates/glucuronides of phenolic aglycones in the *in vitro* system to test biological activities can mimic more closely the *in vivo* situation of glycosides (*51*).

Conclusion

In conclusion, although the bioavailability of glycosides are generally negligible, glycosides become bioavailable after they are transformed into lipophilic aglycones in the intestine. Therefore, glycosides seem to serve as excellent sustained - released natural prodrugs of their aglycones. Based on the physicochemical and pharmacokinetic characteristics, glycosides exhibit promising pharmaceutical properties for oral absorption. The *in vitro* evaluation of glycoside bioactivities based on proper concentration of their putative metabolites is very important for more efficient investigation in this area. Furthermore, a better understanding of glycoside pharmacokinetics can also aid in more rational design of dosage regimen of Chinese medicine in clinical use.

Acknowledgements

Work in the authors' laboratory has been supported in part by grants from National Science Council, Department of Health, R.O.C. and China Medical University.

References

1. Walle, T. *Free Rad. Biol. Med.* **2004**, *36*, 829-837.
2. Zmuidinavicius, D.; Didziapetris, R.; Japertas, P.; Avdeef, A.; Petrauskas, A. *J. Pharm. Sci.* **2003**, *92*, 621-633.
3. Hsiu, S.L.; Huang, T.Y.; Hou, Y.C.; Chin, D.H.; Chao, P.D.L. *Life Sci.* **2002**, *70*, 1481-1489.
4. Yang, C.Y.; Tsai, S.Y.; Chao, P.D.L.; H.F. Yen; Chien, T.M.; Hsiu, S.L. *J. Food Drug Anal.* **2002**, *10*, 143-148.
5. Chen, H.Y.; Wu, T.S.; Wang, J.P.; Kou, S.C.; Chao, P.D.L. *Chin. Pharm. J.* **2001**, *53*, 37-44.
6. Erlund, I.; Kosonen, T.; Alfthan, G.; Maenpaa, J.; Perttunen, K.; Kenraali, J.; Parantainen, J.; Aro, A. *Eur. J. Clin. Pharmacol.* **2000**, *56*, 545-553.

7. Sesink, A.L.A.; O'Leary, K.A.; Hollman, P.C.H. *J. Nutr.* **2001**, *131*, 1938-1941.
8. Setchell, K.D.R.; Brown, N.M.; Zimmer-Nechemias, L.; Brashear, W.T.; Wolfe, B. E.; Kirschner, A.S.; Heubi, J.E. *Am. J. Clin. Nutr.* **2002**, *76*, 447-453.
9. Hollman, P.C.; de Vries, J.H.; van Leeuwen, S.D.; Mengelers, M.J.; Katan, M.B. *Am. J. Clin. Nutr.* **1995**, *62*, 1276-1282.
10. Walle, T.; Otake, Y.; Walle, UK.; Wilson, FA. *Journal of Nutrition.* **2000**, *130*, 2658-2661.
11 Hsiu, S.L.; Lin, Y.T.; Wen, K.C.; Hou, Y.C.; Chao, P.D.L. *Planta Med*, **2003**, *69*, 1113-1118.
12 Ching, H.; Hsiu, S.L.; Hou, Y.C.; Chen, C.C.; Chao P.D.L. *J. Food Drug Anal.* **2001**, *9*, 67-71.
13. Wang, Z.; Kurosaki, Y.; Nakayama, T.; Kimura, T. *Biol. Pharm. Bull.* **1994**, *17*, 1399-1403.
14. Bokkenheuser, V.D.; Shackleton, C.H.L.; Winter, J. *Biochem. J.* **1987**, *248*, 953-956.
15. Schneider, H.; Schwiertz, A.; Collins, M.D.; Blaut, M. *Arch Microbiol.* **1999**, *171*, 81-91.
16. Sesink, A.L.A.; Arts, I.C.W.; Faassen-Peters, M.; Hollman, P.C.H. *J. Nutr.* **2003**, *133*, 773-776.
17. Justesen, U.; Arrigoni, E.; Larsen, B.R.; Amado, R. *Lebensm.-Wiss. u.-Technol.* **2000**, *33*, 424-430.
18. Day, A.J.; Canada, F.J.; Diaz, J.C.; Kroon, P.A.; Mclauchlan, R.; Faulds, C.B.; Plumb, G.W.; Morgan, M.R.; Williamson, G. *FEBS Lett.* **2000**, *468*, 166-170.
19. Nemeth, K.; Plumb, G.W.; Berrin J-G.; Juge, N.; Jacob, R.; Naim, H.Y.; Williamson, G.; Swallow, D.M.; Kroon, P.A. *Eur. J. Nutr.* **2003**, *42*, 29-42.
20. Day, A.J.; Gee, J.M.; DuPont, M.S.; Johnson, I.T.; Williamson, G. *Biochem. Pharmacol.* **2003**, *65*, 1199-206.
21. Mackey, A.D.; Henderson, G.N.; Gregory, J.F. 3rd. *J. Biol. Chem.* **2002**, *277*, 26858-26864.
22. Naim, H.Y.; Lentze, M.J. *J. Biol. Chem.* **1992**, *267*, 25494-25504.
23. Dudley, M.A.; Burrin, D.G.; Quaroni, A.; Rosenberger, J.; Cook, G.; Nichols, B.L.; Reeds, P.J. *Biochem. J.* **1996**, *320*, 735-743.
24. Wilkinson, A.P.; Gee, J.M.; Dupont, M.S.; Needs, P.W.; Mellon, F.A.; Williamson, G.; Johnson, I.T. *Xenobiotica.* **2003**, *33*, 255-264.
25. Schoefer, L.; Mohan, R.; Schwiertz, A.; Braune, A.; Blaut, M. *Appl. Environ. Microbiol.* **2003**, *69*, 5849-5854.
26. Blaut, M.; Schoefer, L.; Braune, A. *Int. J. Vitam. Nutr. Res.* **2003**, *73*, 79-87.

27. Winter, J.; Popoff, M.R.; Grimont, P.; Bokkenheuser, V.D. *Int. J. Syst. Bacteriol.* **1991**, *41*, 355-7.
28. Winter, J.; Moore, L.H.; Dowell, V.R Jr.; Bokkenheuser, V.D. *Appl. Environ. Microbiol.* **1989**, *55*, 1203-1208.
29. Schoefer, L.; Mohan, R.; Braune, A.; Birringer, M.; Blaut, M. *FEMS Microbiol Lett.* **2002**, *208*, 197-202.
30. Braune, A.; Gutschow, M.; Engst, W.; Blaut, M. *Appl. Environ. Microbiol.* **2001**, *67*, 5558-5567.
31. Schneider, H.; Simmering, R.; Hartmann, L.; Pforte, H.; Blaut, M. *J Appl Microbiol.* **2000**, *89*, 1027-1037.
32. Schneider, H.; Blaut, M. *Arch. Microbiol.* **2000**, *173*, 71-75.
33. Schneider, H.; Schwiertz, A.; Collins, MD.; Blaut, M. *Arch. Microbiol.* **1999**, *171*, 81-91.
34. Lin, Y.T.; Hsiu, S.L.; Hou, Y.C.; Chen, H.Y.; Chao, P.D.L. *Biol. Pharm. Bull.* **2003**, *26*, 747-751.
35. Hsiu, S.L.; Tsao, T.W.; Tsai, Y.C.; Ho, H.J.; Chao, P.D.L. *Biol. Pharm. Bull.* **2001**, *24*, 967-969.
36. Lai, M.Y.; Hsiu, S.L.; Hou, Y.C.; Yang, C.Y.; Chao, P.D.L. *J. Pharm. Pharmacol.* **2003**, *55*, 199-209.
37. Lai, M.Y.; Hsiu, S.L.; Chen, C.C.; Hou, Y.C.; Chao, P.D.L. *Biol. Pharm. Bull.* **2003**, *26*, 79-83.
38. Piskula, M.K. *BioFactors.* **2000**, *12*, 175-180.
39. Chen, H.Y.; Wu, T.S.; Hou, Y.C.; Kuo, S.C.; Chao, P.D.L. *J. Food Drug Anal.* **2001**, *9*, 127-131.
40. Bourasset, F.; Cisternino, S.; Temsamani, J.; Scherrmann, J-M. *J. Neurochem.* **2003**, *86*, 1564-1567.
41. Hou, Y.C.; Chao, P.D.L.; Ho, H.J.; Wen, C.C.; Hsiu, S.L. *J. Pharm. Pharmacol.* **2003**, *55*, 199-203.
42. Spencer, J.P.E.; El Mohsen, M.M.A.; Rice-Evans, C. *Arch. Biochem. Biophys.* **2004**, *423*, 148-161.
43. Gauthier, A.; Gulick, PJ.; Ibrahim, RK. *Arch. Biochem. Biophys.* **1998**, *351*, 243-249.
44. Boyle, S.P.; Dobson, V.L.; Duthie, S.J.; Kyle, J.A.M.; Collins, A.R. *Eur. J. Nutr.* **2000**, *39*, 213-223.
45. Hou, Y.C.; Yen, H.F.; Hsiu, S.L; Chen, C.C.; Chao, P.D.L. *J. Chin. Med.* **1999**, *10*, 65-73.
46. Lai, M.Y.; Hou, Y.C.; Hsiu, S.L.; Chen, C.C.; Chao, P.D.L. *J. Food Drug Anal.* **2002**, *10*, 75-80.
47. Hollman, P.C.; van Trijp, J.M.; Buysman, M.N.; van der Gaag, M.S.; Mengelers, M.J.; de Vries, J.H.; Katan, M.B. *FEBS Letters.* **1997**, *418*, 152-156.

48. Hollman, P.C.; Bijsman, M.N.; van Gameren, Y.; Cnossen, E.P.; de Vries, J.H.; Katan, M.B. *Free Rad. Res.* **1999**, *31*, 569-573.
49. Morand, C.; Manach, C.; Crespy, V.; Remesy C. *Bio. Factors*. **2000**, *12*, 169-174.
50. Olthof, M.; Hollman, P.C.H.; Vree, T.B.; Katan, M.B. *J. Nutr.* **2000**, *130*, 1200-1203.
51. Williamson, G. *Phytochem. Rev.* **2002**, *1*, 215-222.

Chapter 17

Moringa, a Novel Plant Rich in Antioxidants, Bioavailable Iron, and Nutrients

Ray-Yu Yang[1,3], Samson C. S. Tsou[1], Tung-Ching Lee[2], Leing-Chung Chang[1], George Kuo[1], and Po-Yong Lai[3]

[1]AVRDC, The World Vegetable Center, Shanhua, Tainan, Taiwan, Republic of China
[2]Department of Food Science and Center for Advanced Food Technology, Rutgers, The State University of New Jersey, 65 Dudley Road, New Brunswick, NJ 08901
[3]Institute of Tropical Agriculture and International Cooperation, National Pingtung University of Science and Technology, Pingtung, Taiwan, Republic of China

A strong food-based approach is critical to alleviate nutritional deficiencies in the tropics. Our survey of over 120 species of Asian indigenous vegetables for nutrient contents, antioxidant activities (AOA), and indigenous knowledge of their medicinal uses indicated that *Moringa oleifera* was among the most promising species. We conducted additional studies to evaluate four Moringa species for AOA, antioxidant contents and nutritional quality, and to investigate *M. oleifera*'s AOA and iron as affected by freezing, boiling, and in vitro digestion. We concluded that the four Moringa species are high in AOA, antioxidant and nutrient contents, low in oxalate content. Boiling in water enhanced aqueous AOA, and the AOA was maintained after simulated digestion. Cooking Moringa increased available iron and raised total available iron of mixtures with mungbean. Moringa, an easily grown perennial, have tremendous potential to improve diets and health.

© 2006 American Chemical Society

Moringa oleifera (L.) is a perennial tree that is widely grown in the tropics (*1*). Its foliage is consumed as a vegetable (*2,3*) and various parts of the tree can be used for industrial oil, medicinal preparations (*4*), and water purification (*5*). Our survey of over 120 species of Asian indigenous vegetables for nutrient contents, antioxidant activities (AOA), and indigenous knowledge of their medicinal uses indicated that *Moringa oleifera* was among the most promising species (*6*). The property of high AOA for Moringa was also reported in other studies (*7-9*). The multiple attributes of Moringa have tremendous potential to improve nutrition and health for the developing world. The wide-spectrum nutrients and phytochemicals it contains are natural, acceptable, affordable, and accessible. Leaves harvested from kitchen gardens and school gardens can be sold in local markets, dried leaves can be stored and taken as supplements during periods of low vegetable and fruit availability and under-nutrition status (*10,11*).

M. oleifera gains more research attentions recently. However, *M. oleifera* is only one of 13 currently known species of Moringa (*4*). Little is known about the nutrient content, AOA and beneficial properties of the other Moringa species. Moringas are clustered into three species groups according to their morphological differences. The first group including four species are slender trees and are principally Asian; the second group including three species are the largest Moringas with the trunks shaped like bottles or elephant feet; the third group including 6 species are tuberous shrubs trees and are confined in a few African countries (*4*). In our study we explored the AOA and nutrition potential of four Moringa species: *M. oleifera* and *M. peregrina* of group 1, and *M. stenopetala* and *M. drouhardii* of group 2. *M. stenopetala* is the most economically important species after *M. oleifera*; Among Moringas, *M. drouhardii* has the most pungent odor similar to mustard oil; and *M. peregrina* has the widest habitat range and the only one of the slender trees extended out of Asia (*4*).

The objectives of this study were to (1) study the variation of AOA and nutrient contents among the four Moringa species; (2) investigate the contribution of antioxidants including vitamin A, C, E and phenolics to AOA; (3) examine the effects of various temperatures and simulated digestion on AOA and available iron of *M. oleifera* leaves.

Materials and Methods

Plant materials and preparation. Seeds of the four Moringa species (Table I) were sown in November 2000 and transplanted to an AVRDC field in April 2001. One to two kg of leaflets from each tree were sampled during 2003 – 2004. No senescing leaflets were included in the study. Fresh leaflets extracted in water or methanol were stored at −70 °C and subjected to AOA analyses. Dried leaflet samples were analyzed for dry matter, protein, calcium and iron contents. Frozen leaflets at −70 °C were used for measuring vitamin antioxidant contents and other nutritional components.

Table I. Four Moringa Species Used in This Study

Species	Plant Age	Part	Group	Origin
oleifera	3 yr	Leaf, stem, seed	Slender tree	India
peregrina	3 yr	Leaf, stem	Slender tree	Arabia, Red Sea area
stenopetala	3 yr	Leaf, stem	Bottle tree	Kenya, Ethiopia
drouhardii	3 yr	Leaf, stem	Bottle tree	Madagascar

AOA methods. The ABTS/H_2O_2/HRP decoloration method was carried out as described in Arnao et al. (*12*) with some modifications (*13*). The capacity of different components to scavenge the ABTS radical cation (ABTS•$^+$) was compared to a standard antioxidant Trolox (0 – 4 mM) in a dose response curve. The DPPH assay was performed as described in Goupy et al. (*14*) with some modifications (*13*). The radical form of DPPH has an absorbance at 517 nm, which disappears upon reduction by H˙ pulled from antioxidant compounds. The SOS was measured by the XA/XOD system as described by Murakami et al. (*15*). Superoxide is generated from the oxidation of xanthine to uric acid by xanthine oxidase. It reacts with NBT to cause a color change from light yellow to dark purple, which can be measured at 560 nm. The AOA against lipid peroxidation by ferric thiocyanate method was tested using linoleic acid and the free radical generator AAPH (*16*).

AO contents. Total soluble phenolics were extracted with methanol from frozen samples and determined using Folin-Ciocalteu reagent (*17*). The determination of total ascorbic acid is on the basis of coupling 2,4-dinitrophenylhydrazine (DNPH) with the ketonic groups of dehydroascorbic acid through the oxidation of ascorbic acid by 2,6-dichlorophenolindophenol (DCPIP) to form yellow-orange color in acidic condition (*18*). Contents of β-Carotene and α-tocopherols were measured by HPLC as described in Hanson et al. (*13*).

Nutritional quality. The protein content was measured with micro-Kjeldahl digestion followed by distillation method (*19*). The determination of calcium and iron contents were performed by ashing procedure, strong acid washing and then detected with Atomic Absorption Spectroscopy (*19*). The contents of oligosaccharides and oxalate were determined using HPLC method.

Simulated digestion. The *in vitro* assay, described by Yang et al. (*20*) involves simulated gastrointestinal digestion using commercial available enzymes including pepsin and pancreatine, with subsequent measurement of the soluble/permeable iron or antioxidant molecules released by the digestion.

Results and Discussion

AOA of the Four Moringa Species

Ranges of AOA. Moringa leaves showed high antioxidant capacity in all four of the antioxidative mechanisms: scavenging of superoxide, ABTS and DPPH radicals and inhibition of lipid peroxidation. The AOA values were the lowest for ILPm (52 TE/g) and highest for ABTSm (411 TE/g) on a dry weight basis, which was around 1,300 − 10,280 TE/100g on a fresh weight basis (Figure 1). These values were 1X − 5X higher compared to ORAC values based on FW of selected high AOA vegetables and fruits including kale (1,770 TE/100g), spinach (1,260 TE/100g), prunes (5,770 TE/100g), raisins (2,830 TE/100g) and blueberries (2,400 TE/100g) *(21)*.

Ranking by averaged AOA. Among the four species, *M. peregina* had the highest AOA values by all the AOA assays except ILPw. The ranking of the four species averaged over AOA (TE/g DW) was: *peregina* (234) >*stenopetala* (121)> *oleifera* (90)= *drouhardii* (90). Although *drouhardii* ranked the last with *M. oleifera*, it showed the second highest value of SOSw next to *peregina*.

AOA methods and W/M extractions. Among the four species, the uppermost two AOA values were found in the methanol extract of *peregina* by ABTS and DPPH methods, which were about 2.1 times higher compared to water extractable AOA. However, ILP and SOS methods measured higher activities from water extracts. The top ILPw and SOSw values were found in *stenopetala* and *peregina*, respectively. These results indicated that water extracted antioxidants of Moringa were superior to methanol extracted antioxidants in superoxide scavenging and inhibition of lipid peroxidation, whereas methanol extracted antioxidants exerted stronger capacity in scavenging of ABTS radical.

Compositions of W/M extracts. Our previous unpublished study on composition of water and methanol extracts of Chinese cedar (*Toona sinensis*) indicated that methanol extract contained four main phytochemical categories based on HPLC peak area and UV-VIS spectra: hydrophilic phenolics (such as gallic acids and chlorogenic acids), less hydrophilic phenolics (such as flavonoids and their glycosides), xanthophylls (such as lutein) and chlorophylls; and water extract contained three main clusters: ascorbate, hydrophilic phenolics, and less hydrophilic phenolics. Despite the high solubility of ascorbate in methanol, the crude methanol extract only contained about 10% of total ascorbate, which was confirmed by using colorimetric methods to measure ascorbate content of water and methanol extracts of Chinese cedar and sweet pepper. Tocopherols were determined from methanol extracts; however, they were not distinguished in the HPLC profile due to their conjugation with fatty acids in general and lower density compared to phenolics. The peak area of less hydrophilic phenolics in water extract was about 10% of that in methanol extract which indicated that a great portion of "less-hydrophilic phenolics" were not water extractable, and thus did not contribute to water extractable AOA.

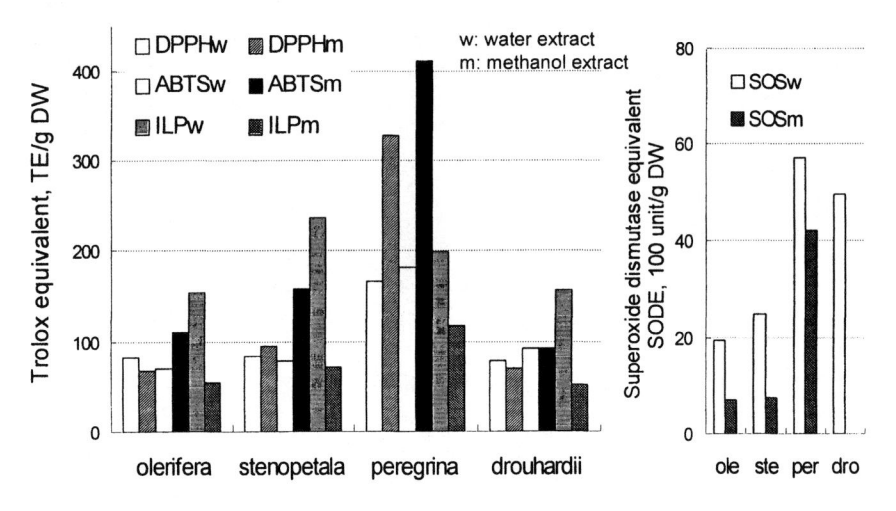

Figure 1. AOA of water and methanol extracts of four Moringa species by four AOA methods. AOA methods followed by subscripts w or m refer to water or methanol extracts, respectively.

Antioxidant Contents of Moringa

Content ranges. Concentrations of four natural antioxidants (total phenolics and antioxidant vitamins A, C and E) were measured for the four species. The content ranges on a dry weight basis were 74 – 210 µmol/g for phenolics, 70 – 100 µmol/g for ascorbate, 1.1 – 2.8 µmol/g for β-carotene and 0.7 – 1.1 µmol/g for α-tocopherol (Figure 2). Antioxidant content of Moringas are high even compared to vegetables and fruits known for high antioxidant contents such as strawberries high in phenolics (330 mg gallic acid equivalent (GAE)/100 g FW, or 190 µmol GAE/g DW) (*22*); hot pepper high in ascorbate (200 mg/100 g FW, or 110 µmol/g DW), carrot high in β-carotene (10 mg/100 g FW, or 1.8 µmol/g DW) and soybean which is high in α-tocopherol (0.85

mg/100 g FW, or 1.8 µmol/g DW) (*23*). Moringas are an excellent source of a wide spectrum of dietary antioxidants.

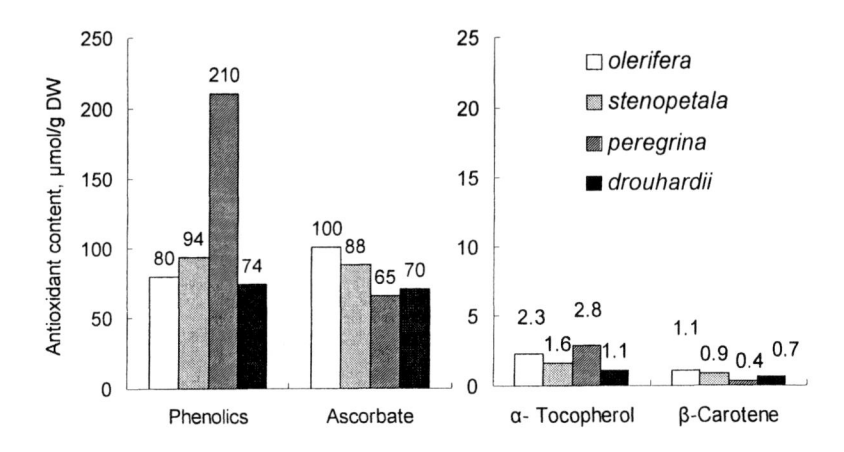

Figure 2. Antioxidant contents of the four Moringa species

Contribution of AO to AOA The concentrations of phenolics and ascorbate were about 25 – 300 times greater than the α-tocopherol and β-carotene contents. From the density point of view, phenolics and ascorbate were the dominant antioxidants in Moringa and they were principally contributive to AOA in both water and methanol extracts. However, only phenolics content showed a linear correlation with methanol extractable AOA (Figure 3). Ascorbate content was independent of either water or methanol extracted AOA. These results suggest that AOA of methanol extracts could be inferred or estimated from phenolics contents, but not from ascorbate content. This conclusion was supported by the evidence that only 10% of ascorbate content was detected in methanol extract. The aqueous phase may contain enzymes and other bioactive molecules, in addition to ascorbate and phenolics, and that may have caused synergisms and/or antagonisms among antioxidants that did not allow prediction of AOA simply by the summing the individual antioxidant contents. However, the crude water extracts of vegetables are more like the food matrix and the actual environments for food intake compared to methanol extracts.

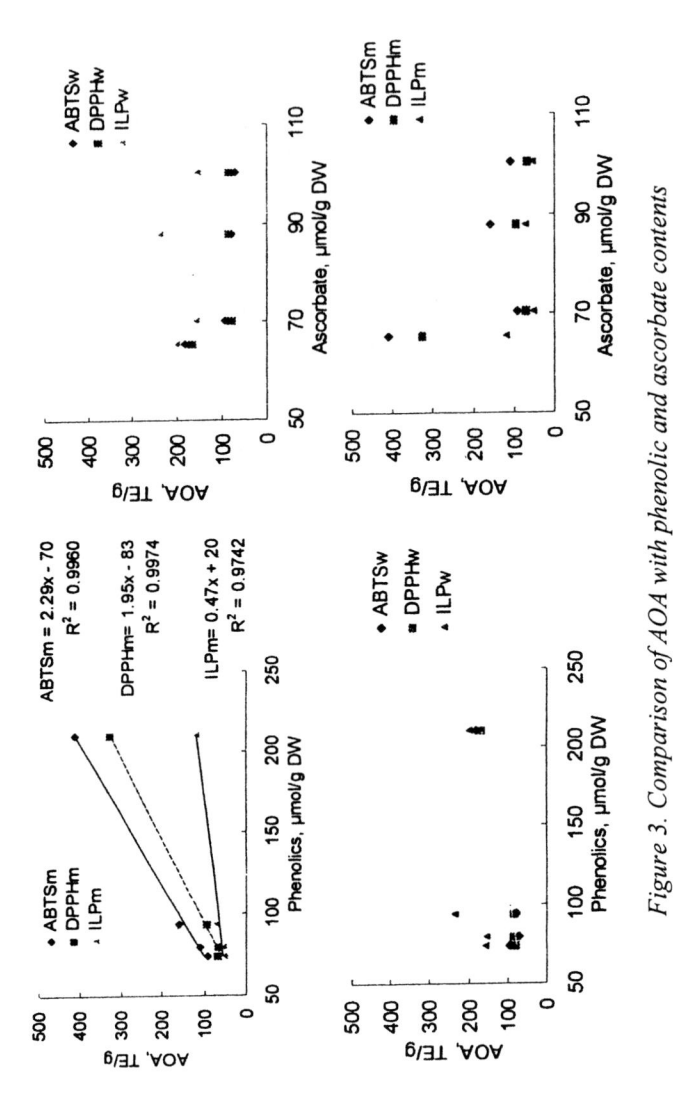

Figure 3. Comparison of AOA with phenolic and ascorbate contents

HPLC Profiles of Phenolics vs. AOA

Antioxidants were extracted from 20 g of mature, fresh leaves of *M. oleifera* in a homogenizer with methanol (1:5, w/v). After heat-acid hydrolysis to breakdown sugar and phenolics linkages, the extract was filtered and then subjected to HPLC analysis. The HPLC eluent was fractionated and collected every minute for 100 minutes, and then freeze-dried for further determination of ABTS scavenging activity. Based on their optical absorption patterns, the methanol extractable antioxidants were thought to be mainly phenolics. The major antioxidants clusters were shown at RT 3-5 min and 22-52 min (Figure 4). After heat-acid treatment the flavonoids such as quercetin (23.6 µmol/g DW) and kaempferol (2.4 µmol/g DW) were detected. Quercetin and its derivatives were the main antioxidants in phenolics in Moringa leaves. The peak profiles were changed after acid treatment and this suggests that most of the phenolic compounds in Moringa were conjugated with sugars.

AOA of Moringa Leaves, Stems and Seeds

Stem extracts showed only 2% – 12% of the AOA values of leaf extracts (Figure 5). Higher AOA in *peregina* stem extract may be due to its higher phenolics content compared with the other three species. Moringa seeds were low in both AOA and phenolics content.

Temperature Effect on AOA

Several 20 g samples of fresh leaves were soaked in 80 ml of prechilled/ preheated water or methanol at –20 °C, 25 °C, 50 °C and 100 °C, respectively, for 10 min, and antioxidants were extracted with the soaking solvent and AOA was measured by ABTS and SOS activities. Antioxidants with ABTSm were stable at temperatures ranging from –20 °C to 100 °C (Figure 6). Boiling even increased ABTSw activity about 2.5 fold and reached the value equal to ABTSm. The increase in AOA due to boiling was also detected by SOSw. Heat treatment may eliminate oxidative reactions in water extracts such as enzymatic oxidations. In methanol extract, generally do not contain bioactive protein. In addition,

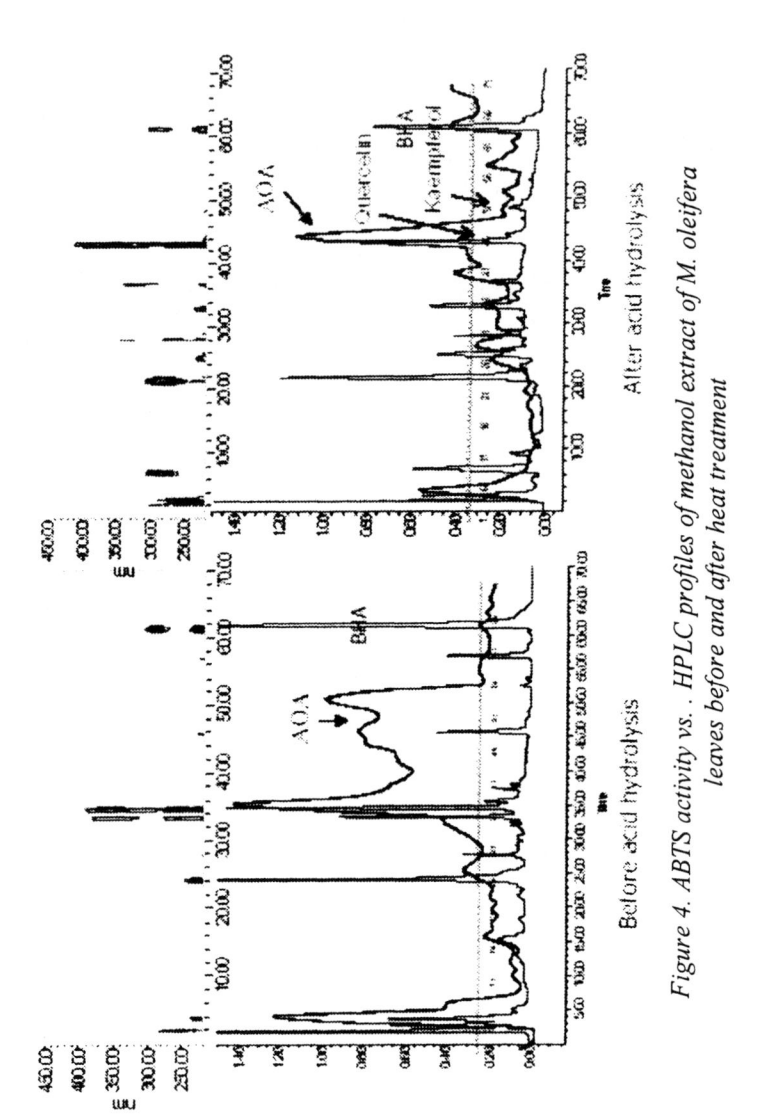

Figure 4. ABTS activity vs. . HPLC profiles of methanol extract of M. oleifera leaves before and after heat treatment

phenolics in methanol extracts are heat stable. Therefore their AOA did not affected by various temperatures. The results suggest that common vegetable processing methods such as freezing, blanching, and boiling would not reduce AOA of Moringa leaves. In fact, higher water extractable AOA could be achieved through blanching and boiling.

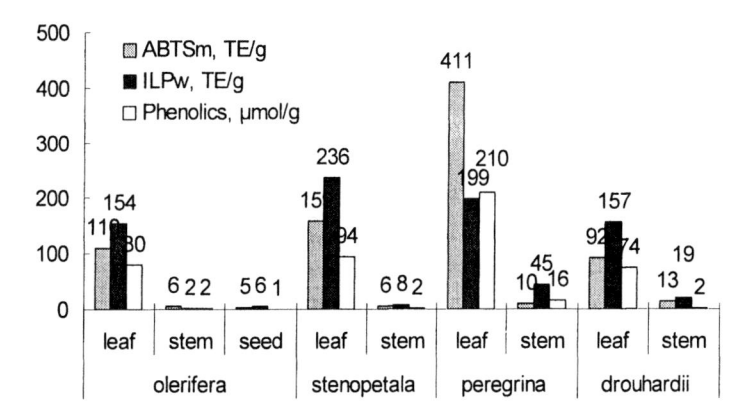

Figure 5. AOA of leaves, steam and seeds of four Moringa species..

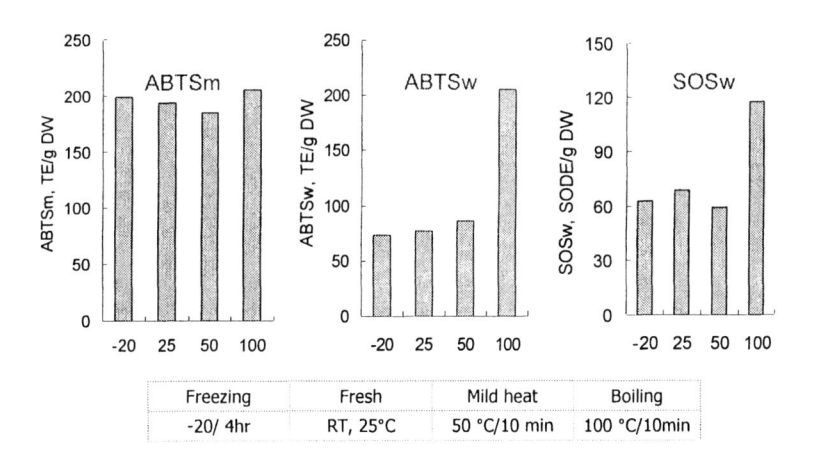

Figure 6. Temperature effect on AOA of M. oleifera

Digestion Effect on AOA

AOA after digestion. The 5 hr in vitro bioavailability assay of antioxidants was performed using fresh leaves with 5% dry matter prepared in water, followed by homogenization, pH2 adjustment, 2 hr of pepsin incubation, addition of dialysis tubing to increase the pH value > 6, and the final stage of 2 hr digestion with intestinal enzymes (porcine bile salts and pancreatine). ABTSw and ILPw of fresh and digested Moringa leaves were measured. The total ABTSw (dialysate and remainder) were increased 1.3-fold after the simulated digestion, whereas the total ILPw dropped 20% (Figure 7). About 30% of the total ABTSw after digestion was contributed from permeable antioxidants (<6000 D). Similarly, 30% of total ILPw was dialyzable.

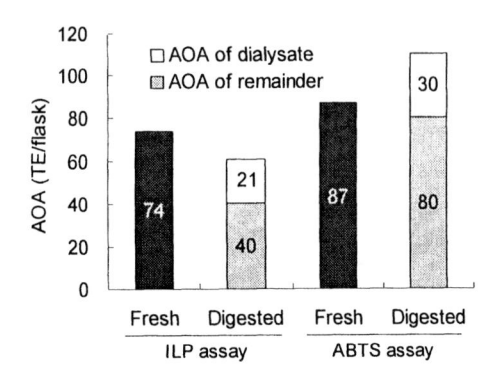

Figure 7. AOA changes before and after simulated digestion

AOA during digestion. ABTSm from each stage of the simulated digestion was monitored (Figure 8). Higher AOA was obtained with acidic conditions after pepsin incubation. The increase may be due to unlocking of the antioxidants from digested protein and partial release of antioxidative phenolic compounds form their derivatives by acid hydrolysis. AOA dropped 10% at pH 6, which implied that the antioxidants were favored by acidic conditions such as in the stomach rather than neutral pH such as found in the intestine. This is supported by the lower AOA found at pH 8 compared to that at pH 2, 4, and 5.5 (the pH value of Moringa meal) (data not shown). During the simulated intestinal digestion, antioxidants gradually penetrated into the dialysis tubing. About 30% of antioxidants were detected in the dialysate, which means they were smaller molecules with molecular weights less than 6000 g/mole and may be easier for uptake by intestines.

Nutritional Quality of the Four Species

Among the four species, *M. oleifera* contained the highest amounts of β-carotene, ascorbate, α-tocopherol and iron, and was the second highest in protein content (Table II). *M. oleifera* grows faster than the other three species under the subtropical low lands in Taiwan, and this species is commonly consumed as a vegetable in South and Southeset Asia. Oligosaccharides and oxalate were reported as anti-nutrient factors in Moringa leaves (*24*). In this study, stachyose and raffinose were not found in mature leaves of the four species. But they were detected in young leaves, but not in mature leaves of *M. stenopetala* harvested at different times (0–14 mg/g). Higher amounts of these two galacto-oligosaccharides (22–98 mg/g) were detected in the seeds of *M. oleifera*. Moringa leaves contained much less oxalate (0.99 ± 0.21 mg/g) than spinach (25–45 mg/g) (*23*). Slightly higher amounts of oxalate (1.37 ± 0.23 mg/g) were measured in young shoots than in mature leaves for ten accessions of *Moringa oleifera*. The data indicated that oxalate and oligosaccharides are not significant anti-nutrient factors in Moringa.

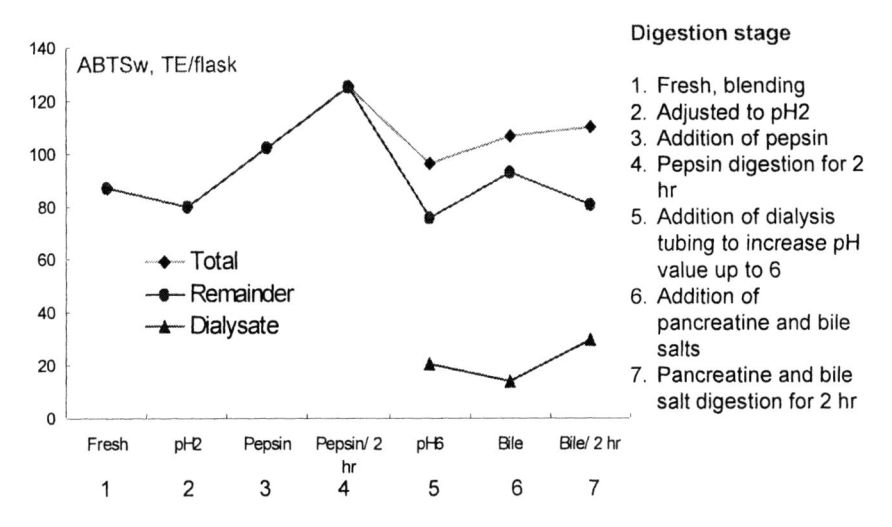

Figure 8. ABTS activity of Moringa leaf meal during simulated digestion

Table II. Nutrient Contents of Moringa Leaves (100 g Fresh Weight)

Specie	DM %	Protein g	β-carotene	Ascorbate	Tocopherol	Iron	Calcium
			----------	---------- mg	----------		
oleifera	24	5.7	15	249	25	9.2	638
stenopetala	24	5.8	13	400	18	5.4	711
peregrina	21	2.9	5	264	28	5.6	458
drouhardii	29	5.0	11	388	14	8.7	745

In Vitro Iron Bioavailability

Cooking effect on Moringa. Iron bioavailability of certain vegetables can be enhanced through cooking (20). Boiling in water enhanced the in vitro iron bioavailability of Moringa fresh leaves and dried powder by 3.5 and 3 times, respectively (**Figure 9**).

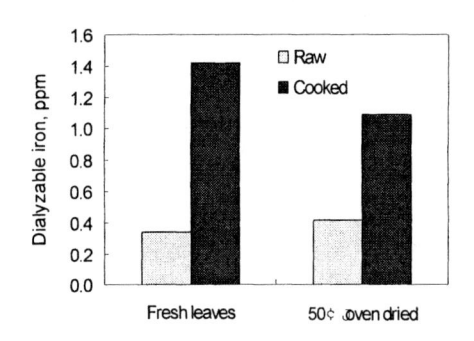

Figure 9. In vitro iron bioavailability of Moringa leaves, raw and cooked

Cooking enhancing effect. The cooking enhancing effect on Moringa could be explained from our previous study modeling how cooking can enhance iron bioavailability from cabbage (25). Iron in plant cells is stored mostly in ferritin, from which iron may be released by proteases or denaturation by heat or low pH. Enzymes such as polyphenol oxidases are compartmentalized in cells until the cells are disrupted by blending or mastication. Almost immediately, the soluble iron becomes bound in iron-polyphenol complexes due to the action of polyphenol oxidases, and is thus rendered unavailable. Heating denatures the polyphenol oxidases preventing their action, but leaves intact a sufficient amount of ascorbic acid to maintain iron in a soluble form through chelation, even at pH 2 in the stomach and pH 6-7 in the intestine. Thus, more available iron can be absorbed from cooked cabbage than from the raw form.

Vegetable combinations Mungbean, a common staple food in South Asia, was cooked with equal amounts (1:1, mungbean:test ingredient, dry matter) of tomato, kale, sweet pepper, and Moringa leaves, and the resulting dialyzable iron was measured. Tomato and Moringa leaves cooked with mungbean raised dialyzable iron compared to the ingredients cooked separately, and the ingredients cooked separately and then mixed (Figure 10). Kale and sweet pepper showed no such enhancing effect. This study suggests mungbean can have their iron bioavailability enhanced by a vegetable, such as tomato and Moringa leaves.

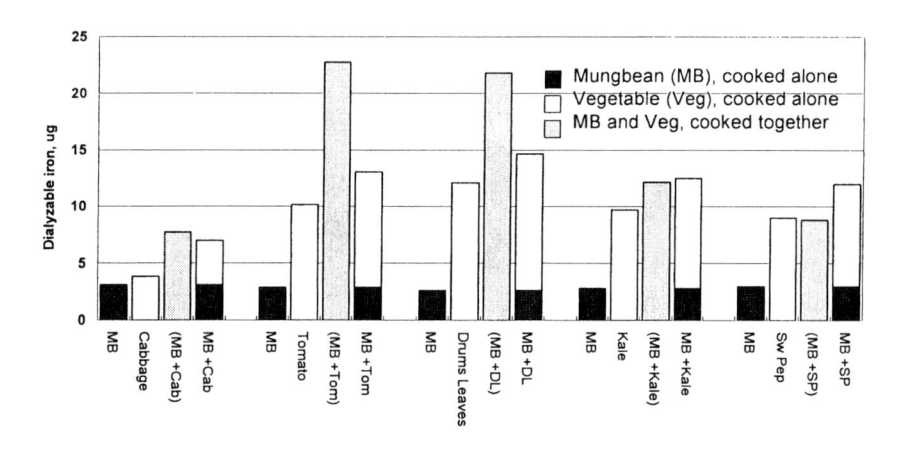

Figure 10. In vitro iron bioavailability of cabbage, tomato, Moringa (drumstick leaves), kale, and sweet peppers, cooked along, or cooked with mungbean (MB)

Conclusion

High AOAs (scavenging of superoxide, ABTS and DPPH radicals and inhibition of lipid peroxidation), antioxidant (β-carotene, ascorbate, α-tocopherols, and phenolics) and nutrient (protein, vitamins A, C, and E and mineral calcium and iron), and low oxalate contents are common features of the four Moringa species. *M. peregina* was the uppermost for AOA and *M. oleifera* has the highest nutrient values among the four. The dominant antioxidants in Moringa leaves were phenolics and ascorbate. Although Moringas contain very high antioxidant vitamins, the α-Tocopherols and β-carotene contents cannot be reflected by the AOA values. Quercetin conjugates were the major antioxidants that contributed to their AOA in methanol extract. Methanol extractable AOA was maintained after freezing and water boiling, whereas water extractable AOA was lower in raw but enhanced after boiling in water. AOA was maintained after simulated digestion; about 30% of AOA were permeable (<6000 D) after simulated digestion. Moringa leaves, high in iron bioavailability after cooked, and helps to unlock iron from mungbean.

Acknowledgments

We thank Council of Agriculture, Taiwan funded this work. (Projects: 93AS-1.1.1-FD-Z3(2) and 92AS-1.1.2-FD-Z2(2))

References

1. Palada, M. C. *HortSci.* **1996**, *31(5)*, 794 – 797.
2. *Prosea – Plant resources of South-East Asia 8 – Vegetables*; Siemonsma J. S.; Kasem P., Eds.; Prosea Foundation: Bogor, Indonesia. 1996.
3. Freiberger, C. E.; Vanderjagt, D. J.; Pastuszyn, A.; Glew, R. S.; Mounkaila, G.; Millson, M.; Glew, R. H. *Plant Foods for Human Nutr.* **1998**, *53*, 57 – 69.
4. Fuglie, L. J. In *The Miracle Tree*, Fuglie, L. J. Ed.; CTA: New York, NY, **2001**; pp 7 – 28.
5. Ndabigengesere, A.; Narasiah, K. S. *Enviro. Tech.* **1996**, *17*, 1103 – 1112.
6. *AVRDC Report 2001*. Kalb, T.; Kuo, G., Eds.; AVRDC publication 02−542; Asian Vegetable Research and Development Center, Shanhua, Tainan. Taiwan, 2002; pp 92 – 94.
7. Lalas, S.; Tsaknis, J. *JAOCS.* **2002**, *79*, 677 – 683.
8. Siddhuraju, P.; Becker, K. *J. Agric. Food Chem.* **2003**, *51*, 2144 – 2155.
9. Bennett, R. N.; Mellon, F. A.; Foidl, N.; Pratt, J. H.; Dupont, M. S.; Perkins, L.; Kroon, P. A. *J. Agric. Food Chem.* **2003**, *51*, 3546 – 3553.
10. Mubarik, A.; Tsou, S. C. S. *Food Policy.* **1997**, *22*, 17−38.
11. Subadra, S.; Monica, J.; Dhabhai, D. *Intl. J. of Food Sci. and Nutr.* **1997**, *48*, 373 – 379.
12. Arnao, M. B.; Cano, A; and Acosta, M. *Food Chem.* **2001**, *73*:239–244.
13. Hanson, P. M.; Yang, R. Y.; Lin, S., Tsou, S. C. S.; Lee. T. C.; Wu, J.; Shieh, J. Gniffke, P.; Ledesma, D. *Plant Genetic Resources− Characterization & Utilization.* **2004**. *2(3)*. (In *press*)
14. Goupy, P.; Hgues, M.; Bovin, P.; Amiot, M. J. *J. Sci. Food Agric.* **1999**, *79*, 1625−1634.

15. Murakami, A.; Ohura, S.; Nakamura, Y.; Koshimizu, K. and Ohigashi, H. *Oncology.* **1996**, *53*, 386 – 391.
16. Mihaljevic, B.; Katusin-Razem, B.; Razem, D. *Free Radic. Biol. Med.* **1996**, *21*, 53 – 63.
17. Singleton, V. L.; Rossi, J. A. *Am. J. Enol. Vitic.* **1965**, *16*, 144–158.
18. Pelletier, O. In *Methods of Vitamin Assay;* Augustin, J.; Klein, B. P.; Becker, D. A.; Venugopal, P. B. Eds.; John Wiley & Sons, Inc. New york. 1985; pp 303-347.
19. *Official Methods of Analysis*; Helrich, K., Eds.; Association of Offical Analytical Chemists: Arlington, VG, 1990.
20. Yang, R. Y.; Tsou, S. C. S.; Lee, T. C. In *Bioactive Compounds in Foods – Effects of processing and storage.* Lee, T.C.; Ho, C.T., Eds.; ACS Symposium 816; American Chemical Society, Washington, D.C., 2002; pp130-142.
21. McBride, J. <http://www.ars.usda.gov/is/pr/1999/990208.htm>
22. Proteggente, A. R; Pannala, A. S.; Paganga, G.; Van Buren, L.; Wagner, E.; Wiseman, S.; Van De Put, F; Dacombe, C.; Rice-Evans, C. A. *Free Radic. Res.* **2002**, *36*, 217–233.
23. USDA National Nutrient Database for Standard Reference. <http://www.nal.usda.gov/fnic/foodcomp/search>
24. Gupta, K; Barat, G. K.; Wagle, D. S.; Chawla, H. K. L. *Food Chem.* **1989**, *31*, 105 – 116.
25. *AVRDC Report 1998.* Stares, J., Eds.; AVRDC publication 99–492; Asian Vegetable Research and Development Center, Shanhua, Tainan. Taiwan, 1999; pp 89 – 92.

Chapter 18

Stability and Transformation of Bioactive Polyphenolic Components of Herbs in Physiological pH

Shiming Li and Chi-Tang Ho

Department of Food Science, Rutgers, The State University of New Jersey, 65 Dudley Road, New Brunswick, NJ 08901–8520

Phenolic compounds occur ubiquitously in plants and are active components in many herbs. Numerous studies have shown that these phenolic compounds possess antioxidant, anti-carcinogenic, anti-inflammatory and anti-viral activity. The recent research results of some new antioxidant formations from (-)-epigallocatechin gallate, a polyphenol in tea catechin, in mild alkaline fluids has led to the metabolic mechanistic study of these biologically active polyphenolic compounds. In our study, chlorogenic acid, a widely spread polyphenolic compound in plants, was incubated in different pH buffer solutions at physiological temperature for various times. It was found that under alkaline conditions some degradation products were formed from chlorogenic acid. The degradation products were isolated and characterized by LC/MS and NMR techniques. Kinetic degradation study of chlorogenic acid has been performed.

Phenolic compounds exist ubiquitously in all terrestrial plants and are found in many food products. Numerous studies have shown that these phenolic compounds possess many health beneficial biological properties (*1-3*). For example, they have strong antioxidant, anti-carcinogenic, anti-mutagenic, anti-inflammatory and anti-viral activity.

It is well known that the major components of tea are catechins (Figure 1) and their galates, particularly the most abundant galate-epigallocatechin gallate (EGCG). Tea is grown in over 30 countries and aside from water, it is the most widely consumed beverage in the world. Increasing scientific evidence shows that tea has great potential for preventing and treating cancer and cardiovascular dieases (*2*). Apple and onion are rich in quercetin and like catechin and its galates, is another widely studied flavonoid as a potential therapeutic agent. One of the most important flavonoids from soy bean is genistein, whose antibody conjugates (B43-genistein and EGF-genistein) are under consideration in clinical development for the treatment of acute lymphoblastic leukaemia and breast cancer, respectively (*1*).

Besides flavonoids, there are other polyphenol nutraceuticals existing in plants. For instance, grapes are rich in resveratrol which is a cancer prevention agent (*4*). Curcumin is another naturally occurring polyphenolic phytochemical isolated from the rhizome of the perennial plant *Curcuma longa*, cultivated throughout the Asia. Curcumin is known for its antioxidant, anti-inflammatory, antimicrobial and anticancer properties. It has great potential use as a cancer prevention agent and as a strong antioxidant (*3,5*).

Coffee, fruits, vegetables and wine are rich in chlorogenic acid. It is estimated that a 200 mL cup of roast and ground coffee can supply 20 mg (weak brew, very dark roast) to 675 mg (strong brew, very pale roast robusta) of chlorogenic acid. Whole apples have been reported to contain 62-385 mg/kg of chlorogenic acid with 5-chlorogenic acid always dominant (*6*). Chlorogenic acid has many health benefits. It can inhibit LDL oxidation and lower the risk of cardiovascular dieases such as stroke. It can also inhibit the formation of carcinogenic and mutagenic *N*-nitroso compounds (*7,8*).

In summary, the biological activities of polyphenols include: anti-inflammatory, heptoprotective, anti-microbial and anti-allergic. Polyphenols also scavenge free radicals, suppress free radical formation and propagation, and inhibit virus-induced cancer, protein kinases and COX-2. In addition, polyphenols have an effect on multidrug resistance.

Structures of some phenolic compounds mentioned are shown in Figure 1.

Degradation Study of Tea Catechins

There is substantial evidence to show that tea catechins exhibit strong antioxidative activity in the plasma, liver and kidney of rats and mice and strong anticarcinogenic activity in rats *in vivo* (*2*). Early investigation of the mechanism

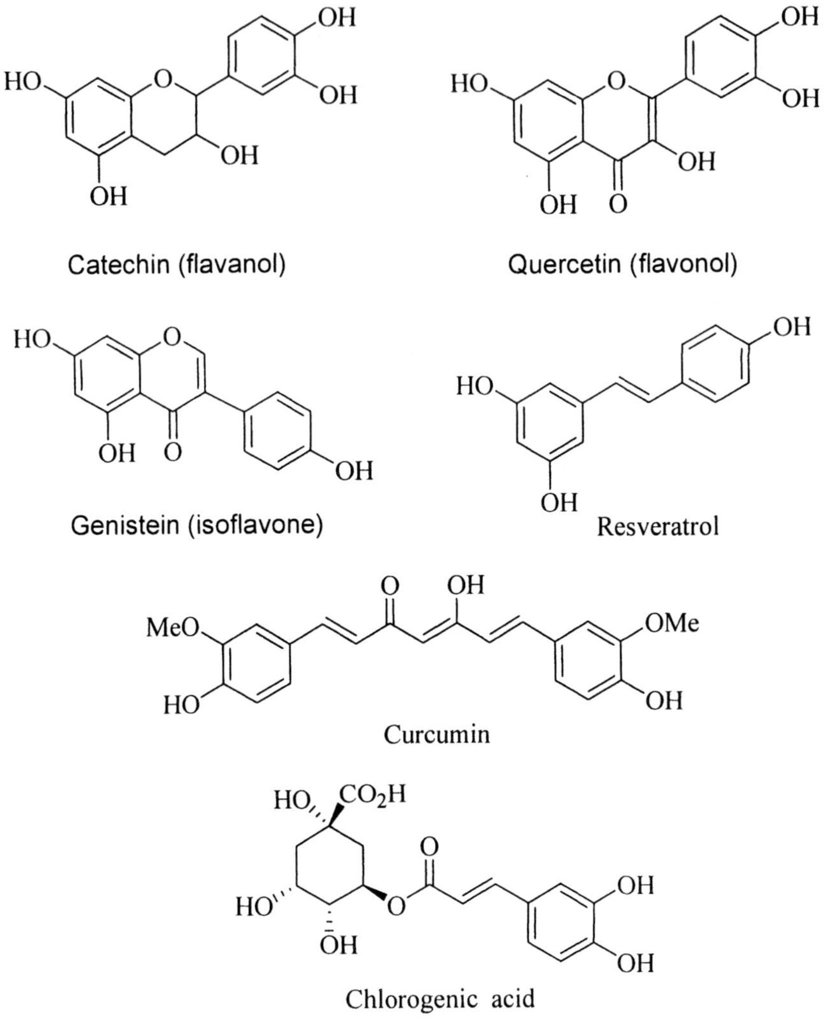

Figure 1. Structures of some polyphenolic compounds

of these effects in animal bodies provided some information about the absorption, distribution and metabolic fate of tea catechins (2,9,10). This research demonstrated that EGCG or other tea catechins administered to rats were detected in the plasma and organs such as liver and kidney. The catechins are conjugated with sulfate and glucuronate by phenol sulfotransferase (EC 2.8.2.1) and UDP (uridinediphosphate)-glucuronyltransferase (EC 2.4.1.17) and are O-methylated by catechol O-methyltransferase (EC 2.1.1.6) or O-methyltransferase (EC 2.1.1.25) in the animals. Recently, it has been reported that there are three new antioxidants being formed from EGCG in mild alkaline solutions (Figure 2): P-1 (theasinensin A), P-2 (a new dimmerized product) and P-3 (theasinensin D) (11). In general, the plasma and bile fluids in mammals as well as human intestinal and pancreatic juices are mildly alkaline. Hence, the formation of these degradation products in the intestinal tract, plasma and bile would be expected.

More interesting findings resulted from both stability study of tea catechins under various pH conditions and antioxidative activity study of EGCG and its dimerization products in authentic intestinal juice and mouse plasma. Compared with EGCG, the compound P-2 has a rapid absorption and a slow metabolization rate. Furthermore, the Fe^{2+} chelating activities of the three products (P-1, P-2 and P-3) and superoxide anion radical-scavenging activity of P-2 was significantly higher than that of EGCG, which indicates that these dimerized products of EGCG are expected to contribute more to *in vivo* antioxidant activity (11).

Degradation Study of Curcumin

There have been a number of reports with regard to the stability and degradation of curcumin (12,13). These studies concluded that curcumin has optimum stability in solutions with pH value less than 7. However, in this pH region, the solubility of curcumin is very low. In the HPLC kinetic study of curcumin degradation by monitoring the changes of curcumin concentration with time, it was observed that curcumin degradation is a first-order reaction. It has also been reported that in buffer solutions at neutral-basic conditions (pH>7) and physiological temperature (37°C), curcumin is extremely unstable and decomposes rapidly (13,14). The major degradation product is *trans*-6-(4'-hydroxy-3'-methoxyphenyl)-2,4-dioxo-5-hexenal and minor ones vanillin, ferulic acid and feruloyl methane (Figure 3). These degradation products were elucidated solely from LC/MS/MS (14). Therefore, further study must be conducted to isolate and confirm the chemical structures of the compounds formed from curcumin degradation.

The activity of curcumin to inhibit mutagenesis, lipid peroxidation and free radical generation has been well documented (3,5,15). The activities of vanillin,

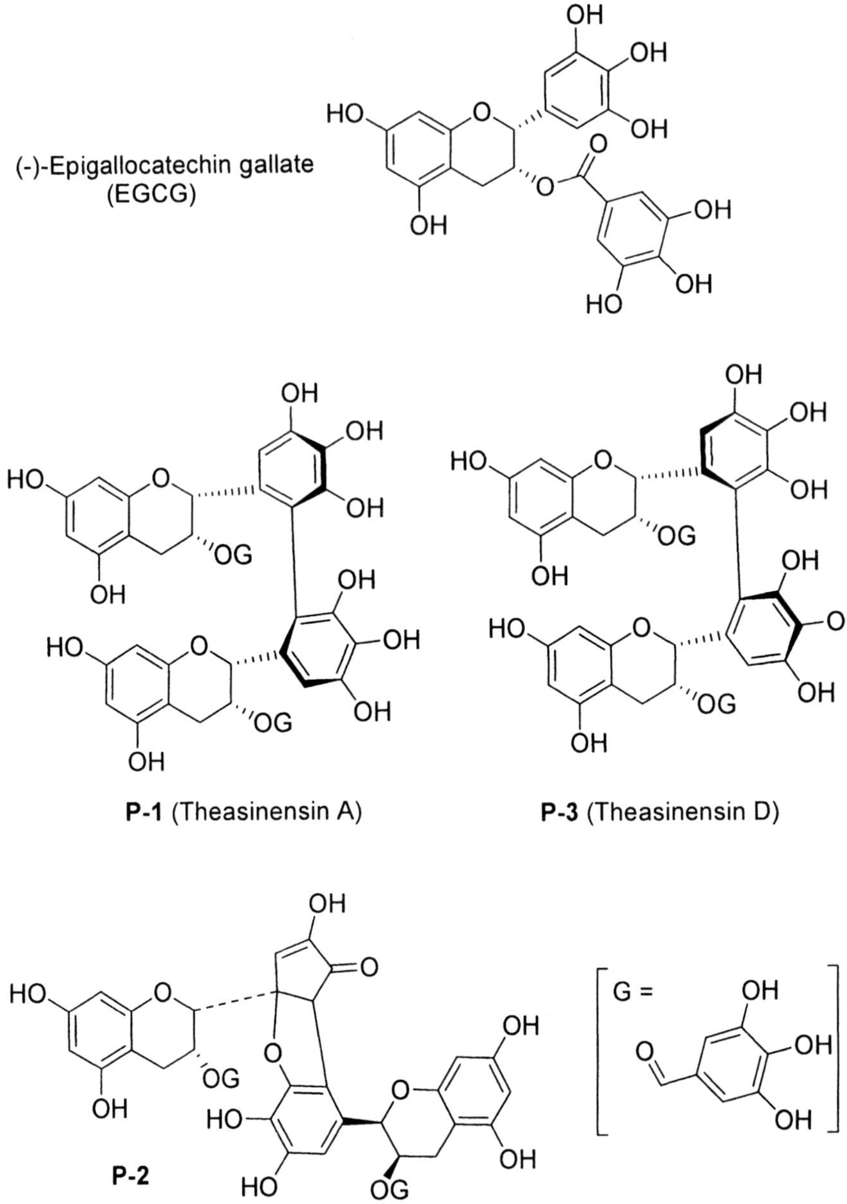

Figure 2. Degradation products of EGCG

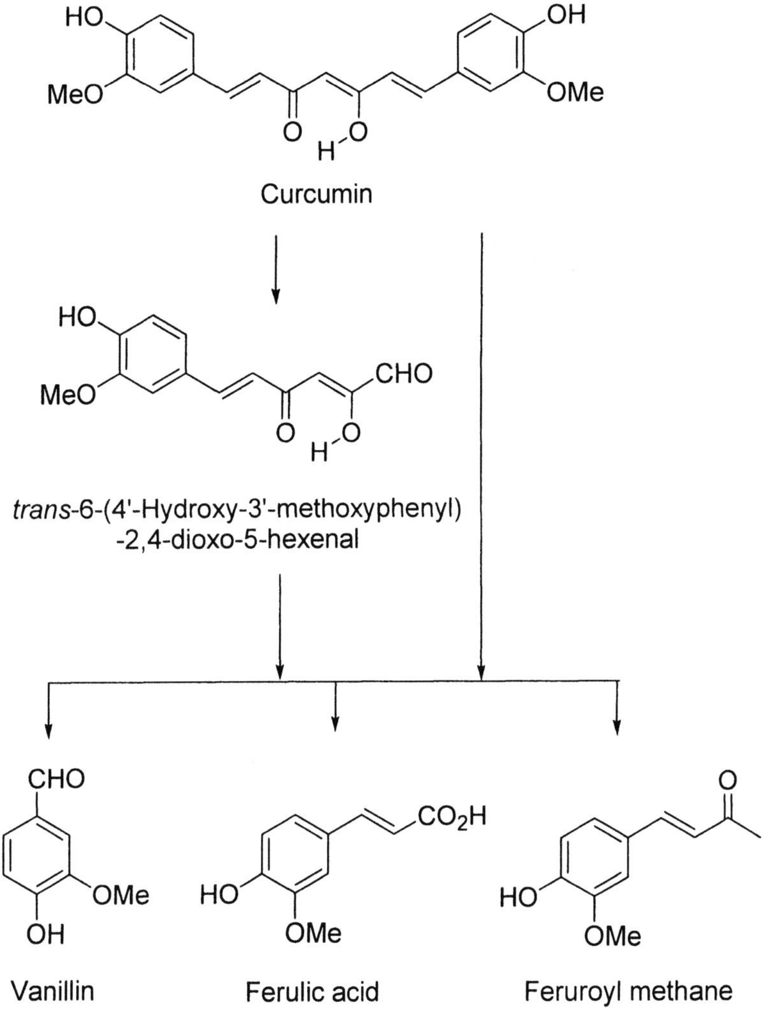

Figure 3. Degradation products of curcumin in a buffer (pH 7.2) at 37 °C.

one of the curcumin degradation products, include the inhibition of mutagenesis in bacterial and mammalian cells and the inhibition of iron-dependent lipid peroxidation in rat brain homogenate, microsomes and mitochondria. Vanillin is also a powerful scavenger of superoxide and hydroxyl radicals. With these aspects, it would be extremely valuable to compare the potency of vanillin and curcumin.

Transformation Study of Chlorogenic Acid

Chlorogenic acid, first introduced in 1846 in describing a coffee bean component, is one of the most widely spread phenolic compounds in plants. A previous degradation study of chlorogenic acid reported by Watanabe in 1998 found the formation of two chlorogenic acid isomers determined by LC/MS from the green pigments formed by mixing chlorogenic acid and glycine under a solution of pH>7 (*16*). The structures of these newly formed isomers were not characterized.

In our study, chlorogenic acid was incubated in different pH buffer solutions, which mimics different physiological pHs, and at physiological temperature and in different time intervals. We observed that in the alkaline conditions, pH 8.5 and 11.2, two transformation products were formed: neochlorogenic acid and cryptochlorogenic acid, characterized by LC/MS and NMR (^{1}H, ^{13}C and NOE). Kinetic study of chlorogenic acid is being performed with an internal standard compound which has been prepared recently by employing organic synthetic methodology.

Experimental Conditions and Procedures

Different buffer solutions of chlorogenic acid were prepared at a concentration of 1 mg/mL and the following five pH conditions: 1.8, 6.4, 7.4, 8.5 and 11.2. These buffer solutions were incubated at 37°C for 0.5, 1, 2 and 4 h. At the end of incubation period the buffer solutions were cooled to 0°C and neutralized. The resultant solution was loaded onto a silica gel plug to remove salts. The collected solution was concentrated *in vaccuo* and the residue was redissolved in 3 ml of water. Analytical samples were taken and the rest of the sample solution was purified by C-$_{18}$ reverse phase HPLC using acetonitrile and water as eluting solvents. The purified fractions were concentrated and the structures were characterized by LC/MS and various NMR techniques.

The kinectic study of chlorogenic acid transformation was done at pH 8.5 and 11.2 with an internal standard synthesized from the methylation of chlorogenic acid. The following flow chart clearly illustrates the experimental procedures (Figure 4).

```
Incubation of chlorogenic acid
            ↓
  Coolness and neutralization
            ↓
  Passing through silica gel plug
            ↓
  HPLC analysis and separation
            ↓
  Aqusition of LC/MS and NMR ($^1$H, $^{13}$C & NOE)
            ↓
      Structure elucidation
            ↓
  Kinetic study in the presence of internal standard
```

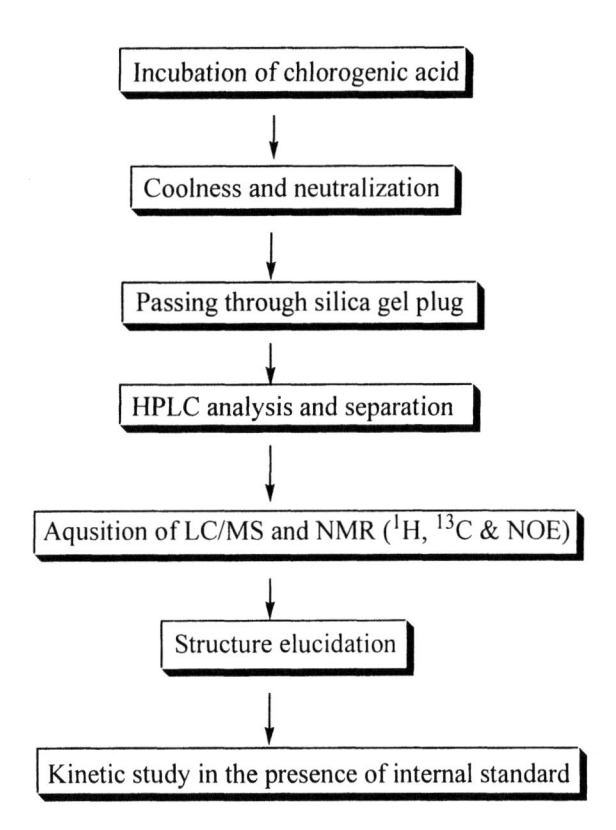

Figure 4. Flow Chart of Experiments.

Separation and Structure Elucidation of Chlorogenic Acid Isomers

Figure 5 is the HPLC traces of chlorogenic acid and its two isomers. The three compounds were separated on reverse phase HPLC. The two isomers of chlorogenic acid are more polar than their parent compound. Isomers a and b were isolated and identified by NOE NMR technique to be neochlorogenic acid and cryptochlorogenic acid respectively.

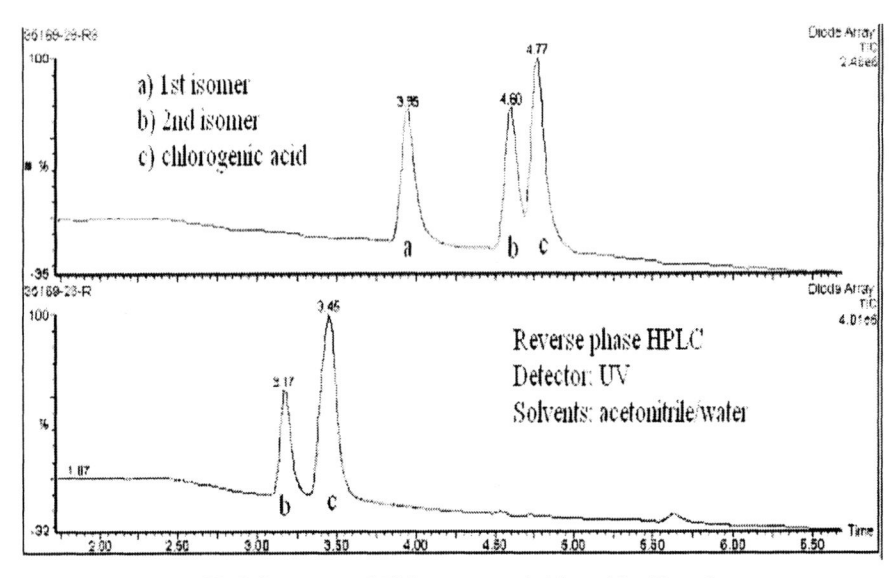

Figure 5. HPLC traces of Chlorogenic Acid and Its Two Isomers

Figure 6 is the proton NMR of chlorogenic acid and the assignment of the three protons on position 3 (Ha), 4 (Hb) and 5(Hc) of the quinic acid ring.

Figures 7 and 8 below show the NOE spectra of neochlorogenic acid and cryptochlorogenic acid (DMSO-$d6$), two isomers formed from the tansformation of chlorogenic acid. In Figure 7, Spectrum I is the proton NMR of neochlorogenic acid. When the proton at 5 position (Hc) was irradiated as shown in spectrum II, it was observed that pronton Hb (4-H) responded from a multiplet to a doublet. Spectrum III shows the singlet of Hb in response changing to Ha decoupling. By irradiating Hb, as shown in spectrum IV, a response was obtained both for Hc going from a multiplet to a quartet and for Ha changing from a multiplet to a triplet. Thus, the NOE experiments confirm the structure of neochlorogenic acid as an isomer a of chlorogenic acid (Figure 5).

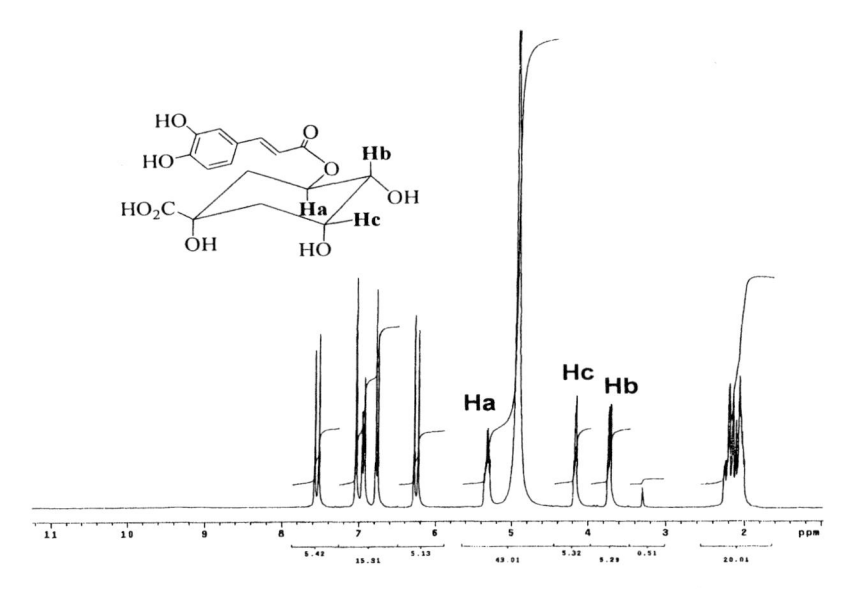

Figure 6. ¹H NMR of Chlorogenic Acid in DMSO-d6

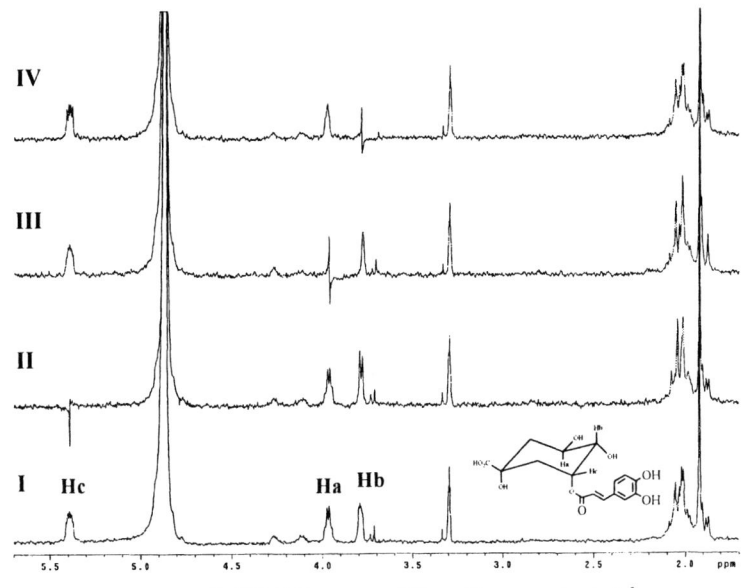

Figure 7. NOE Spectra of Neochlorogenic Acid

Figure 8 shows the ^{1}H NMR (I) and NOE spectra (II, III, IV) of cryptochlorogenic acid (isomer b in Figure 5). Having the same principle and mechanism as the NOE experiment of neochlorogenic acid, when the proton at 5 position (Hc) was irradiated, as shown in spectrum III, it was observed to have the response of pronton Hb. By irradiating Hb, as shown in spectrum II, a response was obtained both for Hc from a multiplet to a quartet and for Ha from a multiplet to a triplet. Thus the structure of cryptochlorogenic acid was confirmed by these NOE experiments.

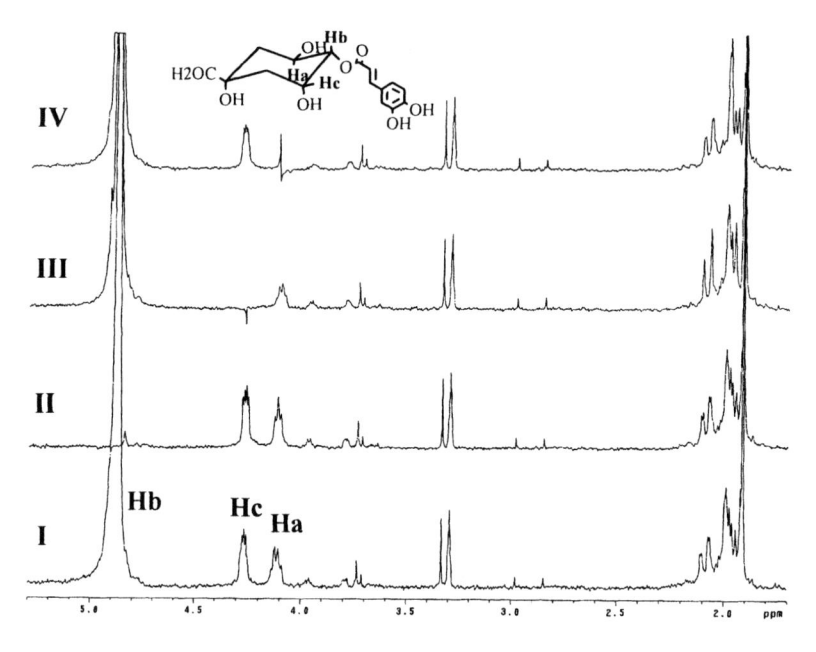

Figure 8. ^{1}H NMR and NOE Spectra of Cryptochlorogenic Acid

Kinetic Study of Chlorogenic Acid Transformation

Chlorogenic acid was incubated with an internal standard, (1S,3R,4R,5R)-3-[(*E*)-3-(3,4-dimethoxy-phenyl)-acryloyloxy]-1,4,5-trihydroxy-cyclohexane-carboxylic acid methyl ester synthesized by the methylation of chlorogenic acid with iodomethane in the presence of potassium carbonate and with DMF as the solvent. The incubation was performed at pH 8.5 and 11.2 at 37°C and different time intervals from 15 min to 4 hours.

At pH 8.5 (Figure 9), the transformation rate of chlorogenic acid was slow and it took about 4 hours to reach a final equilibrium between chlorogenic acid and its two isomers. The gradual percentage increase of the transformation products, namely neochlorogenic acid and cryptochlorogenic acid, and the gradual percentage decrease of chlorogenic acid indicate the slow process at this particular pH. However, calculated data from this experiment showed that the transformation rate was much faster at the first 100 minutes and slower thereafter till a kinetic equilibrium was reached after 4 hours of incubation, when there was approximately one third of each compound: chlorogenic acid and its two transformation products. It is known that human intestinal and pancreatic juices are mildly alkaline with pH values of 8.3 and 7.0 to 8.5, respectively. Therefore, this experiment could be considered as a simplified *in vitro* model and the result could be a hint for a future research topic, that chlorogenic acid would be transferred to its isomers mainly in the first 100 minutes or so after its intake.

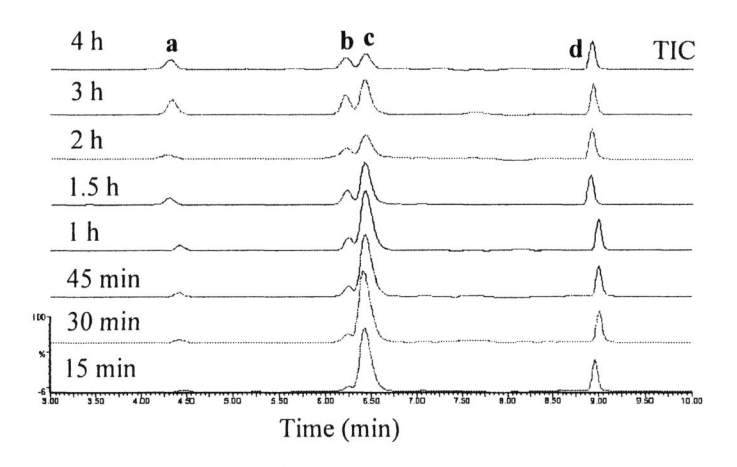

*Figure 9. Transformation of Chlorogenic Acid at pH 8.5 and 37 °C (Note for Figure 9 and Figure 10: **a**, neochlorogenic acid; **b**, cryptochlorogenic acid; **c**, chlorogenic acid; **d**, internal standard; TIC, UV diode array detection)*

However, the transformation of chlorogenic acid at pH 11.2 was shown to be instant. Figure 10 shows that an equilibrium was reached within 30 minutes of incubation and most transformation occurred within 15 minutes.

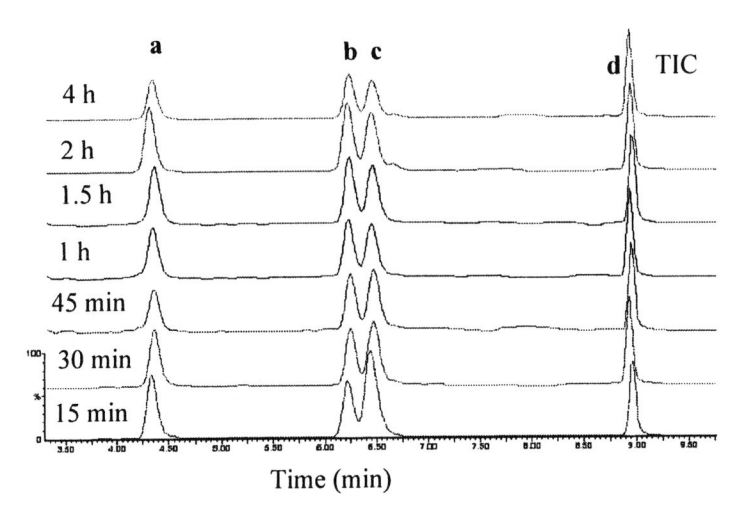

Figure 10. Transformation of Chlorogenic Acid at pH11.2 and 37 °C

Conclusion

In summary, polyphenolic compounds are ubiquitous in plants and most are active ingredients in many herbs. They have many health benefits and possess strong anti-oxidant, anti-cancer, anti-mutagenic, anti-inflammatory and anti-viral properties. It is of great interest that the degradation of tea catechins leads to a stronger biological activity and slower metabolization rate of degradation products of EGCG than their parent compound.

Our current study in the transformation of chlorogenic acid at different physiological pHs showed that chlorogenic acid was readily transformed to its two isomers, neochlorogenic acid and cryptochlorogenic acid, in alkaline buffer solutions and at physiological temperature. It has been observed that upon the establishment of an equilibrium between chlorogenic acid and its two isomers, approximately 100 minutes after the incubation at ph 8.5 and 37 °C, there was an equal distribution among the three compounds. This research result provides very useful information in explanating the potential *in vivo* transformation, absorption and metabolization of chlorogenic acid in alkaline conditions. Further study should be conducted in exploring the mechanism of chlorogenic acid transformation and in evaluating the biological activities of the transformation products.

Acknowledgement

We wish to thank Ted Lambros and Gino Sasso of Hoffmann-La Roche Inc. for their great assistance in HPLC and NMR analysis.

References

1. Lopez-Lazaro, M. *Curr. Med. Chem. – Anti-Cancer Agents*, **2002**, *2*, 691-714.
2. Higdon, J. V.; Frei, B. *Crit. Rev. Food Sci. Nutr.*, **2003**, *43*, 89-143.
3. Aggarwal, B. B.; Kumar, A.; Bharti, A. C. *Anticancer Res.*, **2003**, *23*, 363-398.
4. Wolter, F; Stein, J. *Drugs Fut.*, **2002**, *27*, 949-959.
5. Jayaprakasha, G. K.; Rao, L. J. M.; Sakariah, K. K. *J. Agric. Food Chem.*, **2002**, *50*, 3668-3672.
6. Cliford, M. N. *J. Sci. Food Agric.*, **1999**, *79*, 362-372.
7. Cliford, M. N. *J. Sci. Food Agric.* **2000**, *80*, 1033-1043.
8. Olthof, M. R.; Hollman, P. C. H.; Katan, M. B. *J. Nutr.*, **2001**, *131*, 66-71.
9. Chen, L.; Lee, M.-J.; Li, H.; Yang, C. S. *Drug Metab. Dispos.*, **1997**, *25*, 1045-1050.
10. Nakagawa, K.; Miyazawa, T. *J. Nutr. Sci. Vitaminol.*, **1997**, *43*, 679-684.
11. Yoshino, K.; Suzuki, M.; Sasaki, K.; Miyase, T.; Sano, M. *J. Nutr. Biochem.*, **1999**, *10*, 223-229.
12. Tonnesen, H. H.; Masson, M.; Loftsson, T. *Int. J. Pharm.*, **2002**, *244*, 127-135.
13. Tonnesen, H. H.; Jan, K.; *Z. Lebensm.-Unters.-Forsch.*, **1985**, *180*, 402-404.
14. Wang, Y.-J.; Pan, M.-H.; Cheng, A.-L.; Lin, L.-I.; Ho, Y.-S.; Hsieh, C.-Y.; Lin, J.-K. *J. Pharm. Biomed. Anal.*, **1997**, *15*, 1867-1876.
15. Joe, B.; Lokesh, B. R. *Biochim. Biophys. Acta.*, **1994**, *1224*, 255-263.
16. Watanabe, S.; Naoko, M. *Kiyo-Seitoku Eiyo Tanki Daigaku*, **1998**, *29*, 33-36.

Chapter 19

Bioavailability and Synergistic Effects of Tea Catechins as Antioxidants in the Human Diet

John Shi[1] and Yukio Kakuda[2]

[1]Food Research Center, Agriculture and Agri-Food Canada, Guelph, Ontario N1G 5C9, Canada
[2]Department of Food Science, University of Guelph, Guelph, Ontario N1G 2W1, Canada

Catechins belong to a large group of polyphenolic compounds that are ubiquitous in tea leaves and are characterized by a benzo-y-pyrone structure. Besides their relevance in tea plant, they are also important for humans, because of their high pharmacological activity. Recent interest in these substances has been stimulated by the potential health benefits arising from their antioxidant activity. The beneficial effects on human health has been shown in a number of epidemiological studies that suggest a protective effect against cardiovascular diseases, cancers, and other age related diseases. Their antioxidant activity is enhanced by the synergistic action between catechins, e.g. EGCG, EGC, ECG, EC, pheophytins a and b, and other components in tea leaves. Catechins may complex with proteins, iron and other food compenents and reduce the bioavailability of both compounds or interact with drugs and interfere with the cancer-preventive action of some cancer-fighting medications. Knowledge of the bioavailability and synergistic activities of tea constitutents would provide a clearer picture of the mechanisms by which tea polyphenols exert their therapeutic activity.

Introduction

Interest in the physiological and pharmacological functions of catechins are rapid expanding due to the epidemiological studies showing their versatile health benefits. The levels of these active compounds are directly associated with the daily dietary intake of total antioxidants, thus it is important to evaluate the food you eat as a potential source of catechin. Catechins have been shown to have antioxidative activity, free radical scavenging capacity, preventive effects from coronary heart disease and anticancer activity (*1,2*). Catechins are the most prevalent group of polyphenols found in the tea plant, occurring in virtually all photosynthesizing tea plant cells and is an important source of catechins in both human and animal diets (*3,4*). Catechins cannot be biosynthesized *in situ* by humans and animals (*5*) and must be consumed from plants as feed or food.

Tea (*camellia sinensis*, Theaceae) is one of the most popular beverage consumed throughout the world. An estimated 2.5 million metric tons of dried teas are manufactured annually. The majority of tea beverage is prepared from three types of processed tea: green tea, oolong tea and black tea. Although tea leaves contain more than 2000 components, most attention has been focused on the catechins. Catechins constitute about 30-42% of the dry weight of green tea and 9% of the dry weight of black tea. The major catechins in fresh tea leaves are epigallocatechin gallate (EGCG), epigallocatechin (EGC), epicatechin gallate (ECG) and epicatechin (EC). The composition varies depending on the location where the tea plant is grown, variety of plant, season of harvest, and manufacturing procedures. Catechins are colorless, water soluble compounds that impart bitterness and astringency to tea infusions. The usual composition is 10-15% EGCG, 6-10% EGC, 2-3% ECG, and 2% EC (*6*). The structures of the major catechins are shown in Figure 1. EGCG is the most abundant catechins and has received by far the most attention in clinical studies. In addition, flavonols such as quercetin and their glycosides, caffeine, theobromine, theophylline, and phenolic acids such as gallic acid, are also present as minor constituents.

The possible cancer-preventive activity of tea has received a great deal of attention. Although there is increasing interest in the potential health benefits of tea catechins as a cancer-preventive agent, the bioavailability, absorption, metabolism, tissue distribution, and the biological activities of tea catechins as well as their interaction with other food compounds and medical drugs are not completely known. A better understanding of the association of tea consumption and cancer prevention requires quantitative data on the bioavailability, interactions and disposition of tea catechins.

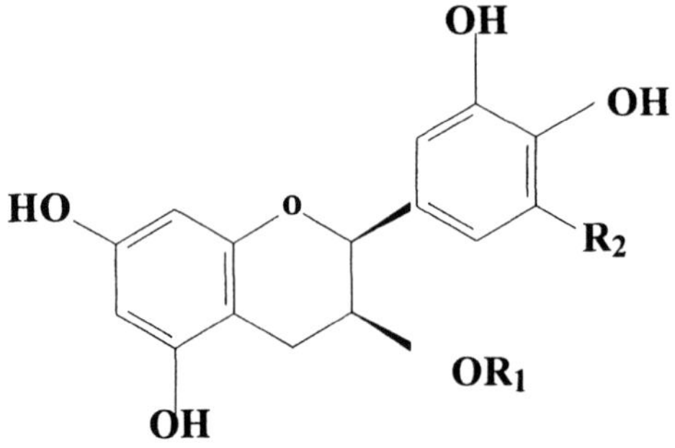

	R_1	R_2
Epigallocatechin gallate	EGCG	Gallate	OH
Epigallocatechin	EGC	H	OH
Epicatechin gallate	ECG	Gallate	H
Epicatechin	EC	H	H

Figure. 1. Major Tea Catechins in Tea

Physiological and Synergistic Effects of Tea Catechins

Catechins have received considerable attention because of their beneficial effects as antioxidants in the prevention of human diseases such as cancer and cardiovascular diseases, and some pathological disorders of gastric and duodenal ulcers, allergies, vascular fragility, and viral and bacterial infections (7). Catechins have exhibited a wide spectrum of pharmacological properties, including antiallergic, anti-inflammatory, antidiabetic, hepato- and gastro-protective, antiviral and antineoplastic activities (7). Some epidemiological studies suggest that regular tea consumption reduces the risk of cancer in humans (8). Catechins have been shown to have a variety of physiological functions. They have *in vitro* and *in vivo* antioxidative activity which was closely related to their preventive effects on various diseases including liver injury, arteriosclerosis and inflammation caused by lipid peroxidation and excessive free radical production.

Catechins may act as antioxidants to inhibit free radical mediated cytotoxicity and lipid peroxidation, as anti-proliferative agents to inhibit tumor growth or as weak estrogen agonists or antagonists to modulate endogenous hormone activity (9). Catechins may also confer protection against chronic diseases such as atherosclerosis and cancer, and assist in the management of menopausal symptoms. Some early studies have revealed that tea polyphenols may be related to a capillary-strengthening property, an antioxidative property responsible for the radioprotective effect, and some antimicrobial property. Hara (10) showed that drinking teas regularly may prevent cardiovascular diseases by increasing plasma antioxidant capacity in humans (11,12). In recent years, considerable attention has been paid to their abilities to inhibit the cell cycle, cell proliferation, and oxidative stress, and to induce deoxification enzymes, apoptosis, and the immune system (13). Multiple mechanisms have been identified for the anti-neoplastic effects of catechins, including antioxidant, anti-inflammatory, anti-proliferative activities, inhibition of bioactivating enzymes, and induction of detoxifying enzymes. Catechin is relatively resistant to heat, oxygen, dryness, and moderate degree of acidity (14).

Antioxidant Activity

Catechins may protect against cancers through inhibition of oxidative damage. Catechins have been labeled as 'high level' natural antioxidants based on their abilities to scavenge free radicals and active oxygen species (15-17). Catechins have been reported to chelate iron and copper and this may partly explain their antioxidant effects. Bors et al. (18) suggested that the antioxidant mechanisms might include synergistic effects. The antioxidant activity of catechins usually increased with an increase in the number of hydroxyl groups and a decrease in glycosylation. EC and ECG with a vicinal diphenol structure in the B ring and a saturated C ring exhibit the strongest effects (Figure 1) (19). EGCG with two triphenol components in its structure, one from the B ring and one from the gallate attachment, has been found to strongly and dose-dependently inhibit histamine release from rat basophilic leukemia cells (20). Salah et al. (21) showed that the total antioxidative activity and the order of effectiveness of tea catechins as radical scavengers were ECG > EGCG > EGC > gallic acid > EC. However, the oxidation of low density lipoproteins is inhibited by EGC, EC, ECG and EGCG to a similar degree, but to a lesser degree in the presence of EGC or gallic acid. These results suggest that the triphenol structure plays an important role in the activities of tea polyphenols (22).

Metabolism and Clinical Effects

Catechins are absorbed by the gastrointestinal tracts of humans, and are excreted either unchanged or as their metabolites in the urine and faeces (5). Measurement of plasma and urine antioxidant power after ingestion of green tea has shown that absorption of tea catechins is rapid (23). catechins enter the systemic circulation soon after ingestion and cause a significant increase in plasma antioxidant status. Benzie *et al.* (*23*) and Feng *et al.* (*24*) suggested that this increase might decrease oxidative damage to DNA and thus reduce the risk of cancer. Catechins have profound effects on the functions of immune and inflammatory cells (*25*). In animal studies, two EGCG methyl ethers isolated from tea significantly inhibited mice allergic reactions (*26*).

A possible reason for the inferred protective effects of catechins against heart disease is their ability to prevent the oxidation of low density lipoproteins to an atherogenic form, although anti-platelet aggregation activity and vasodilatory properties are also reported (*27,28*). Catechin intake may reduce the risk of death from coronary heart disease (*29-32*). Differences in catechin intake in different countries may partly explain the differences in coronary heart disease mortality across populations (*30*). The habitual intake of catechins from tea may lead to a lower risk of atherosclerosis and coronary heart disease, and also protect against stroke (*33,34*). This seems reasonable since tea catechins can reduce blood coagulability, increase fibrinolysis, prevent platelet adhesion and aggregation, and decrease the cholesterol content in aortic walls *in vivo* (*35*). Also green and black teas are able to protect against nitric oxide toxicity, which may be another reason for the beneficial effects observed with tea catechins (*36*).

Some studies have focused on the different functionality of green tea and black tea in cancer-prevention, because of the differences in component composition and variety between green tea and black tea. The results have shown that the cancer prevention activity of green tea was stronger than that of black tea. In the studies on the inhibition of cancer formation by tea in animal models, the effective components were catechins, but black tea contains much lower catechins than green tea (*37*). On the basis of some recent studies, cancer chemopreventive effects of green tea are mediated by EGCG, which is the major catechin constituent of green tea. Catechins are reduced by 85% during black tea manufacturing, only 10% can be accounted for as theaflavins and theaflavic acids. The reduction of cancer risks and chronic diseases by catechins depends on the levels, the composition of catechins, their bioavailability and bioactivity (*38*).

The inhibiting effects of tea components may significantly reduce the risk of several important types of cancers in the world (*39*). Histopathological examinations revealed that both green and black teas were able to inhibit tumor cell proliferation in animal models (*40*). EGCG, EGC and ECG inhibited soybean lipoxygenase, a carcinogen and tumor promoter, most effectively at lower doses (*41*). Tea is a significant source of catechins, with a suggested role in the prevention of cancer (*37*). Polyphenols present in green tea show cancer chemopreventive effects against tumor initiation (*42*) and promotion stages of multistage carcinogenesis in many animal models (*43,44*). Green tea may protect against cancer by causing cell cycle arrest and inducing apoptosis (45), while black tea can produce an inhibitory effect on tumor promotion (46).

Synergistic Effects of EGCG, ECG, EGC, EC, and Pheophytins *a* and *b*

Some catechins such as EC show synergistic effects with other catechins and caffeine (47). According to the study of Suganuma et al. (47), EC can significantly enhance EGCG's potential cancer-preventive activity, but ECG, EGC have only a slight effect on EGCG's inhibitory activity. In one study of bioavailability where decaffeinated green tea and pure EGCG were administered to rats intravenously, the results showed that other components in decaffeinated tea also affected the plasma concentration of EGCG (48). According to Chen et al. (48), EGC and EC seemed to be absorbed faster than EGCG, and EGCG had a much lower bioavailability in terms of absorbed. The low bioavailability of EGCG occurred when given either in decaffeinated green tea or pure EGCG. It seems that EGCG is better absorbed when given through tea infusion rather than in pure form. Synergistic effects on cancer-preventive activity in tea catechin-rich extract shows great potential. EGCC combined with pheophytins *a* and *b* exhibited considerable suppressive activities against the autoxidation of linoleic acid at concentration of 0.2 mg per mL (49). As a control experiment, the same amounts of pure EGCG showed relatively low activities.

Interaction of Catechins with Foods and Medical Cmponents

Some positive associations between intake of catechins and intake of fiber, vitamin C and β-carotene are shown in Table I. In children dietary fiber has a strong association with catechin absorption while catechin intake is inversely associated with intake of alcohol in elderly (50). Catechins also reduces the degradation of α-tocopherol and allow it to scavenge aqueous peroxyl radicals near the membrane surface (51).

Tea catechins have strong affinities for proline rich proteins due to their open extended structure on the one hand, and the high content of proline residues on the other. Hydrophobic interactions and hydrogen bonding result in the formation of protein-catechins complexes, and reduce the absorption of tea catechins and digestibility of protein (52).

Tea catechins have a strong affinity for Fe^{2+}, Fe^{3+} and readily form insoluble complexes. This interaction reduces the absorption efficiency of both the tea catechins and the transition metal (53). The reduction of nonheme-iron absorption in the presence of tea suggests that the chelation of iron may be one of the mechanisms of antioxidant action in the human digestive system (54).

Table I. Partial Rank-order Correlation Coefficients Between Total Catechin Intake and Other Dietary Factors (Arts et al., 2001).

	Children (1-18) (n=1539)	Adults (19-64) (n=3954)	Elderly (>65) (n=707)
Alcohol	-0.01	-0.08	-0.015
Saturated fatty acids	0.5	-0.03	0.02
Polyunsaturated fatty acids	-0.02	-0.02	0.01
Fiber	0.21	0.20	0.13
Vitamin C	0.07	0.17	0.11
Vitamin E	0.01	0.05	0.07
β-carotene	0.06	0.10	0.11

Catechins can enhance cancer-fighting function and reduce toxicity of some medication such as Sulindac and Tamoxifen (47). The mixture of Sulindac + EGCG and Tamoxifen + EGCG synergistically inhibited cell growth of mouse colon cancer, more effectively than either Sulindac or Tamoxifen alone (Figure 2). Tea catechins can act as natural synergistic agents to enhance the cancer-preventive activities and reduce adverse side effects of some medication

CONCLUSION

Catechins are present in tea in large amounts, and are generally nontoxic and can manifest a diverse range of beneficial biological activities. There is much evidence relating catechin consumption and the inhibition of carcinogenesis. Many mechanisms have been proposed on how catechins help to prevent steroid hormone-dependent cancers and several have been validated *in vitro* and in animal models. These include modulation of steroid hormones, inhibition of proliferation, and anticarcinogenic and antioxidative activities. Numerous *in vitro* and *in vivo* studies have been done to try and elucidate the beneficial effects caused by catechins. More information on the health benefits of catechins has become available through ongoing epidemiological studies. Recent research on the health benefits of catechins focused on the occurrences associated with dietary intake, theoretical studies (e.g. mechanisms of the functions of catechins to human) incorporating and/or integrating data from basic research, and *in vitro* and *in vivo* investigations relating individual or a combination of catechins to a specific health problem.

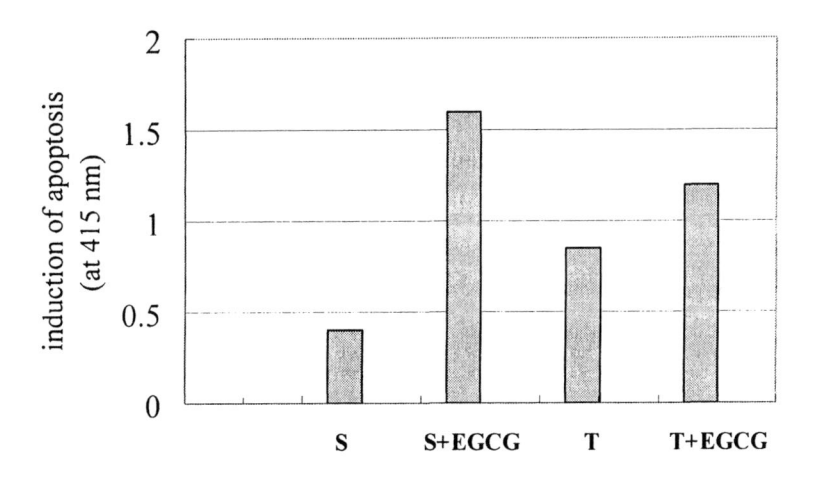

Figure. 2. Synergistic effects of EGCG (75 μm) with Sulindac (S) and Tamoxifen (T) on induction of apoptosis (absorbance at 415 nm) (modified from Suganuma et al., 1999).

The preventive effect may involve the synergistic action between EGCG, EGC, ECG, EC, pheophytins *a* and *b*, and other bioactive components in tea leaves. Catechins have been shown to enhance the cancer-preventive activity of some cancer-fighting medication. On the other hand, protein, iron, and other food components can complex with tea catechins and reduce the bioavailability of both the catechins and the food components. Research on catechins should focus on the bioavailability of catechins and the objective measurement of oxidative damages *in vivo*. More research is needed to develop an authentic and convincing model or system for precisely assessing the human intake and metabolism of catechins, as well as their alleged health benefits.

References

1. Yang, C.S.; Landau, J.M. *J. Nutr.* **2000**, *130*, 2409-2412.
2. Yang, C.S.; Maliakal, P.; Meng, X. *Annu. Rev. Pharmacol. Toxicol.* **2002**, *42*, 25-54.
3. Harborne, J.B.; Turner, B.L. *Plant Chemosystematics*. Academic Press, London, 1984.
4. Clifford, A.H.; Cuppett, S.L. *J. Sci. Food Agric.* **2000**, *80*, 1063-1072.
5. Cook, N.C.; Samman, S. *J. Nutr. Biochem.* **2000**, *7*, 66-76.

6. Harbowy, M.E.; Balentine, D.A. *Crit. Rev. Plant Sci.* **1997**, *16*, 415-480.
7. Zand, R.S.R.; Jenkins, D.J.A.; Diamandis, E.P. *J. Chromatography B*, **2002**, *777*, 219-232.
8. Bushman, J.L. *Nutr. Cancer* **1998**, *31*, 151-159.
9. Lyons-Wall, P.M.; Samman, S. *Nutr. Soc.* Aust. **1997**, *21*, 106-114.
10. Hara, Y. *Prev. Med.* **1992**, *21*, 333.
11. Nakagawa, K.; Ninomiya, M.; Okubo, T.; Aoi, N.; Juneja, L.R.; Kim, M.; Yamanaka, K.; Miyazawa, T. *J. Agric. Food Chem.* **1999**, *47*, 3967-3973.
12. Duthie, G.G.; Duthie, S.J.; Kyle, J.A.M. *Nutr. Res. Rev.* **2000**, *13*, 79-106.
13. Birt, D.F.; Shull, J.D.; Yaktine, A. 1999. In *Modern Nutrition in Health and Disease*, Shils, M.E.; Olson, J.E.; Shike, M.; Ross, A.C., Eds., Williams and Wilkins, Baltimore, 1999.
14. Kuhnau, J. *World Rev. Nutr. Diet* **1976**, *24*, 117-119.
15. Fukumoto, L.R.; Mazza, G. *J. Agric. Food Chem.* **2000**, *48*, 3597-3604.
16. Unno, T.; Sugimoto, A.; Kakuda, T. *J. Sci. Food Agric.* **2000**, *80*, 601-606.
17. Klahorst, S. *World of Food Ingredients* **2002** (April/May), 54-59.
18. Bors, W.; Heller, W.; Michel, C.; Stettmaier, K. In *Handbook of Antioxidants*, Cadenas, E.; Packer, L., Eds., Marcel Dekker Inc., New York, 1996, pp 409-468.
19. Frankel, S.; Robinson, G.E.; Berenbaum, M.R. *J. Apic. Res.* **1998**, *37*, 27-31.
20. Matsuo, N.; Yamada, K.; Shoji, K.; Mori, M.; Sugano, M. *Allergy* **1997**, *52*, 58-64.
21. Salah, N.; Miller, N.J.; Paganga, G.; Tijburg, L.; Bolwell, G.P.; Rice-Evans, C.A. *Arch. Biochem. Biophys.* **1995**, *322*, 339-346.
22. Pannala, A.S.; Chan, T.S.; O'Brien, P.J.; Rice-Evans, C.A. *Biochem. Biophys. Res. Comm.* **2001**, *282*, 1161-1168.
23. Benzie, I.F.F.; Szeto, Y.T.; Strain, J.J.; Tomlinson, B. *Nutr. Cancer* **1999**, *34*, 83-87.
24. Feng, Q.; Torii, Y.; Uchida, K.; Nakamura, Y.; Hara, Y.; Osawa, T. *J. Agric. Food Chem.* **2002**, *50*, 213-220.
25. Middleton, E.; Kandaswami, C. *Biochem. Pharmacol.* **1992**, *43*, 1167-1179.
26. Sano, M.; Suzuki, M.; Miyase, T.; Yoshino, K.; Maeda-Yamamoto, M. *J. Agric. Food Chem.* **1999**, *47*, 1906-1910.
27. Muldoon, M.F.; Kritchevsky, S.B. *Br. Med. J.* **1996**, *312*, 458-59.
28. Chen, C.; Tang, H.R.; Sutcliffe, L.H.; Belton, P.S. *J. Agric. Food Chem.* **2000**, *48*, 5710-5714.

29. Hertog, M.G.; Hollman, P.C.H.; Katan, M.B. *Nutr. Cancer* **1993**, *20*, 21-29.
30. Hertog, M.G.; Kromhout, D.; Aravanis, C.; Blackburn, H.; Buzina, R.; Fidanza, F.; Giampaoli, S.; Jansen, A.; Menotti, A.; Nedeljkovic, S. *Arch. Internal Med.* **1995**, *155*, 381-386.
31. Knekt, P.; Reunanen, A.; Jarvinen, R.; Maatela, J. *Br. Med. J.* **1996**, *312*, 478–481.
32. Yochum, L.A.; Kushi, L.H.; Meyer, K.; Folsom, A.R. *Am. J. Epidemiol.* **2000**, *149*, 943–949.
33. Weisburger, J.H. In *Handbook of Antioxidants*. Cadenas, E.; Packer, L., Eds., Marcel Dekker Inc., New York, 1996, pp 469–486.
34. Tijburg, L.B.M.; Mattern, T.; Folts, J.D.; Weisgerber, U.M.; Katan, M.B. *Crit. Rev. Food Sci. Nutr.* **1997**, *37*, 771–785.
35. Lou, F.Q.; Zhang, M.F.; Zhang, X.G.; Liu, J.M.; Yuan, W.L. *Prev. Med.* **1992**, *21*, 333.
36. Paquay, J.B.G.; Haenen, G.R.M.M.; Stender, G.; Wiseman, S.A.; Tijburg, L.B.M.; Bast, A. *J. Agric. Food Chem.* **2000**, *48*, 5768–5772.
37. Wiseman, S.A.; Balentine, D.A.; Frei, B. *Crit. Rev. Food Sci. Nutr.* **1997**, *37*, 705-718.
38. Bell, J.R.C.; Donovan, J.L.; Wong, R.; Waterhouse, A.L.; German, J.B.; Walzem, R.L.; Kasim-Karakas, S.E. *Am. J. Clin. Nutr.* **2000**, *71*, 103-108.
39. Weisburger, J.H. *Prev. Med.* **1992**, *21*, 331.
40. Chen, Y.C.; Liang, Y.C.; Lin-Shiau, S.Y.; Ho, C.-T.; Lin, J.K. *J. Agric. Food Chem.* **1999**, *47*, 1416–1421.
41. Ho, C.-T.; Lee, C.Y.; Huang, M.T. *Phenolic Compounds in Food and Their Effects on Health I. Analysis, Occurrence & Chemistry*. American Chemical Society, Washington, D.C., 1992.
42. Gensler, H.L.; Timmermann, B.N.; Valcic, S.; Wachter, G.A.; Dorr, R.; Dvorakova, K.; Alberts, D.S. *Nutr. Cancer* **1996**, *26*, 325–335.
43. Dreosti, I.E.; Wargovich, M.J.; Yang, C.S. *Crit. Rev. Food Sci. Nutr.* **1997**, *37*, 761–770.
44. Kivits, G.A.A.; Van der Sman, F.J.P.; Tijburg, L.M.B. *Int. J. Food Sci. Nutr.* **1997**, *48*, 387–392.
45. Ahmad, N.; Feyes, D.K.; Nieminen, A.L.; Agarwal, R.; Mukhtar, H. *J. Nat. Cancer Inst.* **1997**, *89*, 1881–1886.
46. Nakamura, Y.; Harada, S.; Kawase, I.; Matsuda, M.; Tomita, I. *Prev. Med.* **1992**, *21*, 332.
47. Suganuma, M.; Okabe, S.; Kai, Y.; Sueoka, N.; Sueoka, E.; Fujiki, H. *Cancer Res.* **1999**, *59*, 44-47.
48. Chen, L.; Lee, M.J.; Li, H.; Yang, C.S. *Drug Metabolism and Disposition* **1997**, *25*, 1045-1050.

49. Higashi-Okai, K.; Taniguchi, M.; Okai, Y. *J. Sci. Food Agric.* **2000**, *80*, 117-120.
50. Arts, I.C.W.; Hollman, P.C.H.; Feskens E J M, Bueno de Mesquita HB, Krombout D. *Eur. J. Clin. Nutr.* **2001**, *55*, 76-81.
51. Pietta, P.G.; Simonetti, P. *Biochem. Mol. Biol. Int.* **1998**, *44*, 1069-1074.
52. Haslam, E. *J. Nat. Prod.* **1996**, *59*, 205-215.
53. South, P.K.; William, M.S.; House, A. *Nutr. Res.* **1997**, *8*, 1303-1310.
54. Samman, S.; Sandstrom, B.; Toft, M.; Bukhave, K.; Jensen, M.; Sorensen, S.; Hansen, M. *Am. J. Clin. Nutr.* **2001**, *73*, 607-612.

Biological Activities

Chapter 20

Targeting Inflammation Using Asian Herbs

Haiqing Yu, Mohamed M. Rafi, and Chi-Tang Ho

Department of Food Science, Rutgers, The State University of New Jersey, 65 Dudley Road, New Brunswick, NJ 08901–8520

The use of herbal therapy or alternative medicine for the treatment of various inflammatory disorders has been in practice in Asia for centuries, yet only recently it is becoming an increasingly attractive approach over the world. Significant research efforts in the laboratory and in the clinic are ongoing to understand the critical role of certain herbs in the regulation of inflammation. This article has reviewed few selected Asian herbs and their possible mechanism of action in inflammation. The selected herbs include: *Scutellaria baicalensis (Huang Qin), Rheum palmatum, Andrographis paniculata, Polygonum hypoleucum Ohwi,* and *Tripterygium wilfordii Hook f.* Furthermore, possible targets in inflammation are summarized, which will give insights to the development of new anti-inflammatory agents.

Introduction

In Asian countries, medicinal herbs have been used for prevention and treatment of various diseases for thousands of years. Convinced by the efficacy and long history of safe use, western countries started to join in the efforts of exploring health-promoting properties and constituents of these herbs. Asian herbs having anti-inflammatory properties are among those most intensively investigated.

Anti-inflammatory properties of different herbs are mediated through the inhibition of production of cytokines (IL-1β, TNF-α, IL-6, IL-12, IFN-γ), nitric oxide (NO), prostaglandins and leukotriens. The proinflammatory mediators NO and prostaglandins are produced by actions of cyclooxygenases (COX-2) and inducible nitric oxide synthase (iNOS), respectively. These inflammatory mediators are soluble, diffusible molecules that act local and distant sites of tissue damage and infection. The COX-2 and iNOS are important enzymes that mediate most of the inflammatory processes. Improper up-regulation of COX-2 and/or iNOS has been associated with pathophysiology of certain types of human cancers as well as inflammatory disorders (*1,2*). Since inflammation has been shown as one of the factors causing certain types of cancer, herbs with potent anti-inflammatory activities are thought to inhibit carcinogenesis.

Several studies have shown that nuclear factor-kappa B (NF-κB) eukaryotic transcription factor, is involved in regulation of COX-2 and iNOS expression, and some phytochemicals have been shown to inhibit COX-2 and iNOS expression by blocking improper NF-κB activation. Similarly, the inhibition of proinflammatory cytokine by these herbs has also been demonstrated in various inflammatory conditions (*2,3*). The most possible mechanisms underlying inhibition of NFκB activation is by inhibition of Inhibitory Kappa B kinase (IKK), which prevents the degradation of Inhibitory-kappa B (IκB) and thereby hampers subsequent nuclear translocation of the functionally active subunit of NF-κB (*4*).

Activator Protein-1 (AP-1) is another important transcription factor that may be involved in the transcriptional regulation of cytokines and mediators. Many herbs or their constituents affect the formation and activation of AP-1 proteins. Following activation, AP-1 binds to specific recognition sequences in the promoter regions of target genes causing modulation of gene transcription (*5*). COX-2 gene promoter region has binding site for AP-1 (*6*), indicating that COX-2 gene transcription is not only regulated by NFκB, but also by AP-1. There are also reports that elevation of both AP-1 and NFκB may result in greater inflammation than that if either transcription factor alone is activated (*5*).

In this article, literature will be reviewed for anti-inflammatory properties of selected Asian herbs along with their mode of action and elucidated active constituents. The herbs of interest include: *Scutellaria baicalensis* (Huang Qin), *Rheum palmatum*, *Andrographis paniculata*, *Polygonum hypoleucum* Ohwi, and *Tripterygium wilfordii* Hook f.

Scutellaria baicalensis Georgi (Huangqin)

Scutellaria baicalensis is mainly distributed in Baikal Lake area (Russia), northeast Mongolia, and northeast China. It has been extensively used as an anti-inflammatory and anti-cancer therapy in ancient China, where it was called *Huangqin*, and Japan, where it was called *Wogon*. The plant is typically used in herbal combinations for the treatment of inflammatory skin conditions, allergies, high cholesterol, and high blood pressure.

Animal studies have shown that root extract has extremely low toxicity (7,8). It shows topical anti-inflammatory activity in mouse ear edema model, and its potency against arachidonic acid induced edema is comparable to that of reference drug Dexamethasone (9). Ye et al. (10) reported the root extract suppressed PGE2 release in a dose dependent manner, suggesting that COX-2 inhibition might be involved.

In the last 4 years, rigorous studies have been carried out to elucidate the action mechanism of the root extract. The role of ROS in the inflammatory process has been well known. Free radicals liberated from phagocyte cells are implicated in the activation of nuclear factor κB, which serves as a transcription factor that induces transcription of proinflammatory cytokines and COX-2. Thus, Shao et al elucidated the mechanism of its extract regarding the antioxidant activity, showing it acts in cells by scavenging or detoxifying ROS directly, rather than inhibiting generation of intracellular oxidase systems (11). However, a surprising finding illustrated pro-oxidant activity in the Fe^+-EDTA-H_2O_2 system (12), suggesting anti-oxidant activity is not sufficient to explain its inflammatory action. Besides the inhibition effect on both rat liver microsome and red blood cell lipid peroxidation, the root extract was demonstrated to be able to inhibit aminopyrin N-demethylase and xanthine oxidase activities (12).

Studies on the major bioactive constituents of *Scutellaria baicalensis* demonstrated multiple anti-inflammatory pathways. The major active constituents are flavones, including baicalein, wogonin, scullcapflavone I and II, oroxylin A and baicalin (Figure 1) (13,14).

Baicalein has shown to dose-dependently inhibit IL-1β and TNFα induced endothelial leukocyte adhesion molecule-1 (ELAM-1), intercellular adhesion molecule-1 (ICAM-1) expression, and the downstream protein kinase C inhibitor H7 (15). The latest study reported baicalein attenuated NO production in LPS activated mouse microglial cell by suppression of inducible NO synthase (iNOS) induction (16). Furthermore, Baicalein inhibited LPS induced nuclear factor κB (NFκB) activity in BV-2 cells (17).

Wogonin was reported to downregulate PGE2 production and iNOS expression in macrophages (17,18). Kim et al confirmed the action in C6 rat glial cell model induced by LPS, TNFα, and IFNγ. They also observed suppression of transcription factor NFκB activity (19), suggesting NFκB regulated iNOS induction is involved. Wogonin inhibits monocyte chemotactic protein-1 (MCP-1) induction in human umbilical vein endothelial cells; this inhibition is mediated by reducing AP-1 transcriptional activity via the attenuation of ERK1/2 and JNK signal transduction pathways (20).

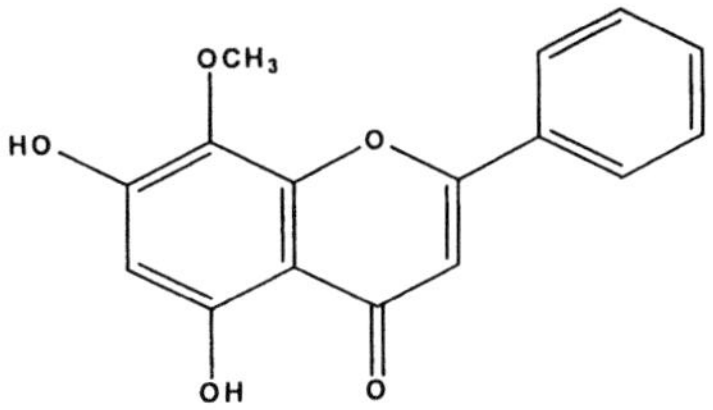

Wogonin

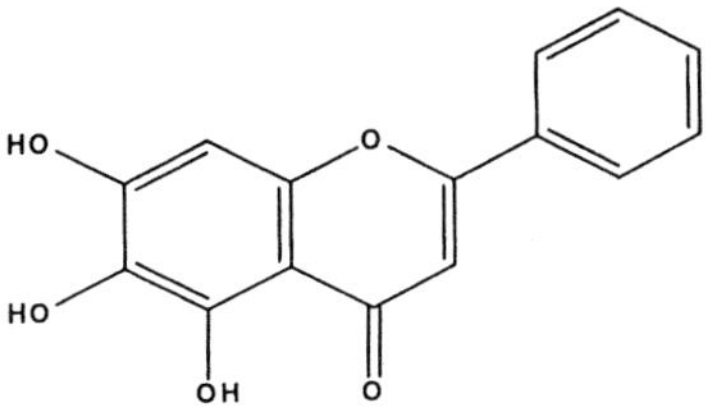

Baicalein

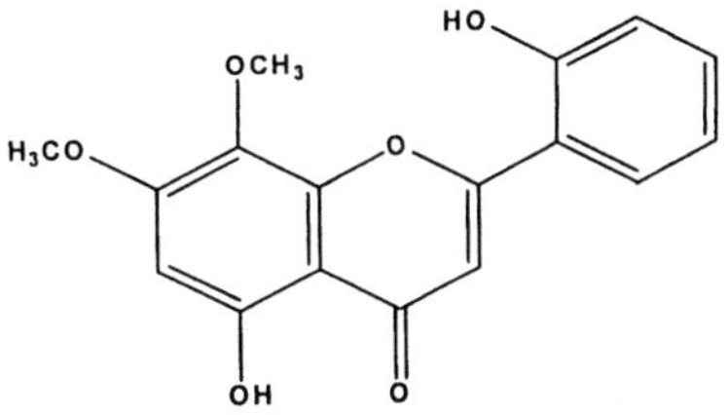

Skullcapflavone I

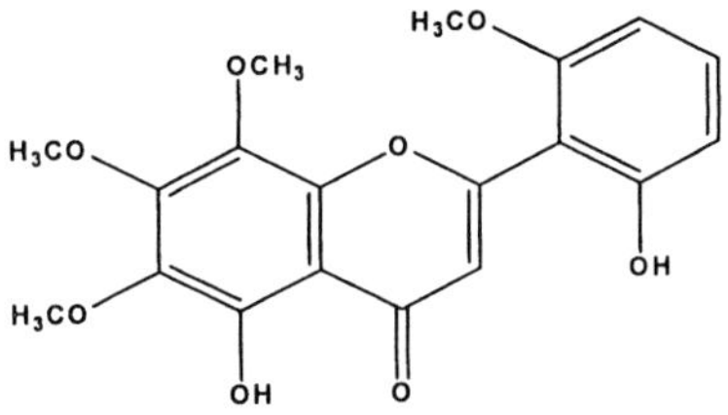

Skullcapflavone II

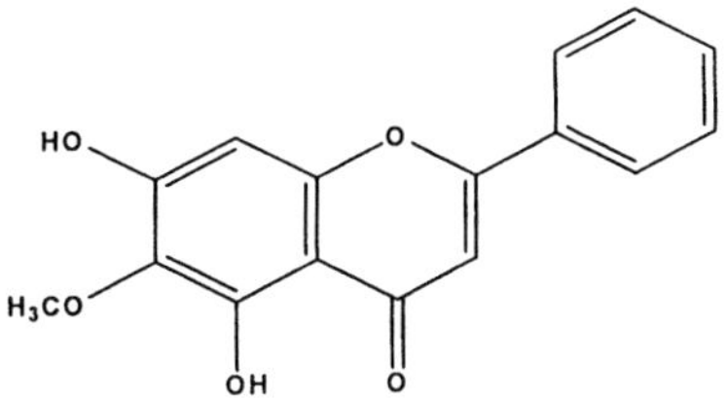

Oroxylin A

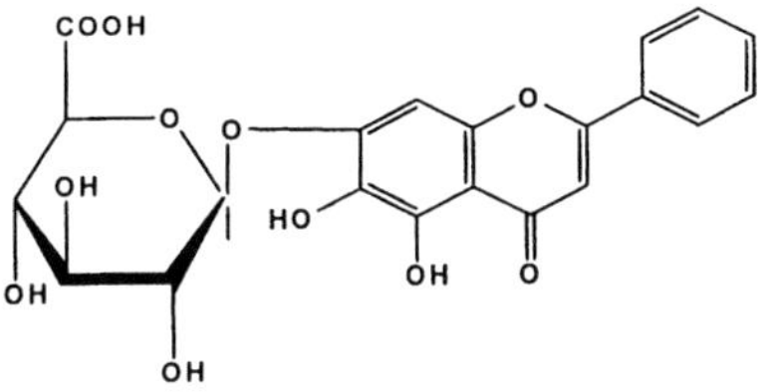

Baicalin

Figure 1. Major bioactive constituents in Scutellaria baicalensis Georgi.

Baicalin has shown to inhibit carrageenan induced paw edema in rats (*17*). In 2000, Li *et al.* (*21*) first reported Baicalin was able to bind a number of chemokines, such as stromal cell-derived factor (SDF)-1α, IL-8, macrophage inflammatory protein (MIP)-1β, monocytes chemotactic protein (MCP)-2, etc. Thus, it was blocking the binding and activation of chemokine receptors, resulting in the arrest of immune/autoimmune response events. Additional study on baicalein showed it also downregulates the expression of proinflammatory cytokines, such as IL-1β, IL-6, TNF, IFN γ, which were induced by superantigens *in vitro*. (*22*).

Oroxylin A is another active flavone constituent shown to inhibit iNOS and COX-2 production in LPS induced macrophages. Further study revealed that NFκB transcription regulation by Oroxylin A was responsible for these effects (*23*).

In summary, most mechanistic studies regarding constituents were done *in vitro*. They act either by direct suppression of ROS species, suppression of chemokines and pro-inflammatory cytokines, or important enzymes involved in modulation of inflammation; some of them are able to regulate transcription factors up-stream.

Rheum palmatum

Rheum palmatum (Polygonaceae), common name Rhubarb, is an important traditional Chinese medicinal herb with demonstrated antibacterial and antipyretic properties. Still widely distributed in Gansu, Tibet, and inner Mongolia regions in China, it has a surprisingly long history of medicinal usage by ancient Chinese people dating back to 2700 BC.

Western societies started to explore *R. palmatum* for its anti-inflammatory activity only recently. In 2001, Cuellar et al. reported its anti-inflammatory action in TPA induced mouse ear edema, especially in acute TPA model (*9*). Additionally, Song *et al.* (*24*) found suppression of IL-6 production in human chronic renal failure patients after treatment with R. Palmatum.

Scientific reports on the actions of Rhubarb constituents remain to be limited. The major known active constituents are a group of anthraquinone derivatives, including aloe-emodin, chrysophanol, emodin, physcion, rhein, and sennosides A and B (Figure 2). Their effects on iNOS and COX-2 production of macrophages were astudied. Emodin and rhein turned out to be most potent in inhibiting NO production with little cytotoxicity. Interestingly, both iNOS and COX-2 protein synthesis were inhibited by emodin, while rhein only inhibited iNOS protein expression. Moreover, synergistic effect between the 2 compounds was observed (*25*). On transcriptional level, NF-κB activity was inhibited by emodin in umbelical vein endothelial cells (*26*).

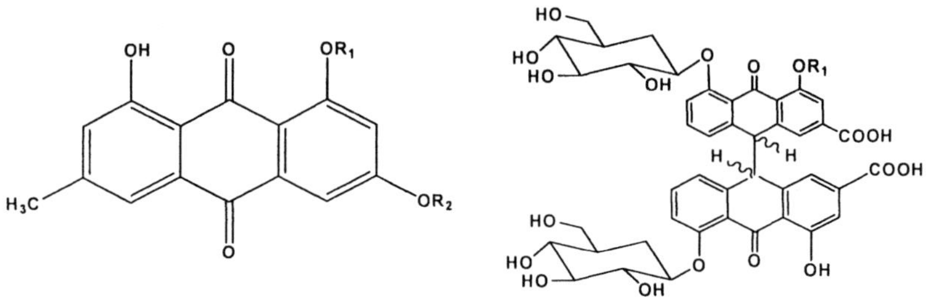

Aloe-emodin	R=CH₂OH	R'=H
Chrysophanol	R=CH₃	R'=H
Emodin	R=CH₃	R'=OH
Physcion	R=CH₃	R'=OCH₃

Sennoside A — *threo*
Sennoside B — *trythro*

Figure 2. Major Anthraquinones in Rheum palmatum.

Polygonum Hypoleucum Ohwi

Polygonum Hypoleucum Ohwi is a vine-like plant that is mainly distributed in mountainous regions in Taiwan. Traditionally, it has been used in China as a therapy for arthritis, rheumarthritis, cough, etc.

P. Hypoleucum Ohwi has the most potent anti-proliferatory activity among 15 Chinese herbs tested by Kuo et al. for modulation effects on human mesangila cell proliferation. On protein level studies indicated that IL-1β and TNFα protein production was suppressed by the herb extract. While on transcriptional level, TNFα mRNA expression was also suppressed (*27*). Their results indicate that gene expression and protein synthesis of proinflammatory cytokine are possible targets for its action.

Four anthraquinones (Fig. 3) were isolated as bioactive constituents. They are emodin, physcion, and their β-D-Glucoside (*28*). Among them, emodin showed most potent suppressing activity on the human mesangial cells proliferation activated by IL-1β and IL-6. Moreover, IL-1β, IL-6, and TNFα production was inhibited; while on mRNA level, IL-1β and TNFα expression was impaired by emodin (*29*).

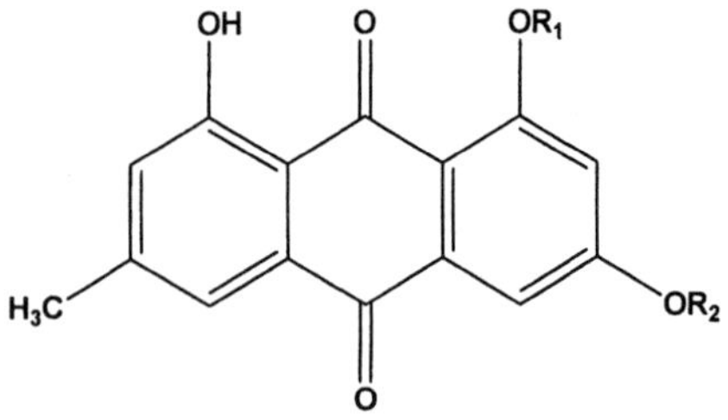

Emodin: R1=R2=H;
Physcion: R1=H, R2=CH3;
Emodin 1-β-O-glucoside: R1=β-D-glucoside, R2=H
Physcion 1-β-O-glucoside: R1=β-D-glucoside, R2=CH3

Figure 3. Major Anthraquinones in Polygonum Hypoleucum Ohwi.

The same research group also evaluated the effects of the 4 anthraquinones on activated T lymphocyte proliferation. Again, emodin has shown to have the highest activity in suppressing cell proliferation. Cytokine production and IL-2 mRNA expression were suppressed in T cells (*30*). The inhibiting effects of emodin on iNOS and COX-2 proteins, as well as transcription factor NF-κB were reported along with herb *Rheum palmatum*.

In summary, anti-inflammatory activity of *P. Hypoleucum Ohwi* has been assessed in human mesangila cell and activated T cell models, along with it major active constituents, while no *in vivo* model has been applied on investigation of this herb. Emodin showed to be an effective anti-inflammatory agent by blocking several pathways in various cell models. Again, further *in vivo* demonstration is needed to establish its anti-inflammatory profile *in vivo* before it can be considered as a neutraceutical candidate.

Andrographis paniculata

Andrographis paniculata (Acanthaceae) is also known as "King of Bitters", since every part of the plant is extremely bitter. Distributed extensively in China (Chuan-Chin-Lian) and India, it has been used by local people as an herbal remedy for arresting diarrhea, alleviating upper respiratory infections, tonsillitis, pharyngitis, laryngitis, pneumonia, tuberculosis, and pyelonephritis.

Along with *in vivo* evidence (*31*), the anti-inflammatory action of *Andrographis paniculata* was shown by human clinical trials. A double blind, placebo controlled clinical study showed that *Kan Jang*, the *Andrographis paniculata Nees* extract fixed combination, is effective in relieving the inflammatory symptoms of sinusitis associated with acute upper respiratory tract infections (*32*).

Diterpenoids were elucidated as the principals for anti-inflammatory activity. The major constituent, andrographolide (Figure 4), is able to suppress inflammation in multiple pathways. First of all, it has demonstrated effects at arresting reactive oxygen and nitrogen species, which play a key role in the inflammation process. Chiou *et al.* (*33*) reported that andrographolide is able to arrest NO production by suppressing iNOS protein in LPS-induced RAW 264.7 cells. Surprisingly, no NO suppression was observed in *in vivo* model (*34*). Additionaly, by modulating protein kinase C-dependent pathway, andrographolide reduced reactive oxygen species H_2O_2 and $O_2\bullet$ production in neutrophils (*35,36*). Although the effect of andrographolide on production of pro-inflammatory cytokines has not been reported, inhibition of intercellular adhesion molecule-1 (ICAM-1) and endothelial-monocyte adhesion (EMA) was observed *in vitro.* ICAM-1 and EMA are upregulated by pro-inflammatory cytokines, which is the key step in development of inflammation (37).

Neoandrographolide (Figure 4), another important diterpene lactone constituent in *Andrographis paniculata,* suppressed the NO production both *in vitro* and *in vivo* (*34*).

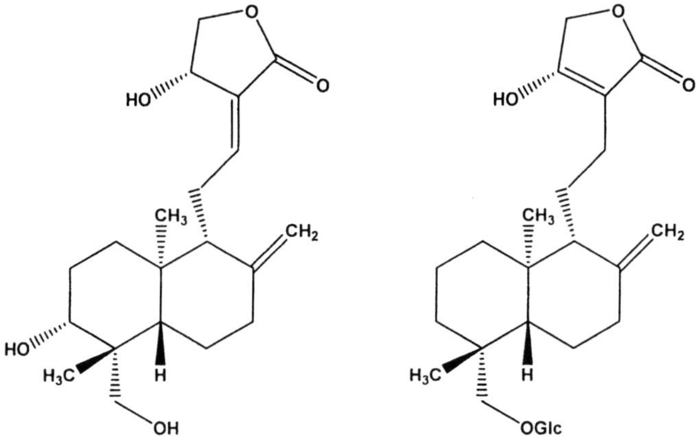

A: Andrographolide **B: Neoandrographolide**

Figure 4. Andrographolide (A) and Neoandrographolide (B) in Andrographis paniculata.

Despite of the promising *in vitro* activities of andrographolide and neoandrographolide, more research has to be done in order to elucidate their action mechanism profile, such as their effects on pro-inflammatory cytokines, transcription factors, etc.

Tripterygium wilfordii Hook f.

Tripterygium wilfordii Hook f. is a vine plant originating in South China and Taiwan, where it is known as *Lei Gong Teng*. It is used as an herbal remedy for inflammation, proteinuric renal disease, autoimmune diseases, etc. Though effective in combating diseases, its toxic reactions, especially associated with long-term use, has also been documented.

Recently, the anti-inflammatory effect was confirmed in numerous *in vitro* and *in vivo* models and in human clinical study. Extracts of *Tripterygium wilfordii Hook f.* significantly inhibited immune responses in all immunocytes, including pro-inflammatory cytokine (IL-1, IL6, IL-8, TNFα, PGE2) secretion from monocytes, IgG secretion from B cells, both cytokine (IL-2, IL-4) secretion and phargocytosis of bacteria from lymphocytes (*38-40*). Li *et al.* (*41*) demonstrated the potent suppression effects on COX-1, COX-2 and 5-LO enzymes *in vitro*, which indicates that it is a nonselective COX inhibitor. Its *in vivo* action were conformed by differenct animal models, such as carrageenan-induced inflammation rats model, acute and subacute inflammation adjuvant arthritis, and type IV and III allergic reaction (*42,43*). Tao et al proved the efficacy of ethanol/ethyl acetate extract on human rheumatoid arthritis in a double-blind, placebo-controlled clinical trial, and its tolerated dosage level were documented in this study (*44*).

Unlike other herbs mentioned above, mechanistic studies and pathway investigation were carried out on the herbal extract itself. Reports so far suggest its capability of modulating COX-2, pro-inflammatory cytokines, intercellular adhesion molecules and Matrix Metalloproteinase. At the transcription level, both NF-κB and AP-1 were inhibited by the herbal extract. The suppression effect on PGE2 release was investigated *in vitro*, and found to be due to suppression of COX-2 mRNA (*40,45*), where NF-κB but not AP-1 was suggested to be involved (*45*). Pro-inflammatory cytokines production such as IL-1, TNFα, IL-6 and IL-8 are suppressed by treatment with extract of the herb. (*46*). Additionally, Intercellular adhesion molecules such as ICAM-1 and VCAM-1 were shown to be the molecular targets of the herb extract, since both secretion and expression of these molecules were inhibited *in vitro* (*47*). Subsequent research showed the herb extract completely inhibited both mRNA and protein expression of Matrix Metalloproteinase MMP-3 and MMP-13 induced by pro-inflammatory cytokines *in vitro*. Mechanistic studies revealed that this action is associated with partially inhibited DNA binding activity of AP-1 and NFκB (*48*) although the exact molecular targets for those 2 transcription factors still remain evasive.

A group of terpenoids such as triptonide, troptolide tropdiolide, tritplidenol, 16-hydroxyl-triptolide, tripchlorolide, triptriolide, tripdioltonide, and 13,14-epoxide-9,11,12-trihydroxyltriptolide are isolated from *Tripterygium wilfordii Hook f.* and believed to be the major effective anti-inflammatory agents (*49-53*). Among those terpenoids, triptolide and tripdiolide (Figure 5) (*54,55*), two diterpenoid triepoxides, were most extensively studied.

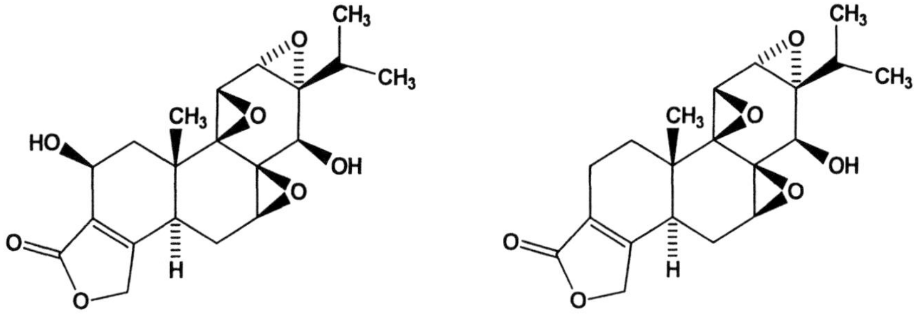

Figure 5. Triptolide (A) and Trotodiolide (B) in Tripterygium wilfordii Hook f.

Triptolide inhibits PGE2 production by selectively suppressing the gene expression and production of COX-2 over COX-1 in human synovial cells and mouse macrophages (*56*). The same research group later demonstrated the inhibition effects of triptolide on gene expression of various pro-inflammatory cytokines, such as IL-1α, IL-1β, TNFα, and IL-6. Furthermore, in human synovial fibroblasts, both the mRNA and protein levels of proMMPs (Pro-Matrix Metalloproteinase) 1 and 3 were downregulated by treatment with triptolides, while the gene expression and production of TIMPs (tissue inhibitors of metalloproteinases) were upregulated (*56*). It is known that pro-inflammatory cytokine stimulated production of matrix metalloproteinases (MMPs) are important markers for rheumatoid arthritis. Therefore the balance and regulation of proMMPs and TIMPs can be monitored for characterization of anti-inflammatory agents. Thus, these results strongly support the anti-inflammatory activity of triptolide on rheumatoid arthritis. On transcription level, NF-κB activity and its binding activity were inhibited by triptolide in several studies (*57,58*). Recently, Jiang et al reported the inhibition on AP-1 transcriptional activity in gastric cancer cells (*59*). Despite all these exciting *in vitro* data, it is surprising to notice that there is a lack of report correlating them to *in vivo* efficacy.

Conclusion

Infection or injury triggers an immediate cascade of inflammatory signals. Various factors are released that recruit inflammatory leukocytes, such as neutrophils and macrophages. The infiltrating leukocytes are activated to express iNOS and COX-2, leading to the production of nitric oxide and prostaglandin PGE_2, respectively. The infiltrated leucocytes at sites of inflammation also produce proinflammatory cytokines such as TNF-α, IL-1β, and IL-12, which altogether worsen the situation. It is widely accepted that NF-κB is one of the master regulators and has a central role in the onset of inflammation. This transcription factor is composed of homo- or heterodimers of Rel-family proteins (p65, p50, p52 cRel, RelB), and the combination of Rel proteins might, in part, determine the pattern of gene expression. NF-κB is held in an inactive state in the cytosol by association with IκB. During inflammation, TNF-α and IL-β signaling in leukocytes activate IκB kinase (IKK), leading to the phosphorylation and degradation of IκB. Once the NF-κB is separated from IκB, it gets translocated to the nucleus and induces expression of multiple genes, including iNOS and COX-2 (Figure 6).

It seems that the attempts of targeting inflammation using Asian herbs is now undergoing a transition from experience-based application to scientific-based application. The investigation has progressed to the cellular and even molecular level. Studies concerning Asian herbs showed the anti-inflammatory effect of herbal extract in *in vitro* and *in vivo* models, some even with clinical studies.

The desire to optimize nutraceuticals, including reducing toxicity and improving effectiveness, necessitates better understanding of their active constituents. Most of the action principals are elucidated and their specific effects towards inflammatory mediators confirmed *in vitro*. Among the active constituents studied, terpenoids, flavonoids, and anthraquinones are of most interest. Their anti-inflammatory effects appear to be associated with their ability to inhibit ROS, NO, prostaglandins, cytokines, and up-stream transcription factors. However, development of effective *in vivo* models for examination of anti-inflammatory agents seems necessary, as there is a lack of effective *in vivo* models compared to the variety of *in vitro* models. Furthermore, for most constituents, upon the establishment of toxicity profile, human trials should be envisioned.

It is worthwhile to mention that much of the attention has been focused on efficacy of the herbs. Numerous data has been obtained on mode of action and their active constituents. However, very few information regarding the metabolic fate of the herbs and their constituents has been reported. Essentially, the development of safe and effective anti-inflammatory neutraceuticals necessitates an in-depth understanding of availability, biotransformation, and disposition of active constituents.

The activation of NFκB begins with stimulation of specific receptor families at the cell surface and recruitment of adaptor proteins, and leads to specific pathways of transduction controlled by various kinases. These pathways converge upon the IKK complex that, in turn, promotes the phosphorylation of IκB and ultimately degradation. As a consequence, the NF-κ B inhibitory protein is removed and free NF-κB is rapidly translocated to the nucleus where it binds to specific promoter regions of various genes encoding, for example, iNOS, COX-2 and inflammatory cytokines. IKK, I κB kinase; LPS, lipopolysaccharide; MEKK1, mitogen-activated protein kinase/extracellular response kinase (MAPK/ERK) kinase kinase 1; MyD88, myeloid differentiation factor; IRAK, interleukin 1 receptor associated kinase; NIK, NF-κB-inducing kinase; TLR, Toll-like receptor; TRADD, TNF-receptor-associated death domain protein; TRAF, TNF-receptor-associated factors.

References

1. Surh, Y. *Mutat. Res.* **1999**, *428(1-2)*, 305-327.
2. Surh, Y.J.; Chun, K.S.; Cha, H.H.; Han, S.S.; Keum, Y.S.; Park, K.K.; Le, S.S. *Mutat. Res.* **2001**, *480-481*, 243-268.
3. Sugimoto, K.; Hanai, H.; Tozawa, K.; Aoshi, T.; Uchijima, M.; Nagata, T.; Koide, Y. *Gastroenterology* **2002**, *123(6)*, 1912-1922.
4. Bremner, P.; Heinrich, M. *J. Pharm. Pharmacol.* **2002**, *54(4)*, 453-472.
5. Adcock, I. M. *Monaldi. Arch. Chest. Dis.* **1997**, *52(2)*, 178-186.\
6. Adderley, S. R.; Fitzgerald, D. J. *J. Biol. Chem.* **1999**, *274(8)*, 5038-5046.
7. Havsteen, B. *Biochem. Pharmacol.* **1983**, *32(7)*, 1141-1148.
8. Kitamura, K.; Honda, M.; Yoshizaki, H.; Yamamoto, S.; Nakane, H.; Fukushima, M.; Ono, K.; Tokunaga, T. *Antiviral Res.* **1998**, *37(2)*, 131-140.
9. Cuellar, M.J.; Giner, R.M.; Recio, M.C.; Manez, S.; Rios, J.L. *Fitoterapia* **2001**, *72(3)*, 221-229.
10. Ye, F.; Xui, L.; Yi, J.; Zhang, W.; Zhang, D.Y. *J. Altern. Complement. Med.* **2002**, *8(5)*, 567-572.
11. Shao, Z.H.; Li, C.Q.; Vanden Hoek, T.L.; Becker, L.B.; Schumacker, P.T.; Wu, J.A.; Attele, A.S.; Yuan, C.S. *J. Mol. Cell Cardiol.* **1999**, *31(10)*, 1885-1895.

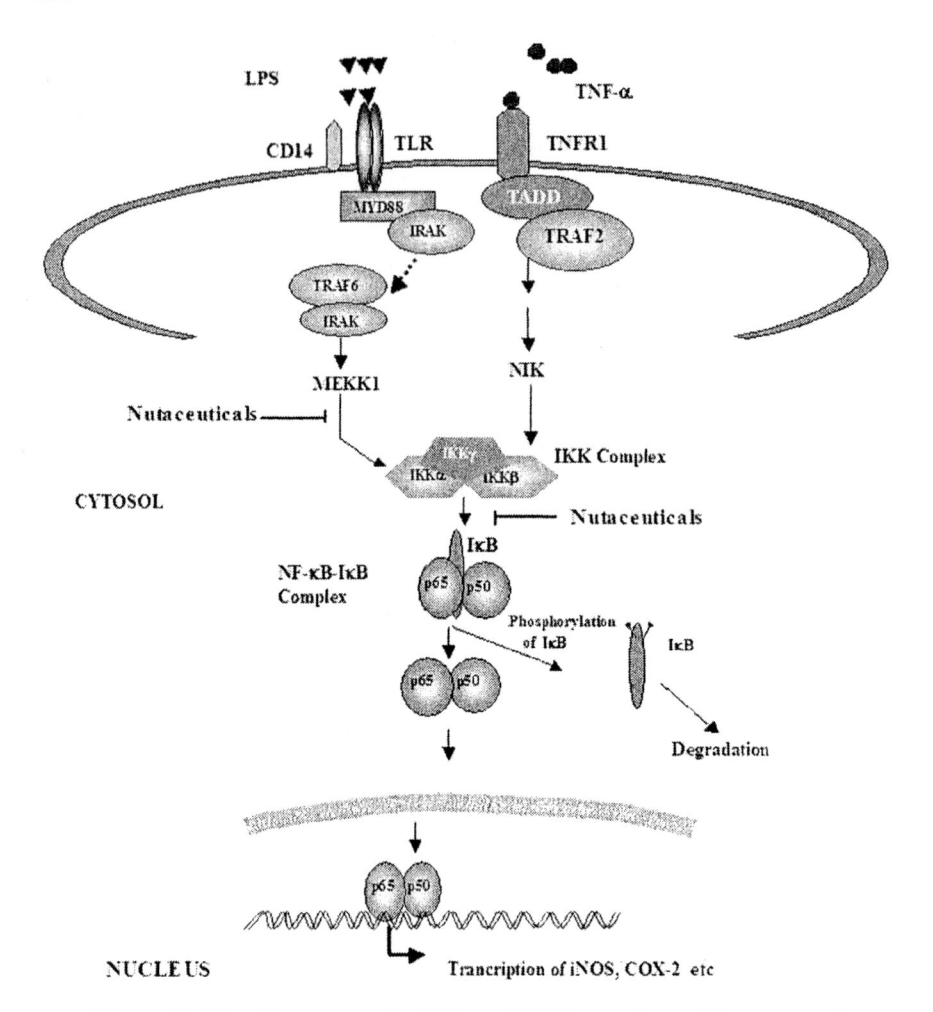

Figure 6. Signaling events for LPS and TNF-α and possible targets of Nutraceuticals.

12. Schinella, G.R.; Tournier, H.A.; Prieto, J.M.; Mordujovich, D.; Rios, J.L. *Life Sci.* **2002**, *70(9)*, 1023-1033.
13. Kimura, Y., Michinori, K., Kimiyo, T., Tani, T., Higashino, M., Arichi, S., Okuda, H., *Chem. Pharm. Bull.* **1982**, *30(1)*, 219-222.

14. Takagi, S.; Yamaki, M.; Inoue, K.; *Yakugaku Zasshi* **1980**, *100*, 1220-1224.
15. Kimura, Y.; Matsushita, N.; Okuda, H. *J. Ethnopharmacol.* **1997**, *57(1)*, 63-67.
16. Suk, K.; Lee, H.; Kang, S.S.; Cho, G.J.; Choi, W.S. *J. Pharmacol. Exp. Ther.* **2003**, *305(2)*, 638-645.
17. Wakabayashi, I. *Pharmacol. Toxicol.* **1999**, *84(6)*, 288-291.
18. Wakabayashi, I.; Yasui, K. *Eur. J. Pharmacol.* **2000**,*.406(3)*, 477-481.
19. Kim, H.; Kim, Y.S.; Kim, S.Y.; Suk, K. *Neurosci. Lett.* **2001**, *309(1)*, 67-71.
20. Chang, Y.L.; Shen, J.J.; Wung, B.S.; Cheng, J.J.; Wang, D.L. *Mol. Pharmacol.* **2001**, *60(3)*, 507-513.
21. Li, B.Q.; Fu, T.; Gong, W.H.; Dunlop, N.; Kung, H.; Yan, Y.; Kang, J.; Wang, J.M. *Immunopharmacology* **2000**, *49(3)*, 295-306.
22. Krakauer, T.; Li, B.Q.; Young, H.A. *FEBS Lett.* **2001**, *500(1-2)*, 52-55.
23. Chen, Y.; Yang, L.; Lee, T.J. *Biochem. Pharmacol.* **2000**, *59(11)*, 1445-1457.
24. Song, H.; Wang, Z.; Zhang, F. *Zhongguo Zhong Xi Yi Jie He Za Zhi* **2000**, *20(2)*, 107-109.
25. Wang, C.C.; Huang, Y.J.; Chen, L.G.; Lee, L.T.; Yang, L.L. *Planta Med.* **2002**, *68(10)*, 869-874.
26. Kumar, A.; Dhawan, S.; Aggarwal, B.B. *Oncogene* **1998**, *17(7)*, 913-918.
27. Kuo, Y.C.; Sun, C.M.; Tsai, W.J.; Ou, J.C.; Chen, W.P.; Lin, C.Y. *J. Lab. Clin. Med.* **1998**, *132(1)*, 76-85.
28. Kuo, Y.C.; Sun, C.M.; Ou, J.C.; Tsai, W.J. *Life Sci.* **1997**, *61(23)*, 2335-2344.
29. Kuo, Y.C.; Tsai, W.J.; Meng, H.C.; Chen, W.P.; Yang, L.Y.; Lin, C.Y. *Life Sci.* **2001**, *68(11)*, 1271-1286.
30. Kuo, Y.C.; Meng, H.C.; Tsai, W.J. *Inflamm. Res.* **2001**, *50(2)*, 73-82.
31. Madav, S.; Tandan S. K.; Lal, J., Tripathi, H.C. *Fitoterapia* **1996**, *67*, 452-458.
32. Gabrielian, E.S.; Shukarian, A.K.; Goukasova, G.I.; Chandanian, G.L.; Panossian, A.G.; Wikman, G.; Wagner, H. *Phytomedicine* **2002**, *9(7)*, 589-597.
33. Chiou, W.F.; Lin, J.J.; Chen, C.F. *Br. J. Pharmacol.* **1998**, *125(2)*, 327-334.
34. Batkhuu, J.; Hattori, K.; Takano, F.; Fushiya, S.; Oshiman, K.; Fujimiya, Y. *Biol. Pharm. Bull.* **2002**, *25(9)*, 1169-1174.
35. Shen, Y.C.; Chen, C.F.; Chiou, W.F. *Planta Med.* **2000**, *66(4)*, 314-317.
36. Shen, Y.C.; Chen, C.F.; Chiou, W.F. *Br. J. Pharmacol.* **2002**, *135(2)*, 399-406.
37. Habtemariam, S. *Phytotherapy Res.* **1998**, *12(1)*, 37-40.
38. Tao, X.L. *Zhongguo Yi Xue Ke Xue Yuan Xue Bao* **1989**, *11(1)*, 36-40.

39. Chang, D.M.; Chang, W.Y.; Kuo, S.Y.; Chang, M.L. *J. Rheumatol.* **1997**, *24(3)*, 436-441.
40. Tao, X.; Schulze-Koops, H.; Ma, L.; Cai, J.; Mao, Y.; Lipsky, P.E. *Arthritis Rheum.* **1998**, *41(1)*, 130-138.
41. Li, R.W.; Lin, G.D.; Myers, S.P.; Leach, D.N. *J. Ethnopharmacol.* **2003**, *85(1)*, 61-67.
42. Tao, X.; Ma, L.; Mao, Y.; Lipsky, P.E. *Inflamm. Res.* **1999**, *48(3)*, 139-148.
43. Li, L.F. *J. Dermatol.* **2000**, *27(7)*, 478-81.
44. Tao, X.; Younger, J.; Fan, F.Z.; Wang, B.; Lipsky, P.E. *Arthritis Rheum.* **2002**, *46(7)*, 1735-1743.
45. Maekawa, K.; Yoshikawa, N.; Du, J.; Nishida, S.; Kitasato, H.; Okamoto, K.; Tanaka, H.; Mizushima, Y.; Kawai, S. *Inflamm. Res.* **1999**, *48(11)*, 575-581.
46. Chang, D.M.; Chang, W.Y.; Kuo, S.Y.; Chang, M.L. *J. Rheumatol.* **1997**, *24(3)*, 436-441.
47. Chang, D.M.; Kuo, S.Y.; Lai, J.H.; Chang, M.L. *Ann. Rheum. Dis.* **1999**, *58(6)*, 366-371.
48. Sylvester, J.; Liacini, A.; Li, W.Q.; Dehnade, F.; Zafarullah, M. *Mol. Pharmacol.* **2001**, *59(5)*, 1196-1205.
49. Lu, X. *Zhongguo Yi Xue Ke Xue Yuan Xue Bao* **1990**, *12(3)*, 157-161.
50. Zhang, L.X.; Yu, F.K.; Zheng, Q.Y.; Fang, Z.; Pan, D. *J. Yao Xue Xue Bao* **1990**, *25(8)*, 573-7.
51. Ma, P.C.; Lu, X.Y.; Yang, J.J.; Zheng, Q.T. *Yao Xue Xue Bao* **1991**, *26(10)*, 759-63.
52. Zhang, D.M.; Yu, D.Q.; Xie, F.Z. *Yao Xue Xue Bao* **1991**, *26(5)*, 341-344.
53. Zhang, C.P.; Lu, X.Y.; Ma, P.C.; Chen, Y.; Zhang, Y.G.; Yan, Z.; Chen, G.F.; Zheng, Q.T.; He, C.H.; Yu, D.Q. *Yao Xue Xue Bao* **1993**, *28(2)*, 110-115.
54. Kupchan, S.M.; Court, W.A.; Dailey, Jr., R.G.; Gilmore, C.J.; Bryan, R.F. *J. Am. Chem. Soc.* **1972**, *94(2*0), 7194-7195.
55. Gu, W.Z.; Chen, R.; Brandwein, S.; McAlpine, J.; Burres, N. *Int. J. Immunopharmacol.* **1995**, *17(5)*, 351-356.
56. Lin, N.; Sato, T.; Ito, A. *Arthritis Rheum.* **2001**, *44(9)*, 2193-2200.
57. Lee, K.Y.; Chang, W.; Qiu, D.; Kao, P.N.; Rosen, G.D. *J. Biol. Chem.* **1999**, *274(19)*, 13451-13455.
58. Zhao, G.; Vaszar, L.T.; Qiu, D.; Shi, L.; Kao, P.N. Am. *J. Physiol. Lung Cell Mol. Physiol.* **2000**, *279(5)*, L958-966.
59. Jiang, X.H.; Wong, B.C.; Lin, M.C.; Zhu, G.H.; Kung, H.F.; Jiang, S.H.; Yang, D.; Lam, S.K. *Oncogene* **2001**, *20(55)*, 8009-8018.

Chapter 21

Induction of Apoptosis by *Ligusticum chuanxiong* in HSC-T6 Stellate Cells

Yun-Lian Lin[1], Ting-Fang Lee[2], Young-Ji Shiao[1], and Yi-Tsau Huang[2]

[1]National Research Institute of Chinese Medicine and [2]Institute of Traditional Medicine, National Yang-Ming University, Taipei, Taiwan

Activation of hepatic stellate cells (HSCs) is a key feature of liver fibrosis. During the development of hepatic fibrosis, activated HSCs are the primary source of extracellular matrix. Induction of HSC apoptosis has been proposed as one of therapeutic strategies for this desease. *Ligusticum chuanxiong* (LC) is a traditional Chinese herb that has been used in the treatment of cardiovascular diseases and to facilitate blood circulation. The present study showed that LC attenuated HSC-T6 growth. By MTT cell viability and lactate dehydrogenase release assay, LC reduced the cell viability in a dose-dependent manner and without significant LDH release. The apoptotic features were observed by cell arrested in S phase, the appearance of a sub-G1 peak and apoptotic cells. The induction of apoptosis by LC was through the activation of caspase-3 and inducing expression of the cell cycle inhibitory proteins, p21 and p27, with no direct cytotoxicity on primary rat hepatocytes. However, tetramethylpyrazine, an active principle of LC had no effect in this studies. The results indicate that LC had a selective effect on HSCs and induce HSC apoptosis, thereby minimizing fibrogenesis. These provide the theoretical basis for LC used to treat and prevent hepatic fibrogenesis, but the active components of LC need to be explored further.

© 2006 American Chemical Society

Hepatic fibrosis is a wound-healing response to various chronic liver injuries (1,2). Stellate cells play a crucial role in the development of liver fibrosis (3,4). As a result of this process, hepatic stellate cells undergo transformation from vitamin A-storing quiescent cells to myofibroblast-like activated cells (5). Activated HSCs exhibit the expression of highly proliferative activity and the accumulation of extracellular matrix (ECM), including the appearance of α-smooth muscle actin (α-SMA) and elevation in type I collagen, the main pathologic feature of hepatic fibrosis (6,7). Suppression of HSC activation and proliferation, and inducing HSC apoptosis have been proposed as one of the therapeutic strategies for the treatment and prevention of hepatic fibrosis (8). Plant-derived antioxidants may emerge as potential anti-fibrotic agents by either protecting hepatocytes against ROS or inhibiting the activation of hepatic stellate cells (9), such as resveratrol (10), curcumin (11), salvianolic acid B (12), epigallocatechin-3-gallate (EGCG) (13-15). Chinese herbal medicine by facilitating blood circulation and dispersing blood stasis has recently attracted much attention for preventing oxidative stress-related diseases including cancers, cardiovascular diseases and degenerative diseases and opened a new route for anti-hepatic fibrosis drug development (16,17).

Ligusticum chuanxiong Hort (LC), a traditional Chinese herb, has been widely used to treat irregular menstrual cycles, cardiovascular diseases and facilitate blood circulation (18) and is claimed to induce vasodilation. Several reports indicated the beneficial effects of *LC*. Tetramethylpyrazine, an active principle of LC, has been shown as a vasodilator (19), a potassium channel opener to lower calcium influx into cultured aortic smooth muscle cells (20), attenuator of iron-induced oxidative damage and apoptosis in cerebellar granule cells (21), scavenger of superoxide anion and decrease nitric oxide production in human poly-morphonuclear leukocytes (22), antiplatelet aggregation (23), portal hypotensive effect (24) and hepatoprotective and therapeutic effect on econazole-induced liver injury (25). However, the effect of LC on liver fibrosis is still not mentioned.

The aim of this study is to elucidate whether LC has any effect on stellate cells. HSC-T6 cells which are immortalized rat liver stellate cells (26) were used to investigate the mechanism involved. The ethanolic extract of the slices of Rhizoma *L. chuanxiong* (LC) was prepared for this study. Cell cycle was analyzed by flow cytometry and apoptotic cells was investigated by TUNEL-staining. Caspase-3 activity was determined by DEVD-*p*-nitroanilide (pNA) substrate. Western blotting was used for detecting cell cycle proteins. In the present study, we demonstrated, for the first time, that LC induced apoptosis in activated stellate cells, cell cycle arrest at S phase by inducing the expression of cell cycle-dependent kinase inhibitor, p21 and p27, and the activation of caspase-3 *in vitro*.

Effect of LC on HSC-T6 Cell Growth

The doubling time of HSC-T6 cells was about 24 h in FBS-containing medium. To evaluate the effect of LC on HSC-T6 growth, preconfluent cells were treated with different concentrations of LC extract in serum free medium for 24 h and 48 h. Cell viability was determined by MTT reduction analysis. As shown in Figure 1, treatment of HSC-T6 with LC decreased cell viability in a dose-dependent manner with an IC$_{50}$ of 233± 7 µg/mL and 257 ± 5 µg/mL for 24 h and 48 h., respectively. The appearance of cell morphology (not shown) exhibits shrinkage with few extracellular matrix (ECM) production and lasted throughout the whole period of treatment. LC cytotoxicity to HSC-T6 cells was also measured as a percentage of lactate dehydrogenase release (LDH) after 24 h treatment (Table I). Compared with control, no significant difference of LDH leakage was found up to a

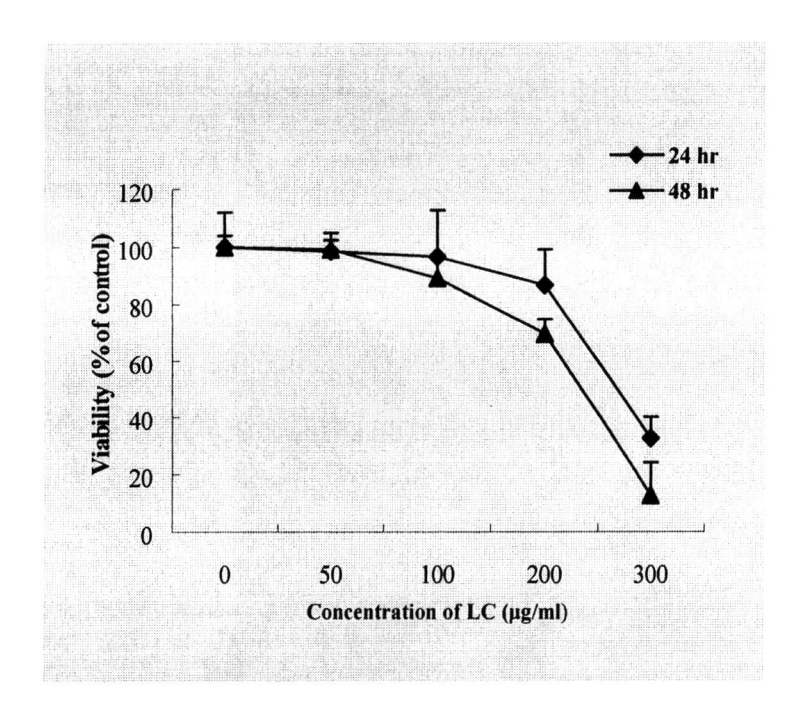

Figure 1. The dosage response of LC in HSC-T6 by MTT assay. 1 x 10^5 HSC-T6 cells were treated with or without various concentrations of LC for 24 and 48 h. MTT reduction assay was performed for cell viability. Each point represents the mean of three independent experiments.

Table I. Lactate Dehydrogenase Release in Cultured HSC-T6 Treated with LC

	\multicolumn{4}{c}{LC (µg/mL)}
	0	50	100	200
Total LDH (%)	31 ± 3	35 ± 3	32 ± 1	33 ± 2

Values are means ± SEM; n=3. Lactate dehydrogenase was calculated as medium LDH/(medium LDH+cellular LDH)

concentration of 200 µg/mL. On the basis of these observations, LC at 200 µg/mL and 24 h treatment was chosen to conduct the following experiments.

Induction of HSC-T6 Apoptosis and the Expression of p21 and p27 by LC

Cytoplasmic and nuclear shrinkage, chromatin condensation, membrane blebbing, internucleosomal degradation of DNA and apoptotic body formation, are characteristic morphological changes of apoptosis (27). Morphological assessment by microscope as well as TUNEL staining and flow cytometric sub-G1 cell analysis were further used to evaluate the effect of LC on HSC-T6 survival. As shown in Figure 2, compared with control, LC significantly increased the number of positive TUNEL-staining of nuclei with fluorescein dye, suggesting an increase in apoptotic cells at 24 h after LC treatment. No significant changes of LDH leakage were previously found in a dose- and time-course studies, suggesting that LC induced DNA damage. LC-induced apoptosis in HSC-T6 cells was also evaluated by sub-G1 cell analysis with propidium iodide (PI) by flow cytometry. The profile of DNA content was obtained by measuring the fluorescence of PI binding to DNA (Figure 2). Cells with lower DNA staining than that of diploid cells were considered as apoptoic. There was an accumulation of subploid population or so-called "sub-G_1 peak after LC-treated cells, when compared with the untreated groups, up to 10% and 22% of total cell counts at 24 h and 48 h, respectively. In addition, LC enhanced the expression of cell cycle inhibitory protein, including p21 and p27 (Figure 3) by about 3.0 and 2.2 fold, respectively. The above results suggest that the reduction of viability after LC treatment might result from apoptosis in HSC-T6 cells.

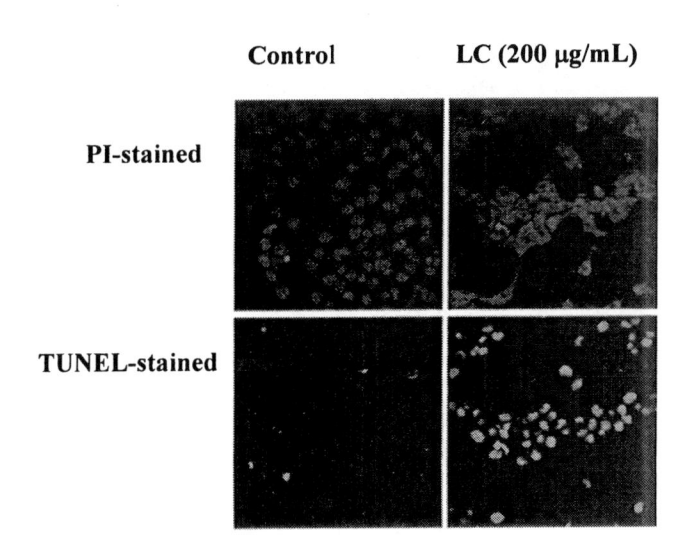

Figure 2. LC induces HSC-T6 apoptosis. Preconfluent cells were treated with or without 200 µg/ml of LC for 24 h. Cells were stained as the protocol provided by the TUNEL kit manufacturer.

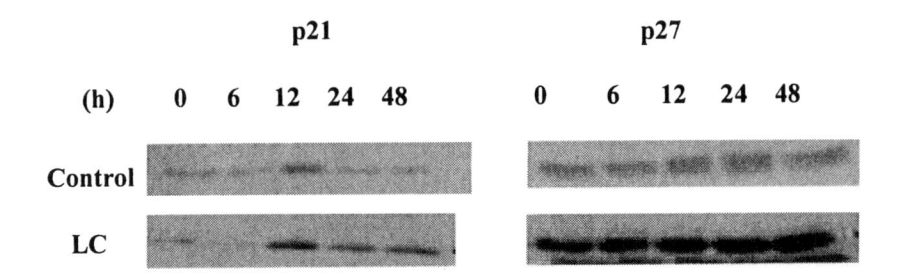

Figure 3. LC alters the expression of cell cycle related proteins in HSC-T6 cells. Cells were treated with or without 200 µg/ml of LC for the indicated time. P21 and p27 were detected by Western blot with antibodies. Each result represented three independent western blotting analyses.

Effects of LC on the Activation of Caspase-3 in HSC-T6 Cells

Caspase-3 is an executive enzyme for cell apoptosis that functions at the late stages of protease casecade. Caspase-3 activity assay was measured by colorimetric assay using acetyl-Asp-Glu-Val-Asp-pNA (Ac-DEVD-pNA) as the specific substrate. Inhibitor of caspase-3 was employed to investigate whether apoptosis was involved in LC-mediated cell death. Ac-DEVD-CHO is the cell-permeable inhibitor for caspase-3. The inhibitor block the activity of caspase-3 but did not interfere with its activation. It was used to rule out non-specific protease activity. Treatment of LC for 24 h, the specific activity of caspase-3 was increased by 2.2 fold of control. Compared with control, the inhibitor of caspase-3 completely abolished the induction of caspase-3 activity (Figure 3). Curcumin was used as a positive control, has been demonstrated as an inducer of HSC apoptosis (*11*), increased caspase-3 activity by about 3.0 fold at 30 μM.

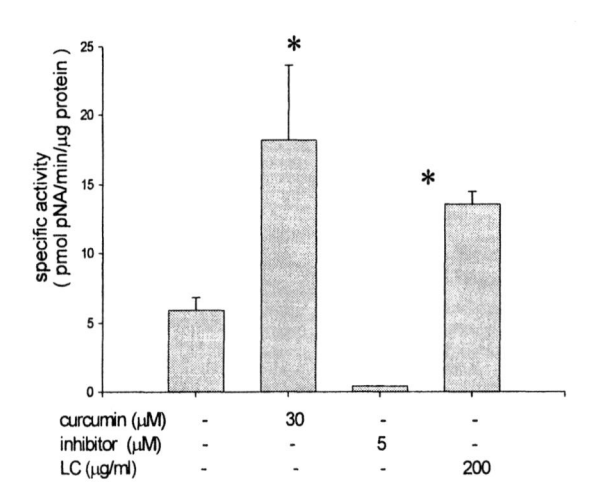

Figure 3. LC induced the activation of Caspase-3 in HSC-T6 cells. Cells were treated with 200 μg/mL of LC for 24 h. Activity of caspase -3 was determined by the cleavage ability of Ac-DEVD-pNA in cell extracts. Results are means ± SEM from three independent experiments. Significant differences between control and drug treated group. *, p<0.05.

Discussion

Activated HSCs are the most relevant cell type responsible for the excessive production and deposition of ECM during the development of hepatic fibrosis (6,7). Therefore, suppression of HSC activation and inducing HSC apoptosis have become a therapeutic strategy in both the treatment and prevention of the hepatic fibrosis (8).

Rhizoma L. chuanxiong (LC) is a popular Chinese herbal medicine, with unique properties of (a) no known adverse effect; (b) low cost; and (c) a long use history in Chinese medicine (18). Tetramethylpyrazine (TMP), an active component of LC has been reported to have cardiovascular effects (19-23). Chang et al. demonstrated its portal hypotensive effect (24). Furthermore, the hepatoprotective and therapeutic effect on econazole-induced liver injury has been investigated by Liu et al (25). However, the effect of LC on liver fibrosis is still not mentioned. In this study, we demonstrated that LC inhibited TSC-T6 growth via an apoptotic mechanism. As evidenced by morphological alterations, characteristic in TUNEL staining (Figure 2) and cell cycle analysis apoptosis is mainly responsible for the effects of LC on inhibition of HSC-T6 growth. But, TMP has no effect in this study even up to 400 µM (data not shown).

Resveratrol inhibits cell growth and induces apoptosis of different cancer cell line by the involvement of S-phase arrest via cyclin A, B1, and β-catenin biomarker reduction (10,28). The E2F transcription factors activate a number of genes responsible for DNA replication and S phase progression (29). Constitutive expression of E2F-1 induces endogenous p21 expression and that p21 is responsible for E2F-1-dependent S phase cell cycle arrest (30). p21 is a potent inhibitor of cyclin-dependent kinase arresting cell cycle via E2F-1 direct interaction with the proximal region of the p21 promoter. p27 binding inhibits the CDK catalytic activity fot its interactions with the N-terminal β-sheet of cdk2 induce conformational changes (31). Overexpression of p27 causes G1 arrest (32). In this study, we found that LC treatment significantly increased p21 and p27 expression to about 4.0 and 2.0 fold, respectively, thereby arresting the cell cycle at S phase.

Apoptosis has been known for mediating HSCs loss during recovery from fibrosis, and control of apoptosis may be a key for regulating fibrosis (31). Yoon and Gores proposed selective induction of apoptosis in activated hepatic stellate cells and differential prevention infected hepatocytes from apoptosis could therapeutically be useful for an ideal antifibrotic therapeutic strategy (8). In this study, we demonstrated LC can be selectively effect activated hepatic stellate cells with no direct cytotoxicity on primary rat hepatocytes (data not shown).

In summary, the present study demonstrated that LC has marked effect on HSC-T6 stellate cells in vitro. It inhibits cell growth and induces cell death by apoptosis through enhanced p21 and p27 expression, cell cycle arrest and caspase-3 activation. These provide the theoretical basis for LC used to treat and prevent hepatic fibrogenesis. But the active components of LC remain to be explored.

Acknowledgements

We gratefully acknowledge the kind provision of HSC-T6 cells by Dr Scott L. Friedman, Division of Liver Diseases, The Mount Sinai School of Medicine, New York. This work was supported in part by a grant of National Science Council of The Republic of China, Taiwan (NSC 92-2320-077-006 and NSC 93-2914-1-077-001-A1).

References

1. Olaso, E.; Friedman, S. L. *J. Hepatol.* **1998**, *29*, 836-847.
2. Friedman, S. L. *J. Biol. Chem.* **2000**, *275*, 2247-2250.
3. Friedman, S. L. *New Eng. J. Med.* **1993**, *328*, 1828-1835.
4. Gressner, A. M. *J. Hepatol* **1995**, *22* (Suppl. 2), 28-36.
5. Friedman, S. L.; Roll, F. J.; Boyles, J.; Bissell, D. M. *Proc. Nat. Acad. Sci. (USA)* **1985**, *82*, 8681-8685.
6. Bataller, R.; Brenner, D. A. *Seminars in Liver Disease* **2001**, *21*, 437-451.
7. Weiner, F. R.; Giambrone, M. A.; Czaja, M. J.; Shah, A.; Annoni, G.; Takahashi, S.; Eghbali, M.; Zern, M. A. *Hepatology* **1990**, *11*, 111-117.
8. Yoon, J. H.; Gores, G. J. *J. Hepatol.* **2002**, *37*, 400-410.
9. Gebhardt, R. *Planta Med.* **2002**, *68*, 289-296.
10. Joe, A. K.; Liu, H.; Suzui, M.; Vural, M. E.; Xiao, D.; Weinstein, I. B. *Clin. Cancer Res.* **2002**, *8*, 893-903.
11. Xu, J.; Fu, Y.; Chen, A. *Am. J. Physiol. Gastrointest. Liver Physiol.* **2003**, *285*, G20-G30.
12. Liu, P.; Hu, Y. Y.; Liu, C.; Zhu, D. Y.; Xue, H. M.; Xu, Z. Q.; Xu, L. M.; Liu, C. H.; Gu, H. T.; Zhang, Z. Q. *World J. Gastroenterol.* **2002**, *8*, 679-685.
13. Chen, A.; Zhang, L.; Xu, J.; Tang, J. *Biochem. J.* **2002**, *368*, 695-704.
14. Chen, A.; Zhang, L. *J. Biol. Chem.* **2003**, *278*, 23381-23389.
15. Sakata, R.; Ueno, T.; Nakamura, T.; Sakamoto, M.; Torimura, T.; Sata, M. *J. Hepatol.* **2004**, *40*, 52-59.
16. Yan, J. H. *Chinese Traditional Drugs*; Chih-Yin Publishing, Taipei, Taiwan, **1991**, pp 456
17. Wasser, S.; Sian Ho, J. M.; Ang, H. K.; Tan, C. E. L. *J. Hepatol.* **1998**, *29*, 760-771.
18. Xiao, P. G. *Modern Chinese Materia Medica*; Chemical Industry Press: Beijing, China, **2002**; Vol. 1, pp 121.
19. Tsai, C. C.; Lai, T. Y.; Huang, W. C.; Yang, T.; Liu, I. M.; Wong, K. L.; Chan, P.; Cheng, J. T. *Planta Med* **2003**, *69*, 557-558.
20. Wong, K. L.; Chan, P.; Huang, W. C.; Yang, T. L.; Liu, I. M.; Lai, T. Y.; Tsai, C. C.; Chang, J. T. *Clin. Exp. Pharmacol. Physiol.* **2003**, *30*, 793-798.
21. Zhang, Z.; Wei, T.; Hou, J.; Li, G.; Yu, S.; Xin, W. *Eur. J. Pharmacol.* **2003**, *467*, 41-47.

22. Zhang, Z.; Wei, T.; Hou, J.; Li, G.; Yu, S.; Xin, W. *Life Sci.* **2003**, *72*, 2465-2472.
23. Sheu, J. R.; Kan, Y. C.; Hung, W. C.; Lin, C. H.; Yen, M. H. *Life Sci.* **2000**, *67*, 937-947.
24. Chang, F. C.; Chen, K. J.; Lin, J. G.; Hong, C. Y.; Huang, Y. T. *Pharm. Pharmacol.* **1998**, *50*, 881-884.
25. Liu, C. F.; Lin, C. C.; Ng, L. T.; Lin, S. C. *Planta Med.* **2002**, *68*, 510-514.
26. Vogel, S.; Piantedosi, R.; Frank, J.; Lalazar, A.; Rockey, D. C.; Friedman, S. L.; Blaner, W. S. *J. Lipid Res.* **2000**, *41*, 882-892.
27. Cohen, J. J. *Immunol. Toda.* **1993**, *14*, 126-130.
28. Radhakrishnan, S. K.; Feliciano, C. S.; Najmabadi, F.; Haegebarth, A.; Kandel, E. S.; Tyner, A. L.; Gaartel, A. L. *Oncogene*, **2004**, 1-4.
29. Nevins, J. R. *Hum. Mol. Genet.* **2001**, *10,* 699-703.
30. Ragione, F. D.; Cucciolla, V.; Borriello, A.; Pietra, V. D.; Racioppi, L.; Soldati, G.; Manna, C.; Galletti, P.; Zappia, V. *Biochem. Biophys. Res. Commun.* **1998**, *250*, 53-58.
31. Russo, A. A.; Jeffrey, P. D.; Patten, A. K.; Massague, J.; Pavletich, N. P. *Nature* **1996**, *382*, 325-331.
32. Toyoshima, H.; Hunter, T. *Cell* **1994**, *78*, 67-74.

Chapter 22

Effect of Combined Use of Isothiocyanate and Black Tea Extract on Dental Caries

Hideki Masuda, Saori Hirooka, and Toshio Inoue

Material R&D Laboratories, Ogawa & Company, Ltd., 15-7, Chidori, Urayasushi, Chiba 279-0032, Japan

Isothiocyanates, the main volatile components in wasabi and horseradish, have been reported to have an inhibitory effect on dental caries. On the other hand, the extracts of tea, such as green tea, oolong tea, and black tea, are well-known to have an anticaries activity. The polyphenols in tea extracts are considered to play an important role in the anticaries activity. Therefore, the combined use of isothiocyanate and tea polyphenols can be expected to have a stronger anticaries activity compared to that of their single use due to a different mechanism of action. In this study, the combination of isothiocyanate and black tea extract was done *in vitro* and *in vivo*, and analyzed using an isobologram.

Introduction

Isothiocyanates, the volatile components in wasabi and horseradish, are well-known to have a variety of biological functions, such as an antimicrobial, anti-platelet aggregatory, anticancer, anthelmintic, and antiasthmatic activities (*1-13*). As for the anticaries activity, the isothiocyanates are expected to be useful as anticaries because of their high antimicrobial activity. The pathway of formation of dental caries is as follows: the mutans streptococci, sucrose, and glucosyltransferases (GTases) which are released from mutans streptococci, form a water-insoluble glucan (*14,15*). The mutans streptococci then multiply in the water-insoluble glucan and produce organic acids, such as lactic acid, by the glycolytic pathway. Finally, the tooth enamel is dissolved by the organic acid that lead to the dental caries. Recently, isothiocyanates have been reported to show the anticaries activity *in vitro* and *in vivo* (*16*). On the other hand, the anticaries activity of green tea extract (*17*), oolong tea (*18-20*), black tea (*21-23*), were reported in detail. In addition, the anti-caries activity of the tea polyphenols, such as catechins and their polymers has also been studied (*24-31*). In this study, we focused on the combined use of isothiocyanates and black tea extract for the following reason: 1) The main mechanism for anticaries by the isothiocyanates is considered to be different from that of the tea extract (*16*). That is to say, the significant antibacterial activity of the isothiocyanates for mutans streptococci and the high inhibition of tea extract for GTases. 2) Taking into account the pungency of the isothiocyanates and the astringency of tea polyphenol, the smaller the dose, the better the practical value.

Experimental

Materials: 5-Methylthiopentyl isothiocyanate was obtained from 4-pentenyl thiocyanate. (*32,33*). The black tea extract (13 g) was obtained from dry leaf in China (50 g) by extraction with 50% ethanol for 24 h at room temperature, followed by filtration, evaporation, and freeze-drying.

Animals and Treatment: Twenty-day-old male Wistar rats were purchased from Shimizu Laboratory Supplies (Kyoto, Japan). The animals were kept in stainless steel cages and housed in an air-conditioned room maintained at 24 ± 2 °C and at a humidity of 55 ± 15 %. All animals were weaned after 20 days from birth. After weaning, the rats were fed a normal diet containing tetracycline (4 mg/g) and drinking water containing penicillin G (4000 unit/L) ad libitum for two days. Three days later, all animals were randomly distributed into three groups. The control: The rats were infected with *Streptococcus sobrinus* (*S. sobrinus*) 6715 and fed the Diet 2000 throughout the experiment period. 5-Methylthiopentyl isothiocyanate and the black tea extract treatment groups: The rats infected with *S. sobrinus* 6715 were fed the Diet 2000 containing 200 ppm of 5-methylthiopentyl isothiocyanate, 2000 ppm of black tea extract, and the combined use of 200 ppm of 5-methylthiopentyl isothiocyanate and 2000 ppm of black tea extract. Diet 2000 was purchased from CLEA Japan, Inc. (Tokyo, Japan). Diet 2000: Wheat powder (6 g), sucrose (56 g), skim milk powder (28 g), the leaf powder of alfalfa (3 g), leber (1 g), beer yeast (4 g), and sodium chloride containing iodine (2 g). Fifty-five days later, the rats were killed and their jaws removed.

Data and Statistical Analysis: The caries score was calculated at 24 fissures of the upper and lower molars (*34*). All data are presented as means ± S.E. The statistical analysis was performed using the Kruskal-Wallis and Steel test. A probability value of less than 1 % was considered significant.

Inhibitory Effect on Sucrose Dependent Adherence by Growing Cells of *S. sobrinus* 6715: BHI broth (4.9 mL) containing *S. sobrinus* 6715 and 1 % sucrose was pipetted into the test tube. *S. sobrinus* 6715 was obtained from the Institute of Physical and Chemical Research (RIKEN), Saitama, Japan. An N,N-dimethylformamide (50 µL) solution of isothiocyanate, the black tea extract, or both of isothiocyanate and black tea extract was added to the test tube, and vigorously mixed. The content of the test tube (200 µL) was pipetted into the microplate followed by sealing with cap. The microplate was allowed to stand for 10 h at 37 °C in the anaerobic jar. The BHI broth in the microplate was rinsed by distilled water (twice). The microplate was soaked in 50 % ethanol overnight to remove the BHI broth and the black tea extract. After drying the microplate, a 1N NaOH aqueous solution (150 µL) was added to it. The microplate was shaken for more than 6 hours. The microplate was neutralized by a 1N HCl aqueous solution. The solution was diluted with distilled water and its diluted solution (250 µL) was added to the test tube. Two hundred and fifty microliters of a 5 % phenol aqueous solution was added and vigorously mixed. Concentrated sulfuric acid (1.25 mL) was dropwise added to it and allowed to stand for 20 min. The transmittance at 490 nm was measured using a Bausch & Lomb Spectronic 20 spectrophotometer.

Data and Statistical Analysis: All tests were run four times. The calculated inhibition value of sucrose dependent adherence by the growing cells of *S. sobrinus* 6715 for each sample was obtained by a probit analysis with SPSS from the observed inhibition value. The calculated inhibition values obtained by the probit analysis are presented as means ± S.E.

Results and Discussion

Taking into account the different mechanism of the isothiocyanate and tea extract, a more effective inhibition effect against the dental caries is expected by their combined use rather than by the single use of one (*16*). Figure 1 shows the caries score of the rats after being fed the diets containing 2000 ppm of black tea extract, 200 ppm of 5-methylthiopentyl isothiocyanate (**1**), and both of them. Figure 2 is a photograph of the dental caries of the rats being fed a diet. A significant anticaries activity was obtained by the combined use of isothiocyanate (**1**) and black tea extract compared to the single use of only black

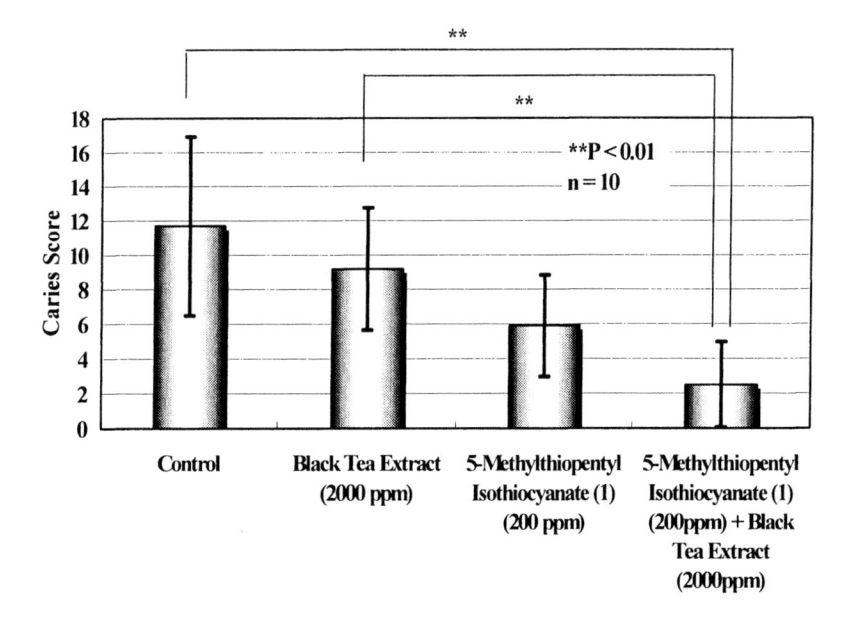

Figure 1. Caries score of rats after being fed a diet containing 5-methylthiopentyl isothiocyanate (1) and/or black tea extract. P: Observed significance level of the test. n: Number of rats. N.S.: No significance.

tea extract or isothiocyanate (**1**) (Figure 1). When considering their practical value, the characteristic flavor of isothiocyanate (**1**), that is, a radish-like and pickle-like flavor (*35*), and the astringent flavor of the black tea extract, it is useful to reduce their concentrations. In order to clarify the anticaries effect for the combined use of the isothiocyanate and the black tea extract, we used an isobolographic analysis (*39-46*). An isobologram, a plot of equally effective doses by their combined use, has never used for the model study of dental plaque. In order to evaluate the interaction of the combination, a continuous study of the dose-response is considered to be more reliable than the simple study. Figures 3 shows the IC_{50} (50% Inhibitory Concentration) isobologram for sucrose dependent adherence by the growing cells of *S. sobrinus* 6715 obtained by a combination of isothiocyanate (**1**) and the black tea extract. These IC_{50} values were calculated from the observed inhibition values by a probit analysis. The solid line connecting the IC_{50} point for the single use of isothiocyanate and that for the black tea extract is the theoretical additive line (*42*). The area surrounded by the dashed lines is considered to show an additive effect by this combination. Therefore, data to the upper right of this area indicate the antagonistic effect,

Control

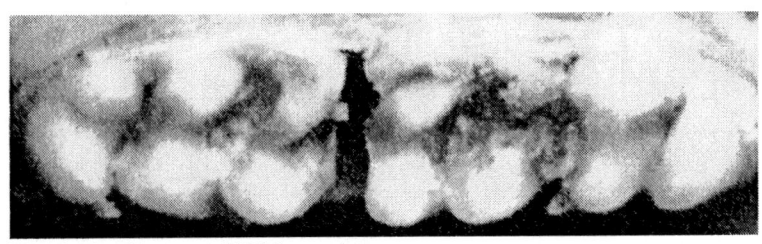

Black Tea Extract (2000 ppm)

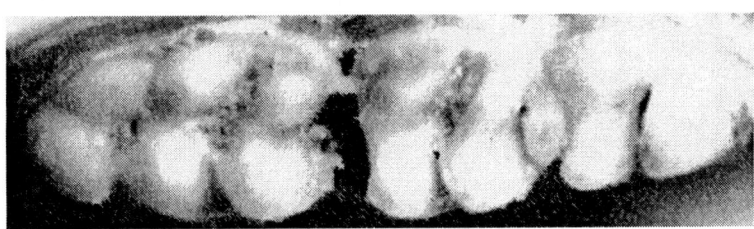

5-Methylthiopentyl Isothiocyanate (1) (200 ppm)

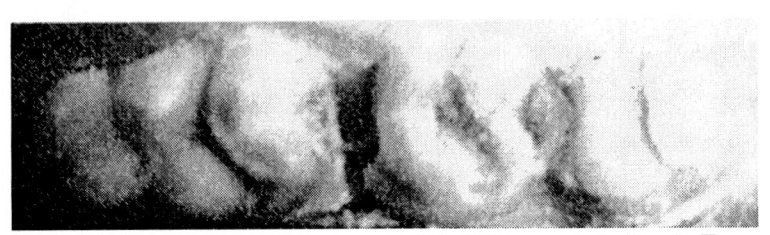

5-Methylthiopentyl Isothiocyanate (1) (200 ppm) + Black Tea Extract (2000 ppm)

Figure 2. A part of rat molars after being fed a diet containing 5-methylthiopentyl isothiocyanate (1) and/or black tea extract compared to that of the control. Control: Being fed diet without isothiocyanate 1 and/or black tea extract.

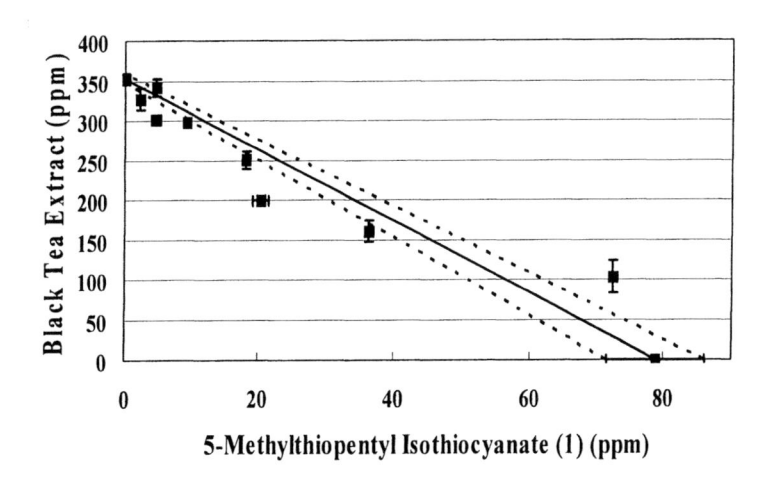

Figure 3. IC$_{50}$ isobologram for sucrose dependent adherence by growing cells of S. sobrinus 6715 obtained by combination of 5-methylthiopentyl isothiocyanate (1) and black tea extract. IC$_{50}$: 50% Inhibitory concentration.

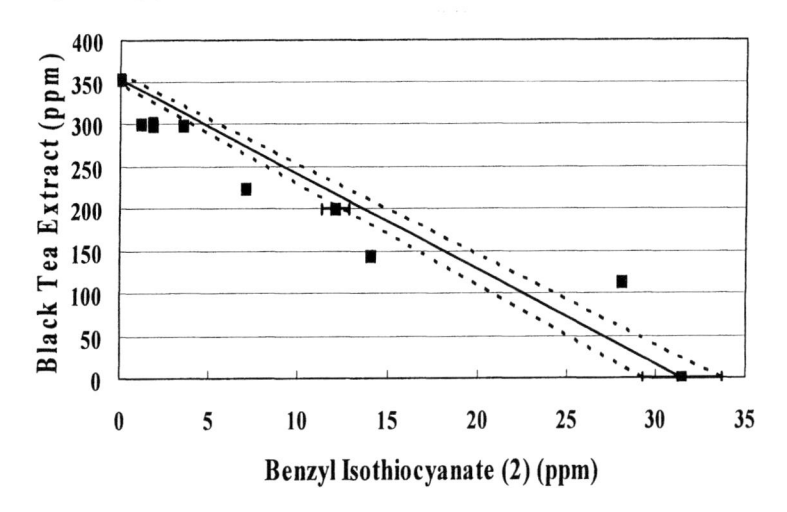

Figure 4. IC$_{50}$ isobologram for sucrose dependent adherence by growing cells of S. sobrinus 6715 obtained by combination of benzyl isothiocyanate (2) and black tea extract. IC$_{50}$: 50% Inhibitory concentration.

while data to the lower left show the synergistic effect. The combined use of isothiocyanate (1) and the black tea extract generally gave a slightly synergistic effect. The inhibition of the sucrose dependent adherence by the growing cells of *S. sobrinus* 6715 is considered to be obtained by the action of the antimicrobial effect and GTase-inhibition effect. The antimicrobial effect for mutans streptococci of the isothiocyanate is higher than that of the black tea extract (*16,23*). On the other hand, the GTase-inhibitory effect of the tea extract is reported to be significant due to the interaction between the protein and the polyphenol (*16,18,24-31*). As for the isothiocyanate, the interaction of the protein was studied using papain, insulin, bovine serum albumin, ovalbumin, or lysozyme (*36-38*). Thus, the differet action mechanism and action strength against the sucrose dependent adherence by the growing cells of *S. sobrinus* 6715 seems to give the combined effect shown in Figure 3. In addition, the combined use of benzyl isothiocyanate (2) and black tea extract provided, in general, a more synergistic effect than the combination with isothiocyanate (1) (Figure 4). The inhibition activity of the benzyl substituent is higher than that of the methylthio substituent.

Conclusion

The combined use of isothiocyanate and black tea extract on the inhibition of the sucrose dependent adherence by the growing cells of mutans streptococci, that is a dental plaque model, was studied using an isobologram. The continuous dose-response data plotted in the isobologram were obtained by using microplate methods. From the isobologram, the combined use of isothiocyanate and black tea extract has been generally found to show a slightly synergistic effect on the sucrose dependent adherence by the growing cells of mutans streptococci.

References

1. Kishimoto, N; Tano, T.; Harada, Y.; Masuda, H. *Journal of the Japan Association of Food Preservation Scientists* **1999**, *25*, 7-13.
2. Masuda, H.; Harada, Y.; Kishimoto, N.; Tano, T. In *Aroma Active Compounds in Foods*; Takeoka, G. R.; Guntert, M.; Engel, K.-H. Ed.; ACS Symposium Series 794; American Chemical Society: Washington, DC, 2001; pp. 229-250.
3. Kumagai, H.; Kashima, N.; Seki, T.; Sakurai, H.; Ishii, K.; Ariga, T. *Biosci. Biotech. Biochem.* **1994**, *58*, 2131-2135.
4. Morimitsu, Y.; Hayashi, K.; Osawa, T. *Nihon Koshinryo Kenkyukai Koen Yoshishu* **1997**, 17-18.

5. Stoner, G. D.; Galati, A. J.; Schmidt, C. J.; Morse, M. J. In *Food Phytochemicals for Cancer Prevention I*; Huang, M.-T.; Osawa, T.; Ho, C.-T.; Rosen, R. T., Eds.; ACS Symposium Series 546; American Chemical Society: Washington, DC, 1994; pp 173-180.
6. Drobnica, L.; Gemeiner, P. In *Protein Structure and Evolution*; Fox, J. L.; Deyl, Z.; Blazej, A., Eds.; Marcel Dekker, Inc., New York, 1976; pp 105-115.
7. Hecht, S. S. *J. Nutr.* **1999**, *129*, 768S-774S.
8. Zhang, Y.; Talalay, P.; Cho, C.-G.; Posner, G. H. *Proc. Natl. Acad. Sci.* **1992**, *89*, 2399-2403.
9. Ono, H.; Adachi, K.; Fuke, Y.; Shinohara, K. *Nippon Shokuhin Kagaku Kogaku Kaishi* **1996**, *43*, 1092-1097.
10. Fuke, Y.; Sawaki, S.; Nomura, T.; Ryoyama, K. *Nippon Shokuhin Kagaku Kogaku Kaishi* **2000**, *47*, 760-766.
11. Hamajima, F.; Yasuhiro, C.; Nishihara, S. *Jap. J. Parasit.* **1969**, *18*, 498-504.
12. Dorsch, W.; Adam, O.; Weber, J.; Ziegeltrum, T. *European Journal of Pharmacology* **1985**, *107*, 17-24.
13. Adcock, J. J.; Garland, L. G. *Br. J. Pharmacol.* **1980**, *69*, 167-169.
14. Hamada, S.; Slade, H. D. *Microbiological Reviews*, **1980**, *44*, 331-384.
15. Loesche W. J. *Microbiological Reviews*, **1986**, *50*, 353-380.
16. Masuda, H.; Inoue, T.; Kobayashi, Y. In *Anticaries Effect of Wasabi Components*; Ho, C.-T.; Lin, J.-K.; Zheng, Q. Y. Ed.; ACS Symposium Series Symposium Series 859; American Chemical Society: Washington, DC, 2003; pp. 142-153.
17. Sakanaka, S.; Shimura, N.; Aizawa, M.; Kim, M.; Yamamoto, T. *Biosci. Biotech. Biochem.* **1992**, *56*, 592-594.
18. Ooshima, T.; Minami, T.; Aono, W.; Izumitani, A.; Sobue, S.; Fujiwara, T.; Kawabata, S.; Hamada, S. *Caries. Res.* **1993**, *27*, 124-129.
19. Ooshima, T.; Minami, T.; Aono, W.; Tamura, Y.; Hamada, S. *Caries. Res.* **1994**, *28*, 146-149
20. Ooshima, T.; Minami, T.; Matsumoto, M.; Fujiwara, T.; Sobue, S.; Hamada, S. *Caries. Res.* **1998**, *32*, 75-80.
21. Linke, H. A. B.; LeGeros, R. Z. *Internat. J. Food Sci. Nutr.* **2003**, *54*, 89-95.
22. Yoshino, K.; Nakamura, Y.; Ikeya, H.; Sei, T.; Inoue, A.; Sano, M.; Tomita, I. *J. Food Hyg. Soc. Japan* **1996**, *37*, 104-108.
23. Sakanaka, S.; Kim, M.; Taniguchi, M.; Yamamoto, T. *Agric. Biol. Chem.* **1989**, *53*, 2307-2311.
24. Sakanaka, S.; Sato, T.; Kim, M.; Yamamoto, T. *Agric. Biol. Chem.* **1990**, *54*, 2925-2929.
25. Otake, S.; Makimura, M.; Kuroki, T.; Nishihara, Y.; Hirasawa, M. *Caries. Res.* **1991**, *25*, 438-443.

26. Hattori, M.; Kusumoto, I. T.; Namba, T.; Ishigami, T.; Hara, Y. *Chem. Pharam. Bull.* **1990**, *38*, 717-720.
27. Nakahara, K.; Kontani, M.; Ono, H.; Kodama, T.; Tanaka, T.; Oooshima, T.; Hamada, S. *Appl. Environ. Microbiol.* **1995**, *61*, 2768-2770.
28. Nakahara, K.; Kawabata, S.; Ono, H.; Ogura, K.; Tanaka, T.; Ooshima, T.; Hamada, S. *Appl. Environ. Microbiol.* **1993**, *59*, 968-973.
29. Hamada, S.; Kontani, M.; Hosono, H.; Ono, H.; Tanaka, T.; Ooshima, T.; Mitsunaga, T.; Abe, I. *FEMA Microbiology Letters* **1996**, *143*, 35-40.
30. Manjo, F. *Nippon Shokuhin Shinsozai Kenkyukaishi* **2000**, *3*, 61-68.
31. Lee, M.-J.; Lambert, J.D.; Prabhu, S.; Meng, X.; Lu, H.; Maliakal, P.; Ho, C.-T.; Yang, C. S. *Cancer Epidem. Biomarkers Preven.* **2004**, *13*, 132-137.
32. Masuda, H.; Tsuda, T.; Tateba, H.; Mihara, S., Japan Patent 90,221,255, 1990.
33. Harada, Y.; Masuda, H.; Kameda, W., Japan Patent 95,215,931, 1995.
34. Kamo, T. *J. Dental Health* **1960**, *17*, 31-39.
35. Masuda, H. Harada, Y.; Tanaka, K.; Nakajima, M.; Tateba, H. In *Biotechnology for Improved Foods and Flavors*; Takeoka, G. R.; Teranishi, R.; Williams, P. J.; Kobayashi, K. Eds.; ACS Symposium Series 637; American Chemical Society: Washington, DC, 1996; pp. 67-78.
36. Tang, C.-S. *J. Food Sci..* **1974**, *39*, 94-96.
37. Tang, C.-S.; Tang, W.-J. *Biochim. Biophys. Acta* **1976**, *452*, 510-520.
38. Kawakishi, S.; Kaneko, T. *J. Agric. Food Chem.* **1987**, *35*, 85-88.
39. Berenbaum, M. C. *J. Clin. Exp. Immunology* **1977**, *28*, 1-18.
40. Tallarida, R. J.; Porreca, F.; Cowan, A. *Life Science* **1989**, *45*, 947-961.
41. Berenbaum, M. C. *Pharmacol. Rev.* **1989**, *41*, 93-141.
42. Gressner, P. K. *Toxicology* **1995**, *105*, 161-179.
43. Hu, W. Y.; Reiffenstein, R. J.; Wong, L. *Alcohol and Drug Research* **1986**, *7*, 107-117.
44. Brunet, B. L.; Reiffenstein, R. J.; Williams, T.; Wong, L. *Alcohol and Drug Research* **1986**, *6*, 341-349.
45. Altenburger, R.; Boedeker, W.; Faust, M.; Grimme, L. H. *Ecotoxicology and Environmental Safety* **1990**, *20*, 98-114.
46. Kim, K. S.; Chun, Y. S.; Chon, S. U.; Suh, J. K. *Anaesthesia* **1998**, *53*, 872-878.

Chapter 23

Cytotoxic Properties of Leaf Essential Oil and Components from Indigenous Cinnamon (*Cinnamomum osmophloeum* Kaneh)

Tzou-Chi Huang[1], Chi-Tang Ho[2], Hui-Yin Fu[3], and Min-Hsiung Pan[4]

[1]Department of Food Science, National Pingtung University, 912 Pingtung, Taiwan
[2]Department of Food Science, Rutgers, The State University of New Jersey, 65 Dudley Road, New Brunswick, NJ 08901–8520
[3]Department of Food Science and Technology, Tajen Institute of Technology, 907, Pingtung, Taiwan
[4]Department of Food Science, National Kao Hsiung Marine University, Kao Hsiung, Taiwan

The essential oils from leaves of three *Cinnamomum osmophloeum* clones (A, B and C) and their major chemical constituents were investigated. GC and GC-MS analyses showed that *Cinnamomum osmophloeum* clones A, B and C contain *trans*-cinnamaldehyde (90.6%), eugenol (91.3%) and linalool (83.7%), respectively, as the major component. The inhibitory effects of the essential oils on the growth of human leukemia cell lines U937 and K562 Hep-1, *in vitro* were investigated. A strong cell proliferation inhibition activity for *Cinnamomum osmophloeum* clone A was observed for cell lines K562, whereas *Cinnamomum osmophloeum* clone B and *Cinnamomum osmophloeum* clone C gave comparatively low inhibition. This trend paralleled the cytotoxic properties of authentic *trans*-cinnamaldehyde, eugenol and linalool, respectively.

Volatile Composition of Cinnamon Essential Oil

Cinnamon, also known as Canela and Sees (Ceylon cinnamon), is native to Sri Lanka and India, and has also been cultivated in Brazil, Mauritius and Jamaica. Cinnamon is used primarily in the flavor and fragrance industries to impart a cinnamon flavor and/or fragrance to various types of foods, beverages, medical products, and perfumes (*1*). *Cinnamomum osmophloeum* Kaneh, also known as pseudocinnamomum, belongs to the Angiospermae subphylum, the *Dicotyledoneae* class, the *Archichlamydeae* subclass, the *Laurceae* family and the *Cinnamomum* genus. Indigenous cinnamon *Cinnamomum osmophloeum* Kaneh is an endemic tree that grows in Taiwan's natural hardwood forest at elevations between 400 and 1500 m (*2*). Both bark and leaves of *Cinnamomum osmophloeum* Kaneh from the primary forests of central Taiwan were found to contain high levels of cinnamaldehyde (*3*). Hu *et al.* (*4*) analyzed the composition of the essential oil of *C. osmophloeum* leaves collected from 21 provenances in central, southern, and eastern Taiwan. On the basis of the chemical composition of different leaf essential oils, *C. osmophloeum* was classified into nine types: cassia, cinnamaldehyde, coumarin, linalool, eugenol, camphor, 4-terpineol, linalool-terpineol and mixed types. Similar to *C. cassia* bark oil, *C. osmophloeum* is a major source of *trans*-cinnamaldehyde which is recognized as a bioactive substance with potential health effects (*5*). Cinnamaldehyde has been reported to have anti-platelet aggregating and vasodilatory action, inhibiting collagen-induced platelet aggregation in a dose dependent manner, and antiproliferative activity against human solid tumor cells (*6*).

The oils obtained from leaves of three types, namely, eugenol (clone A), linalool (clone B), and cinnamaldehyde (clone C) cultivated by farmers in Taiwan were examined by GC and GC/MS. At least 13 compounds were characterized, as shown in Table I. Clones A, B and C contain *trans*-cinnamaldehyde (90.6%), eugenol (91.3%) and linalool (83.7%), respectively as the major component characterized in this experiment. The oils obtained from leaf, stembark and rootbark of *Cinnamomum pauciflorum* Nees (Lauraceae), growing wild in northeast India, were examined by GC (*7*). Cinnamaldehyde was the main component of the oil, with the highest percentage found in the leaf oil (94%), followed by the rootbark oil (92.4%) and stembark oils (85.1%). The oils of four species of Cinnamomum from Madagascar were studied. *C. zeylanicum* contained predominantly (*E*)-cinnamaldehyde together with camphor (15%). *C. camphora* was of the 1,8-cineole type (58-63%). *C. fragrans* contained α- and βpinenes (8% and 10%), 1,8-cineole (28%) and β-caryophyllene (13%). *C. angustifolium* contained mostly α-phellandrene, rho-cymene, α-caryophyllene, 1,8-cineole and α-pinene *(8)*. The leaf and stem bark oils of *Cinnamomum sulphuratum* Nees, growing wild in northeast India, were investigated by GC and GC/MS. Geranial (27.8%) geraniol (23.2%) and neral (17.6%) were the major constituents of the leaf oil. The bark oil was rich in (*E*)-cinnamaldehyde (65.6%) *(9)*.

Table I. Volatile Compounds Identified in the Essential Oil of *Cinnamomum*

Compound	\multicolumn{4}{c}{*Concentration (%)*}
	A	B	C	Stem Bark
α-pinene	0.48	0.54	0.07	0.04
camphene	0.26	0.17	0.12	0.27
benzaldehyde	0.01	0.63	0.08	0.18
β-pinene	0.24	0.62	0.58	0.01
linalool	2.56	87.72	0.91	1.06
benzenepropanal	0.04	0.32	0.45	0.46
methylchavicol	0.01	0.43	0.62	0.32
cis-cinnamaldehyde	0.02	0.01	0.57	0.60
trans-cinnamaldehyde	1.23	3.02	90.58	91.41
bornyl acetate	0.06	1.01	0.36	0.74
eugenol	91.15	2.25	1.21	1.47
D-limonene	0.34	0.42	0.01	0.02
cinnamyl acetate	3.60	2.86	4.44	3.42

The levels of the total essential oil in cinnamon leaf samples were determined by distillation. The fresh cinnamon leaves harvested in August, 2003 from a local farm (Taiwan Cinnamon Co., Pingtung) were used in this study. The cinnamon leaves were dehydrated to a final moisture content of approximately 5%, moisture-free basis or less, as follows: Low temperature air dryer: drying temperature 30±2 °C, and total drying time 20 h.; Freeze-drying: freezing temperature –30 °C, drying temperature 30 °C, vacuum 0.15-0.20 mm Hg, drying time 12 h.; Solar drying: dried under sun shine at temperatures ranging between 30 °C to 35 °C.; Hot air shelf-drying: drying temperature 80±2 °C, and total drying time 16 h. Changes in the recovery of the volatile compounds due to drying are shown in Figure 1. The recovery of total essential oil from fresh leaves in air-drying at low temperature (30 °C), freeze-drying, solar drying and hot air-dried (80 °C), was 94%, 93%, 89%, and 75%, respectively. Freeze drying and air-drying at 30 °C produced quite similar results and caused hardly any loss in volatiles as compared to fresh cinnamon leaf, whereas solar drying and hot air-drying at 80 °C brought about substantial losses in cinnamon leaf essential oil.

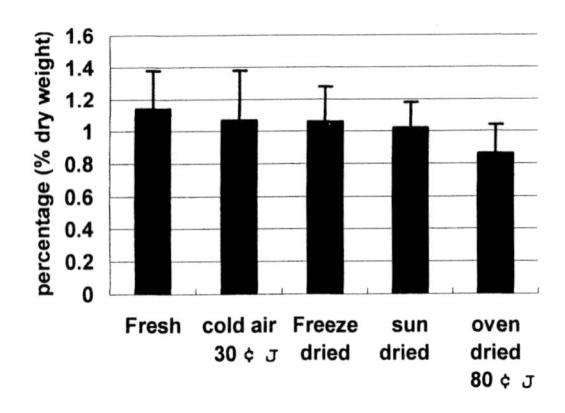

Figure 1. Effect of drying on the essential oil recovery from cinnamon leave.

Principal component analysis (PCA) provides a better understanding of the variation among volatile compositions of samples and has been widely used to evaluate food flavors (*10*). Each principal component of the axis is a linear combination of the original variable. From the GC analysis of 5 different drying methods, 13 major peaks were used as variable for principle component analysis (PCA). In PCA, it can be observed that the first four principal components had eigenvalues greater than 1.0 and accounted for 87.62% of the total variance. Figure 2 shows a plot of the values for the first two principal components, which together account for 68.60% (43.71% + 24.89%) of the variation in the data. From a chemical compositional point of view, the quality of the air-dried samples (30 °C) is similar to that of freeze dried samples, since both were grouped from fresh cinnamon leave. Considering the operation cost, air-drying (30 °C) rather than freeze drying is recommended as a cost-effective method for farmers to store their cinnamon.

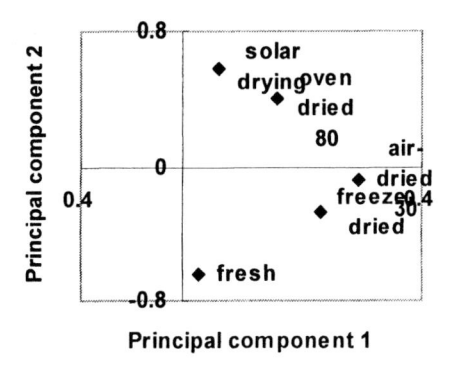

Figure 2. Plot of samples of fresh and dried cinnamon leaf on the coordinate grid defined by principal component 1 and 2.

Antimicrobial and Insecticidal Activity of Cinnamaldehyde

Cinnamaldehyde, the principal of cinnamon flavor, is a potent antimicrobial compound present in cinnamon essential oils. Cinnamaldehyde exhibits strong antibacterial activity against several human food-borne pathogens (*11*), intestinal bacteria (*12*), and meat spoilage organisms (*13*). Cinnamaldehyde revealed potent inhibition against postharvest disease fungi (*14*) and aflatoxin production fungi (*15*). Insecticidal activity of cinnamaldehyde against oak nut weevil (*16*), termite (*17*), and grain storage insects (*18*) was reported as well. These inhibitory activities were attributed, at least in part, to the inactivation of rat intestinal and nephritic Na^+/K^+ ATPase (*19*). *In vitro* studies demonstrated an inhibitory effect on activity of many enzymes, including mushroom tyrosinase (*20*) and nitric oxide synthase (*21*).

Cytotoxic Properties of Leaf Essential Oil

Three human leukemia cell lines were obtained from the American Type Culture Collection (ATCC), namely K562, CEM and U937 cell lines. These cell lines were cultured in RPMI 1640 medium with 10% fetal bovine serum (Gibco, BRL), supplemented with 5 mm L-glutamine and 50 µg/mL of antibiotics (Penicillin/Streptomycin, Gibco, BRL) at 37 °C in a humidified 5% CO_2 incubator (*22*). Cells were seeded in 96-well plates at an initial density of 2.0×10^5/mL in 180 µL RPMi-1640 medium per well. As a control vehicle, 0.1%

of DMSO was added to the cells. Following 48 h of incubation with these test samples, cell viability was assayed with a Luminesscent ATP detection assay kit (Packard BioScience, B.V.). Luminescence was measured on a Multilabel Reader (Mithras LB 940). Cinnamon essential oil showed different inhibition properties on various cell lines. Among the tested cell lines, cinnamaldehyde type essential oil showed the most potent inhibition activity on K562 and U937 cell lines followed by eugenol essential oil, and both were in a dose dependent manner, as shown in Table II. No antiproliferative activity was noted on either K562 or U937 cell lines for linalool oil. The IC_{50} values of cinnamaldehyde type essential oil on human leukemia cell lines K562 and U937 were 15.21 and 35.97µM, respectively. As similar trend was found for authentic cinnamaldehyde, eugenol and linalool. The structures of cinnamaldehyde, eugenol and linalool are illustrated in Figure 3. Qualitative analysis by gas chromatography verified the identity of cinnamaldehyde as the major ingredient present in cinnamon essential oil with a relative abundance, 90-fold greater than any of the other distillated ingredients. The high concentration of *trans*-cinnamaldehyde may account for the high antiproferative activity of cinnamaldehyde type essential oil against both K562 and U937 cell lines. The α, β unsaturated carbonyl group on cinnamaldehyde may be responsible for the potent cell growth inhibition activity. Although eugenol contains an aldehyde group as well, the saturated carbonyl group may be responsible for this activity. Linalool was found to be inactive and this may be attributed to its structural properties.

Table II. The IC_{50} on Human Leukemia Cell Lines K562 and U937

Samples	Cell lines (IC50 µM)	
	K562	U937
Cinnamon oil A	15.2	35.9
Cinnamon oil B	73.4	92.7
Cinnamon oil C	-	-
Cinnamaldehyde	14.7	28.7
Eugenol	69.8	83.9
Linalool	-	-

Cinnamon oil A= Cinnamaldehyde type; Cinnamon oil B=Linalool type; Cinnamon oil C=Eugenol typ
*Means of triplicate trials

Cinnalmaldehyde   Eugenol   Linalool

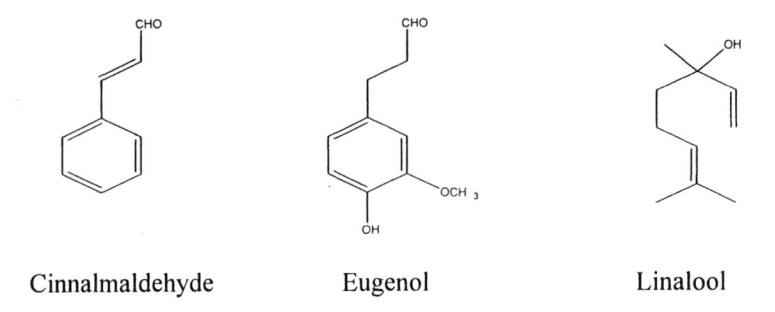

Figure 3. Chemical structures of cinnamaldehyde, eugenol and linalool.

Effects of Cinnamaldehyde on Morphological Changes of K562 Cell

K562 cells treated with 0, 50, 100 and 200 µM of cinnamaldehyde for 48 h were studied. Cells were harvested and washed with ice-cold PBS, and then examined by microscopy. Figure 4 shows the representative morphological changes of K562 cells exposed to cinnamaldehyde (50 µM) for 48 h. Under control conditions, K562 cells appeared normal and the nuclei were round and homogeneous. After being treated with cinnamaldehyde, the cells exhibited characteristic features of apoptosis, including plasma membrane blebbing and cell shrinkage, as shown in Figure 4B. These results indicate that cinnamaldehyde induced apoptosis in K562 cells.

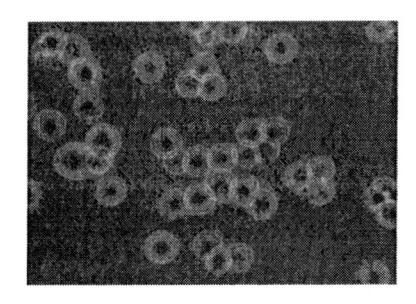

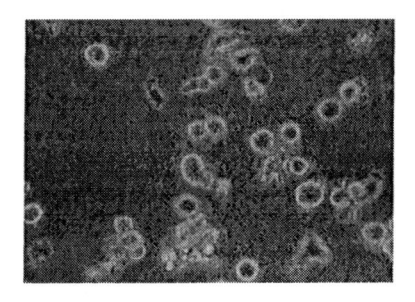

Figure 4. Morphological changes of K562 cell after exposure to trans-cinnamaldehyde (50 µM) for 48 h. (A) untreated cells; (B) cells treated with trans cinnamaldehyde. The condensed chromosomes are seen as spots in the nucleus by acridine orange staining; apoptotic cells are shown with arrowheads, magnification x 200.

Cinnamaldehyde-induced DNA Fragmentation of Leukemic Cell Lines

The effect of cinnamaldehyde on DNA fragmentation, a hallmark of apoptosis, was determined by incubating K562 cells with different concentrations of cinnamaldehyde for 48 h. DNA fragmentation appeared at 200 µM of cinnamaldehyde treatment, and the DNA ladder response was dose-dependent from 50 to 200 µM treatment (Figure 5).

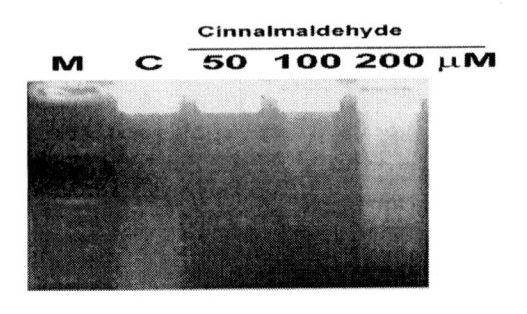

Figure 5. Induction of DNA fragmentation by (A) treated with various cinnamaldehyde as indicated for 12 hr, and (B) HL-60 cells were treated with 60µM flavonoids for indicated time. Internucleosomal DNA fragmentations were analyzed by agarose electrophoresis. M: one hundred-base pairs DNA ladder size maker.

The degree of DNA fragmentation in cinnamaldehyde treated K562 cells was determined by flow cytometry analysis. K562 cells were treated with 0.5 % DMSO as a control or with different concentrations of cinnamaldehyde for 48 h. A fluorometric method using DAPI was applied to quantify the amount of fragmented DNA. As sub G1 peak represents apoptotic cells with a lower DNA content. Cinnamaldehyde showed a dose dependent DNA fragmentation effect on K562 cells. At each time point cells were harvested, washed twice with phosphate-buffered saline (PBS) and fixed in 70% ethanol for at least 2 h at -20 °C. Fixed cells were washed with PBS, incubated with 1 ml PBS containing 0.5 µg/ml RNase A and 0.5% Triton X-100 for 30 min at 37 °C, and then stained with 50 µg/mL propidium iodide. The stained cells were analyzed by a FACScan laser flow cytometer (Becton Dickinson, San Jose, CA) and ModFit LT cell cycle analysis software (Verity Software, Topsham, ME). After 12 h of treatment, a sub G1 (sub 2N) DNA peak, which has been suggested to be the apoptotic DNA, was detected. The percentage of apoptotic cells is shown in Figure 6. Cinnamaldehyde appeared to be a potent apoptosis-inducing agent for

K562 cells and the apoptotic effects were found to be dose-dependent up to 200 µM.

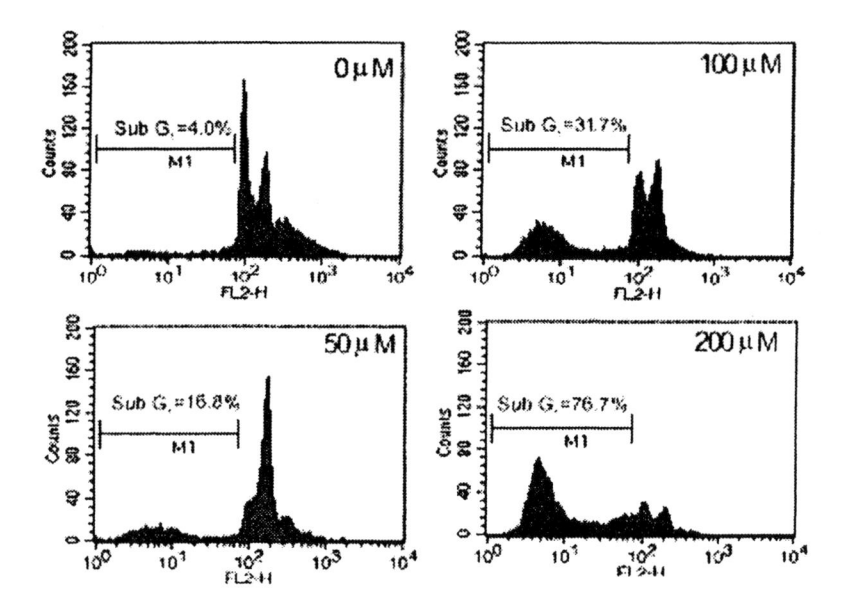

Figure 6. Induction of DNA fragmentation in K562 cells by cinnamaldehyde. K562 cells treated with increasing doses of cinnamaldehyde for 12 h. Determination of sub-G1 cells in cinnamaldehyde-treated K562 cells by flow cytometry.

To access the generation of superoxide radical (H_2O_2), an H_2O_2-sensitive probe DCFH-DA was used. DCFH-DA is deacetylated in a cell to a nonfluorescent compound, 2',7'-dichlorofluorescin, which remains trapped within the cell and is oxidized by H_2O_2 in the presence of endogenous peroxide to a highly fluorescent compound, 2',7'-dichlorofluorescein (DCF) (23). The time course of cinnamaldehyde on intracellular superoxide radical levels were studied. Figure 7 shows the typical histograms of superoxide radical in control and cinnamaldehyde-treated cells. The level of intracellular superoxide radical in cinnamaldehyde-treated cells decreases in a time-dependent manner.

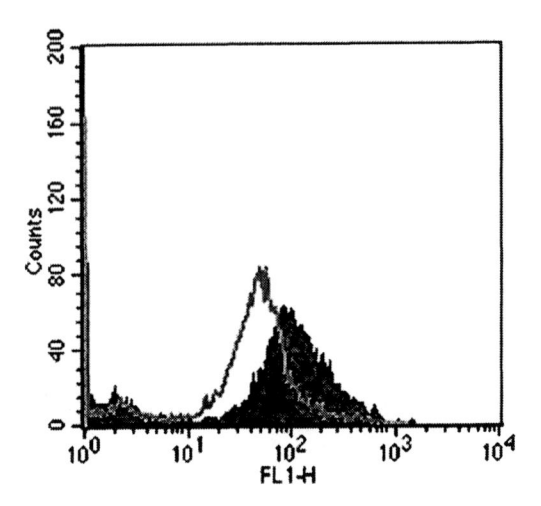

Figure 7. Effect of cinnamaldehyde on the level of intracellular superoxide radicals. Typical histograms showing the decrease of intracellular superoxide radical in cinnamaldehyde-treated K 562 cells. Cells were treated with 60μM cinnamaldehyde for 0 and 6 h, respectively.

The cells were harvested at 0 h, 2 h, 4 h and 6 h after treatment with 60 μM of cinnamaldehyde. Briefly, K562 cells were exposed to cinnamaldehyde, and the mitochondrial transmembrane potential was measured directly using 40nM 3,3'-dihexyloxacarbocyanine (DiOC6) (Molecular Probe Eugene, Oregon). The samples (10^4 events) were analyzed for fluorescence (FL1 detector, filter 430/30 nm band pass) using a FACScan flow cytometer (Becton Dickinson, San Jose, CA). The cytogram in Figure 8 shows a typical result of the increase of ΔΨm in cinnamaldehyde induced apoptosis in K562 cells. During apoptosis in cinnamaldehyde induced apoptosis in K562 cells, dissipation of ΔΨm led to leakage of DiOC6(3) from the mitochondrial matrix, which can be measured by flow cytometry as a decrease in the fluorescent intensity of the dye. The decrease of green fluorescence indicates an increase of mitochondrial membrane potential, and the number in each histogram is the percentage of cells with lower mitochondrial membrane potential. It was found that cinnamaldehyde was able to increase mitochondrial membrane potential in a time-dependent manner. It is well established that the change of mitochondrial membrane potential (ΔΨm) causes the disruption of the outer mitochondrial membrane and contributes to the release of cytochrome *c* (*24*). These observations suggested that an apoptosis-inducing mechanism, via the mitochondria pathway, was triggered by cinnamaldehyde in K562 cells.

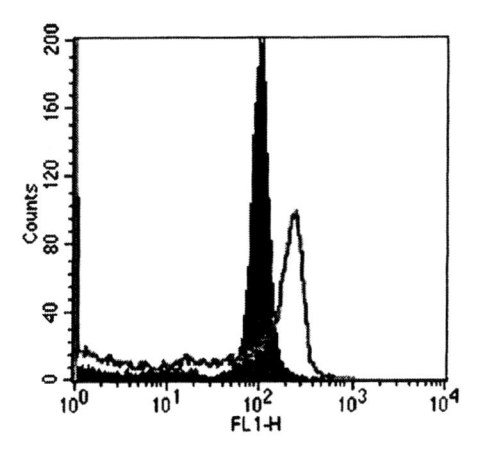

Figure 8. Effect of cinnamaldehyde on the level of intracellular superoxide radicals. Typical histograms showing the decrease of intracellular superoxide radical in cinnamaldehyde-treated K 562 cells. Cells were treated with 60µM cinnamaldehyde for 0 and 6 h respectively.

Cinnalmaldehyde exhibited a concentration-dependent dissipation of the membrane potential producing a 50% increase in the fluorescence intensity. NADH-oxidase (complex I) and succinate dehydrogenase (complex II) constitute, respectively, the main entry site of reducing equivalents NADH and FADH2 into the respiratory chain, where oxidation occurs, leading ultimately to phosphorylation of ADP into ATP (25). Cinnamaldehyde exhibited a concentration-dependent inhibition on NADH oxidation. Among the well established sites where electron transfer occurs coupled with proton release into the cytosol is complex-I (NADH-oxidase). A major consequence of the inhibition of cinnamaldehyde on NADH oxidase is a decrease in ATP production with subsequent accumulation of oxidized dysfuction proteins that will ultimately impair the mitochondrial function leading to cell death (26). The cinnamaldehyde effect on membrane potential and stimulation in ATPase activity are indicative of their possible influence on proton redistribution across the membrane. We postulated that cinnamaldehyde may inhibit NADH oxidase, a proton pumping site, ultimately resulting in a decline in ATP level in the treated K562 cells.

The disruption of mitochondrial membrane potential results from mitochondrial permeability transition due to the permeability transition pore opening. The permeability transition pore is a multi-protein complex. The opening pore has been shown to possess redox-sensitive sites of critical vicinal thiols which are in redox equilibrium with matrix glutathione (27). These thiols are suggested to be a key factor of apoptotic signaling pathways. Therefore, depletion or oxidation of these thiols or glutathione can facilitate the membrane potential transition (28).

To investigate the influence of cinnamaldehyde treatment on the level of intracellular glutathione, K562 cells were treated with cinnamaldehyde (200 µM). Harvested cells were treated with 5% TCA and GSH was determined by an enzymatic method using glutathione reductase. Our results confirmed that cinnamaldehyde is a powerful GSH depletor. A high GSH to GSSG ratio is maintained within cells, which has been shown to be important in the structural integrity and functional processes of membranes, the maintenance and polymerization of microtubules, the conformation of proteins and modulation of their activities (29). Maintaining a high GSH to GSSG ratio contributes to the redox homeostasis of a cell, which provides an antioxidant defense mechanism against ionizing radiation, reactive oxygen species, free radicals and toxic xenobiotics (30).

Structural changes of cells associated with oxidative stress, such as increased vacuolization and membrane blebbing, were found to be closely correlated with glutathione depletion (31). Cinnamaldehyde has been reported to be a potent depletor of hepatic GSH *in vivo* (32). Previous studies demonstrated that cinnamaldehyde displays the novel property of inducing apoptosis in rat Leukemia L1210 cells through rapid depletion of intracellular thiols (33). Cinnamaldehyde is conjugated across the carbon double bond with reduced glutathione (34). The overall findings thus suggest that the mitochondrial permeability transition that resulted from intracellular thiol depletion is a critical event in cinnamaldehyde-induced apoptosis.

Recently, the molecular mechanisms of apoptosis have been actively investigated in various *in vitro* and *in vivo* models, and numerous studies have demonstrated that mitochondria play a pivotal role in transducing a variety of proapoptotic stimuli (35). It seems that various proapoptotic stimuli provoke alteration of the permeability of apoptotic proteins, such as cytochrome c. While cytochrome c has been demonstrated to be involved in the activation of a caspase casade (36), Ka *et al.* (37) postulated that cinnamaldehyde can act as a cytostatic agent for human promyelocytic leukemia HL-60 cells by modulating the mitochondrial permeability transition. The antiproliferative activity effect of cinnamaldehyde was proposed to be the result of a ROS- and caspase-induced apoptosis. The cytotoxicity of cinnamaldehyde was thus suggested to be the consequence of increased covalent binding of cinnamaldehyde to critical proteins. The cytotoxic activity of cinnamaldehyde might follow a similar mechanism. A non specific Schiff base reaction may occur between the unsaturated aldehyde group and either the sulfhydryl or amino group, leading to the inactivation of related enzymes. Thus we postulate that the exerted effects on membrane potential of cinnamaldehyde may be attributed to the release of protons into the matrix or to interaction with sulfhydryl groups of membrane proteins. Other contributing factors such as swelling of mitochondria, calcium mobilization and membrane perturbation, which would influence mitochondrial enzyme activities and function should not be ruled out (Figure 9). In the present study, we further demonstrate that mitochondrial permeability transition resulting from intracellular thiol depletion is a critical event in cinnamaldehyde induced apoptosis.

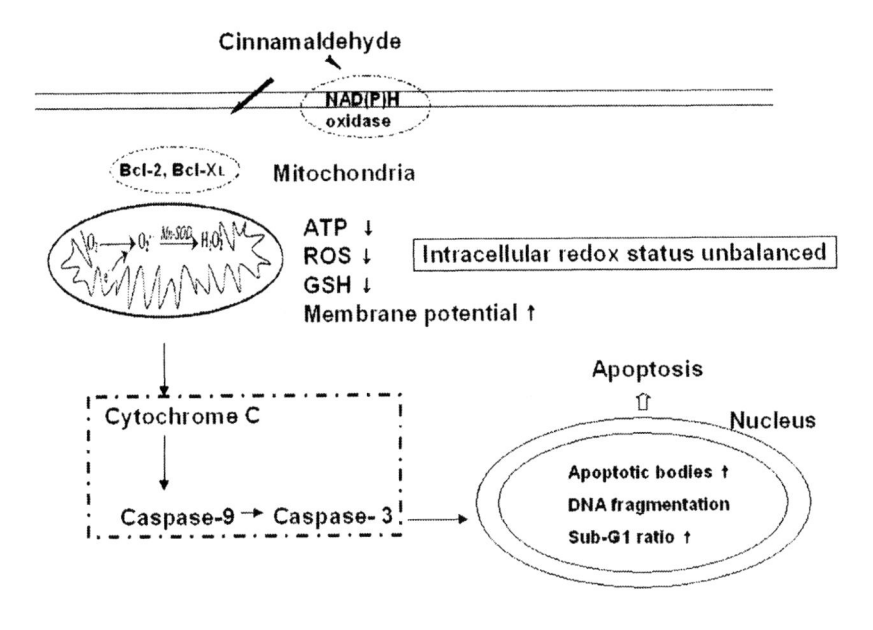

Figure 9. Schematic representation of action mechanism by cinnamaldehyde induced apoptosis in K562.

References

1. Ter Heide, R. *J. Agric. Food Chem.* **1972**, *20*, 747-751.
2. Jayaprakasha, G.K.; Rao, L.J.M.; Sakariah, K.K. *J. Agric. Food Chem.* **2000**, *48*, 4294-4295.
3. Fang, J.M.; Chen, S.A.; Cheng, Y.H. *J. Agric. Food Chem.* **1989**, *37*, 744-746.
4. Hu, T.W.; Lin, Y.T.; Ho, C.K. *Bull. Taiwan For. Res. Inst. Eng.* **1985**. *78*, 18.
5. Hoskins, J.A. *J. Appl. Toxico.* **1984**, *4*, 283-292.
6. Chang, S.T.; Chen, P.F.; Chang, S.T. *J. Ethnopharmacology* **2001**, *77*, 123-127.
7. Nath, S.C.; Hazarika, A.K.; Baruah, A. *J. Essen. Oil Res.* **1996**, *8*, 421-422.
8. Chalchat, J.C.; Valade, I. *J. Essen. Oil Res.* **2000**, *12*, 537-540.
9. Baruah, A.; Nath, S.C.; Leclercq, P.A. *J. Essen. Oil Res.* **1999**, *11*, 194-196.
10. Kawakami, M.; Sachs, R.M.; Shibamoto, T. *J. Agric. Food Chem.* **1990**, *38*, 1657-1661.
11. Chang, S.T.; Chen, P.F.; Chang, S.C. 2001. *J. Ethopharmacology* **2001**, *77*, 123-127.
12. Lee, H. S.; Ahn, Y. J. *J. Agric. Food Chem.* **1998**, *46*, 8-12.
13. Ouattara, B.; Simard, R.E.; Holley, R.A.; Piette, G.J.P.; Begin, A. *J. Food Microbiol.* **1997**, *37*, 155-162.
14. Sivakumar, D.; Wijerstnam, R.S.W.; Wijerstnam, R.L.C.; Abeyesekere, M.. *Crop Protection* **2002**, *21*, 847-852.
15. Singh, H.B.; Srivastava, M.; Singh, A.B.; Srivastava, A.K. *Allergy* **1995**, *50*, 995-999.
16. Park, J.D.; Lee, S.G.; Kim, C.S.; Byun, B.K. *FRI. J. For. Sci.* **1998**, *57*, 151-156.
17. Chang, S.T.; Cheng, S.S. *J. Agric. Food Chem.* **2002**, *50*, 1389-1392.
18. Huang, Y.; Ho, S.H. *J. Stored Pro. Res.* **1998**, *34*, 11-17.
19. Kreydiyyeh, S.I.; Usta, J.; Copti, I. *Food Chem. Toxicol.* **2000**, *38*, 755-762.
20. Lee, S. E.; Kim, M.K.; Lee, S.G.; Ahn, Y.J.; Lee, H.S. *Food Sci. Biotechnol.* **2000**, *9*, 330-333.
21. Lee, H.S.; Kim, B.S.; Kim, M.K. *J. Agric. Food Chem.* **2002**, *50*, 7700-7703.
22. Rong, Y.; Yang, E.B.; Zhang, K.; Mack, P. *Anticancer Res.* **2000**, *20*, 4339-4345.
23. Rota, C.; Chignell, C.F.; Mason, R.P. *Free Rad. Biol. Med.* **1999**, *27*, 873-881.
24. Zhuang, J.; Cohen, G.M. *Toxicol. Lett.*. **1998**, *102-103*, 121-129.

25. Usta, J.; Kreydiyyeh, S.; Bajakian, K.; Nakkash-Chmaisse, H. *Food Chem. Toxicol.* **2002**, *40*, 935-940.
26. Shigenega, M.; Hagen, T.M.; Ames, B. *Proc. Natl. Acad. Sci. U.S.A.* **1994**, *91*, 10771-10778.
27. Yang, C.F.; Shen, H.M.; Ong, C.N. *Arch. Biochem. Biophys.* **2000**, *374*, 142-152.
28. Marchetti, P.; Decaudin, D.; Macho, A. Zamzami, N.; Hirsch, T.; Susin, S.A.; Kroemer, G. *Eur. J. Immnol.* **1997**, *27*, 289-296.
29. Kosower, N.S.; Kosower, E.M. *Int. Rev. Cytol.* **1978**, *54*, 109-160.
30. Arrick, B.A.; Nathan, C.F. 1984. *Cancer Res.* **1984**, *44*, 4224-4232.
31. Ault, J.G.; Laurence, D.A. *Exper. Cell Res.* **2003**, *285*, 9-14.
32. Boyland E.; Chasseaud, L.F. *Biochem. Pharm.* **1970**, *19*, 1526-1528.
33. Moon, K.H.; Pack, M.Y. *Drug Chem. Toxicol.* **1983**, *6*, 521-535.
34. Swales, N.J.; Caldwell, J. *Toxicol. in Vitro.* **1996**, *10*, 37-42.
35. Mignotte, B.; Vayssiere, J.L. *Euro. J. Biochem.* **1998**, *252*, 1-15.
36. Green, D.R.; Kroemer, G. *Trends Cell Biol.* **1998**, *8*, 267-271.
37. Ka, H.; Park, H.J.; Jung, H.J.; Choi, J.W.; Cho, K.S.; Ha, J.; Lee, K.T. *Cancer Lett.* **2003**, *196*, 143-152.

Chapter 24

Effect of Black Tea Theaflavins on 12-*O*-Tetradecanoylphorbol-13-acetate-Induced Inflammation

Expression of Pro-Inflammatory Cytokines and Arachidonic Acid Metabolism in Mouse Ear and Colon Carcinogenesis in Min (Apc+/–) Mice

Mou-Tuan Huang[1], Yue Liu[1], Divya Ramji[2], Shengmin Sang[1,2], Robert T. Rosen[2], Geetha Ghai[2], Chung S. Yang[1], and Chi-Tang Ho[2]

[1]Department of Chemical Biology, Ernest Mario School of Pharmacy, Rutgers the State University of New Jersey, Piscataway, NJ 08854–8020
[2]Department of Food Science and Center for Advanced Food Technology, Rutgers, The State University of New Jersey, 65 Dudley Road, New Brunswick, NJ 08901–8520

We have examined the effects of black tea constituents, theaflavin mixture (TFs), on 12-*O*-tetradecanoylphorbol-13-acetate-induced inflammation, up-expression of pro-inflammatory cytokines, interleukin-1β (IL-1β) and interleukin-6 (IL-6) protein levels and formation of prostaglandin E_2 (PGE_2) and leukotriene B_4 (LTB_4) levels in ears of CD-1 mouse as well as on spontaneous colorectal carcinogenesis in Min (Apc+/-) mice. Two doses of TFs, 0.18 and 0.71 mg, were topically applied to mice ears twice a day for 3.5 days (7 treatments) 20 mins prior to each TPA (0.4 nmol) treatment. These doses of TFs inhibited TPA-induced a) inflammation by 74 and 97%, b) up-expression of IL-1β protein levels by 93 and 99%, and c) up-expression of IL-6 protein levels by 58 and 99%, respectively. In these experiments, TFs (0.18 mg or 0.71 mg) also inhibited TPA

© 2006 American Chemical Society

induced formation of PGE$_2$ by 27-100% and LTB$_4$ by 92-100%. In addition, oral administration of 0.2% or 0.4% TFs (28%) enriched black tea extract in drinking water as sole source of drinking fluid to female Min (Apc+/-) mice for 10 weeks inhibited the formation of the numbers of colorectal tumors per mouse by 39 or 29%, respectively. The percentage of mice bearing colorectal tumors was also inhibited by 39 or 29%, respectively. Furthermore, in these mice there was a reduction in the formation of the number of small intestinal tumors per mouse by 39 or 29% respectively.

Black tea accounts for about 80% of tea consumed worldwide ([1]). Black tea is produced from fresh tea leaves by fermentation through enzymatic oxidation ([1,2]). The hot water extract of black tea contains about a 2-6% mixture of theaflavin polyphenols, and greater than 20% thearubigens ([1-3]). The major theaflavin constituents of black tea are theaflavin (TF), theaflavin-3-gallate (TF-3-G), theaflavin-3'-gallate (TF-3'-G) and theaflavin-3,3'-digallate (TF-3,3'-diG). Black tea and green tea polyphenols have antioxidant activity ([4-6]) and the reported biological effects of black tea may be partially related to its antioxidant properties. Liang *et al.* ([5]) showed that topical application of the major constituents of black tea inhibited TPA-induced mouse ear edema. Additionally, black tea extract also be reported to have anti-inflammatory activity in the carrageenan-induced paw edema model in rats ([5]). Furthermore, the black tea constituent theaflavins were observed to inhibit TPA-induced edema of the mouse ear with the following order of potency where TF-3,3'-diG > TF-3-G = TF-3'-G > TF ([6]). Luceri *et al.* reported that administration of black tea extract equal to 40 mg of black tea polyphenols/kg/rat inhibited AOM-induced expression of COX-2, iNOS, glutathione *S*-transferase (GST), GST-M2 and GST-P in colon tumors ([7]). Pan *et al.* ([8]) reported that black tea polyphenols inhibited the activation of NFκB in activated murine macrophages (RAW 264.7 cell line). Theaflavin-3,3'-digallate (TF-3,3'-diG) is the most potent inhibitor of the activation of NFκB among black tea polyphenols ([9]). Recently, our laboratory demonstrated that topical application of TF and TF-3,3'-diG inhibited TPA-induced up-expression of IL-1β and IL-6 protein levels as well as inhibited TPA-induced increasing levels of PGE$_2$ and LTB$_4$ in mouse ear tissues ([6]). TF-2 (a mixture of TF-3-G and TF-3'-G) has been shown to inhibit COX-2 in cell lines ([10]). These mechanisms add strength to the claim that black tea may be an anti-inflammatory agent by affecting the molecular targets that lead to inflammation. This report deals with the effect of black tea on TPA-induced inflammation, expression of pro-inflammatory cytokines, arachidonic acid metabolism in mouse ear and colon carcinogenesis in C57BL/6J Min (Apc+/-) mice providing further evidence to its mechanism of action.

Materials and Methods

Animals and Chemicals

Female CD-1 mice (3-4 weeks old) were purchased from the Charles River Breeding Laboratories (Kingston, NY). Female C57BL/6J Min (Apc+/-) mice were purchased from Jackson Laboratories (Bar Harbor, ME). Theaflavin related compounds (theaflavin mixture) were prepared as previously described (*11*). 12-*O*-Tetradecanoylphorbol-13-acetate (TPA) was purchased from Sigma Chemicals Company (St. Louis, MO). IL-1β and IL-6 ELISA kits were purchased from BioSource (Camarillo, CA). PGE_2 and LTB_4 EIA kits were purchased from Cayman (Ann Arbor, MI). Phosphate buffered saline was used to homogenize tissue samples.

Animal treatment CD-1 mice

Female CD-1 mice (23-30 days old, obtained from Charles River Breeding Laboratories, Kingston, NY) were divided into 3-4 groups of 5-6 mice each. All test compounds were dissolved in acetone, which served as the solvent (negative) control as it does not induce inflammation. A TPA control group was done to indicate the maximum induction of inflammation in mouse ear. Black tea theaflavin mixtures were evaluated in a dose dependent manner for their inhibitory effect on mouse ear inflammatory model.

Topical application of theaflavins and TPA

Both ears of CD-1 mice were treated topically with 15 µL acetone (solvent control group), TPA (TPA control) or test compound in acetone 20 mins before topical application of 15 µL acetone (solvent control) or 0.4 nmol TPA in acetone on the mouse ear. This treatment was continued twice a day for 3.5 days (7 treatments). Six hours after the last TPA treatment, the mice were sacrificed by cervical dislocation and the ears punches (6-mm in diameter) were taken and weighed. The ear punches from each group were then combined and homogenized with phosphate buffered saline. The resulting homogenate was centrifuged and the supernatant was used to test for the levels of the various inflammatory mediators using the enzyme linked immunosorbent assay (ELISA).

Oral administration of theaflavin

Two doses, 5 and 10 mg of theaflavin were evaluated to examine the dose-response. For these experiments theaflavins were dissolved in water and given by gavages to CD-1 mice prior to TPA application. Each dose was given in two sequences. The 5 mg dose was given twice at 2.5 mg each time. Similarly, the 10 mg dose was given twice at 5 mg each. The total pretreatment time for theaflavin was 60 mins and these studies were design so that the first dose of theaflavin was given 60 mins and the second dose 20 mins prior to TPA application. At 6 hrs after TPA treatment, mice were sacrificed to evaluate ear inflammation.

Female C57BL/6J Min (Apc+/-) mice

For spontaneous colon carcinogenesis study, female C57BL/6J Min mice (APC+/-; 5-6 weeks old; 3 groups with 13 mice per group) were given water (control), 0.2% or 0.4% black tea in drinking water as sole source of drinking fluid for 10 weeks, for these experimental animals. The mice were then sacrificed for colorectal cancer evaluation.

Synthesis of Theaflavin Mixtures

A mixture of theaflavins was synthesized from green tea polyphenols using enzymatic oxidation methods. Specifically, after filtration, the crude green tea polyphenol (1.8 g, commercial sample containing 80% catechins) was loaded directly onto a Sephadex LH-20 column eluted first with 95% ethanol to remove non-catechin flavonoids, and then the column was eluted with acetone to obtain a mixture of tea catechins (1.34 g). The tea catechins were dissolved in a pH-5 buffer (50 mL), which contained 4 mg horseradish peroxidase. While being stirred, 3.0 mL of 3.13% H_2O_2 was added 5 times during 1 hr. The enzymatic reaction solution containing catechins and crude peroxidase had turned into a reddish solution during oxidation reaction. The reaction mixture was extracted by ethyl acetate (50 mL × 3). After concentration, the residue (0.97 g) was subjected to Sephadex LH 20 column eluted with acetone-water solvent system (from 35% to 50%). 350 mg of a theaflavin mixture was obtained (TF1: TF2: TF3 is 1:1:1; TF1 is theaflavin, TF2 is the mixture of theaflavin monogallates, TF3 is theaflavin 3,3'-digallate).

Preparation of Ear Homogenates

Tissues were homogenized in a phosphate buffered saline solution containing 0.4 M NaCl, 0.05% Tween-20, 0.5% bovine serum albumin, 0.1 mM phenylmethylsulphonyl fluoride, 0.1 mM benzethonium, 10 mM EDTA and 20 mM KI aprotinin per mL. The homogenates were centrifuged at 12,000 x g for 60 mins at 4 °C. The supernatant was used for the determination of cytokine levels.

A two-site sandwich Enzyme-Linked ImmunoSorbant Assay (ELISA) was used to assay for cytokines.

ELISA Assay Procedure

The IL-1β and IL-6 ELISA kits follow the same basic procedure. The capture antibody, diluted with PBS, was used to coat a 96-well plate overnight at room temperature. The plate was then washed, blocked (1% BSA, 5% sucrose in PBS with 0.05% NaN$_3$), and washed again. The standards were added to the plate leaving at least one zero concentration well and one blank well. The diluted samples (1:3-1:8) were then added to the plate. After incubating for 2 hrs the plates were washed and the detection antibody added. After incubating for another 2 hours the plates were washed and Streptavidin-HRP was added. After 20 mins incubating, the plates were washed and substrate (H$_2$O$_2$ and tetramethylbenzidine) was added. After another 20 mins incubating, the stop solution (2 N H$_2$SO$_4$) was added and the plates were read with a microplate reader at a wavelength of 450 nm.

The LTB$_4$ and PGE$_2$ ELISA kits follow the same basic procedure. The well plates are pre-coated with goat polyclonal anti-mouse IgG and blocking proteins. The standards and samples are added to the wells and incubated for an hour with tracer and anti-serum. After washing, Ellman's Reagent is added for color development. After incubating in the dark, the plate is read by microplate reader at a wavelength of 420 nm.

Results

Effect of Topical Application of Theaflavin Mixture (TFs) on TPA-induced Ear Inflammation Mouse Model:

Topical application of the two concentrations of TFs (either 0.18 or 0.71 mg) to both ears of CD-1 mice 20 mins before each TPA (0.4 nmol) treatment inhibited TPA induced ear inflammation by 74 and 97%, respectively (Panel A, Figure 1). In these experiments TPA-induced up-regulation and increased expression of IL-1β protein levels were inhibited by 93 or 99%, respectively (Panel B, Figure 1), and inhibited TPA-induced up-expression of IL-6 protein levels by 58 or 99%, respectively (Panel C, Figure 1). The results indicated that TFs strongly inhibited TPA-induced inflammation, up-expression of IL-1 and IL-6 protein levels in mouse ears.

Effects of Topical Application of TFs on TPA-induced Arachidonic Acid Metabolism in Ears of CD-1 Mice

Effects of TFs on TPA-induced formation of arachidonic acid metabolites, PGE_2 and LTB_4 were evaluated in the mouse ears. PGE2 and LTB4 are formed by TPA-induced increase in metabolism of arachidonic acid (AA) via cyclooxygenase and lipoxygenase pathways, respectively. Results showed that there was a concentration dependent inhibition of formation of these metabolites when TF was topically applied to mice ears. PGE_2 levels were reduced by 27 and 100% (Panel A, Figure 2), while LTB_4 levels were down by 92 and 100%, respectively (Panel B, Figure 2). Although, TFs inhibited TPA-induced formation of AA metabolites, PGE2 and LTB4 from both cyclooxygenase and lipoxygenase pathways, our data suggests that TFs had a stronger inhibitory effect on the formation of AA metabolites from lipoxygenase pathway than the cyclooxygenase pathway.

Inhibitory Effect of Oral Administration of TFs on TPA-induced Edema of Mouse Ears

Oral intubation of 2.5 or 5.0 mg of TFs to female CD-1 mice at 60 and 20 mins before topical application of TPA (1.5 nmol) inhibited TPA-induced edema of CD-1 mouse ears by 45 and 56%, respectively. Aspirin (2.5 mg), which is a known anti-inflammatory agent served as positive inhibitor control, inhibited TPA-induced edema of mouse ear by 35%. The inhibitory effect of both low dose (5 mg) and high dose (10 mg) of TFs and aspirin (2.5 mg) were statistically different from the positive control TPA (group 2).

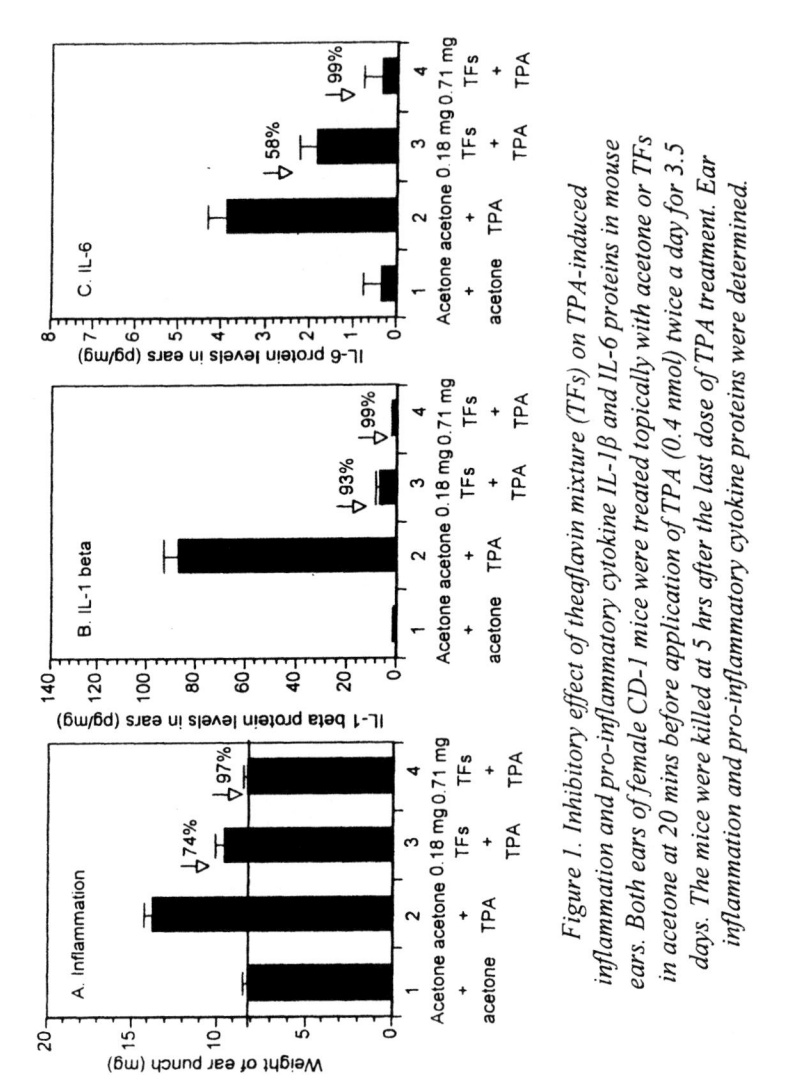

Figure 1. Inhibitory effect of theaflavin mixture (TFs) on TPA-induced inflammation and pro-inflammatory cytokine IL-1β and IL-6 proteins in mouse ears. Both ears of female CD-1 mice were treated topically with acetone or TFs in acetone at 20 mins before application of TPA (0.4 nmol) twice a day for 3.5 days. The mice were killed at 5 hrs after the last dose of TPA treatment. Ear inflammation and pro-inflammatory cytokine proteins were determined.

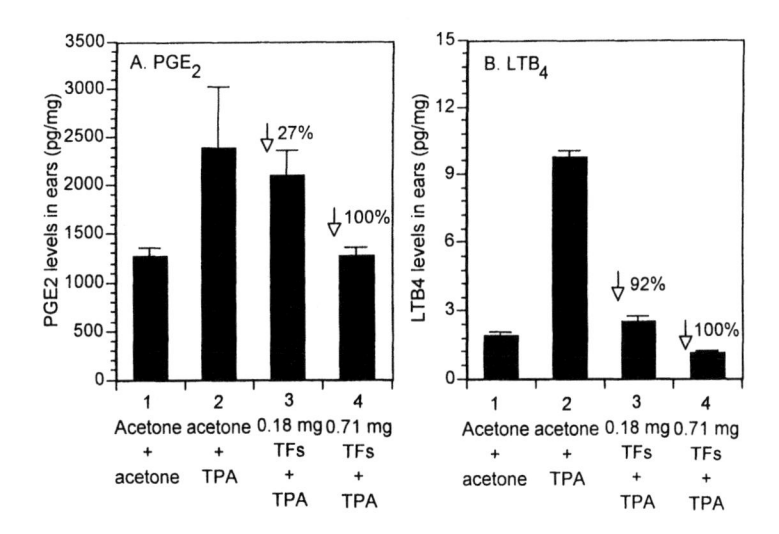

Figure 2. Inhibitory effect of theaflavins (TFs) on TPA-induced formation of PGE_2 and LBT_4 levels in mouse ears. Both ears of female CD-1 mice were treated topically with acetone or TFs in acetone at 20 mins before application of TPA (0.4 nmol) twice a day for 3.5 days. The mice were killed at 5 hrs after the last dose of TPA treatment. PGE_2 and LBT_4 proteins were determined.

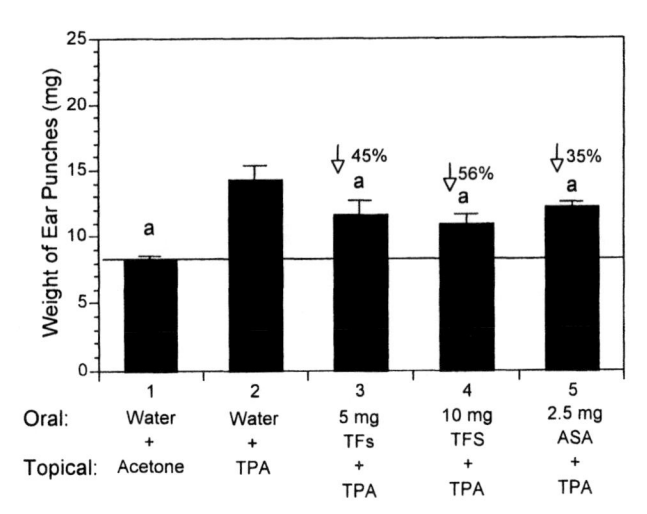

[a] Statistically different from group 2 TPA (P<0.05) as determined by the Student's t test.

Figure 3. Inhibitory effects of oral administration of theaflavin mixture (TFs) on TPA-induced ear inflammation in CD-1 mice, Female CD-1 mice were given 2.5 mg and 5.0 mg of TFs or aspirin (ASA, 1.25 mg) in 1 mL water by oral gavage at 60 and 20 mins before topical application of acetone or TPA (1.3 nmol) on both ears. The mice were killed at 6 hrs after TPA treatment. Ear punches (6-mm in diameter) were taken and weighed. Data are mean±SE from 12 ears average.

Inhibitory Effect of Oral Administration of Black Tea Extract (Enriched with Theaflavin Mixture) on Colorectal Carcinogenesis in Min (Apc+/-) Mice Model

Oral administration of 0.2% and 0.4% of black tea extract (28% TFs) as sole source of drinking fluid to female C57BL/6J Min (Apc+/-) mice inhibited the formation of the numbers of colorectal tumors per mouse by 59 or 34%, respectively. The percent of mice bearing colorectal tumors was inhibited by 56 or 29%, respectively (Figure 4). Oral administration of 0.2% and 0.4% of black tea extract enriched with theaflavin in water, the sole source of drinking fluid to female Min mice (Apc+/-) for 10 weeks inhibited formation of the numbers of small intestinal tumors per mouse by 39 and 29%, respectively. The total numbers of small and large intestinal tumors per mouse were inhibited by 39 and 29%, respectively (Figure 5).

There was no change in the body weight (water control group vs. black tea group.)

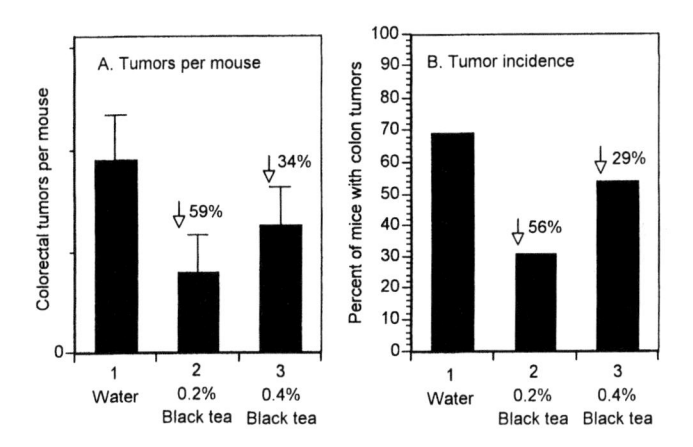

Figure 4. Effect of oral administration of black tea extract (enriched with theaflavins) on formation of colorectal tumors in C57BL/6J Min mouse (Apc+/-). Female Min mouse (Apc+/-; 5-6 week old; 13 mice per group) were orally administered with water, 0.2% black tea and 0.4% black tea for 10 weeks. The mice were killed and corectal tumors were examined and counted. Data are expressed as the mean±SE as determined by Student's t test.

Discussion

Present results demonstrated that topical application of theaflavin mixture (TFs) strongly inhibited TPA-induced inflammation, up-expression of IL-1β and IL-6 protein levels in ears of CD-1 mice. Additionally, TFs also strongly inhibited TPA-induced formation of PGE_2 and LTB_4 levels in mouse ears, however, the inhibition of formation of LTB_4 was greater than that of PGE_2. These data suggested that TFs inhibited arachidonic acid metabolism by blocking the lipoxygenase pathway to a greater extent than the cyclooxygenase pathway. Thus, combination of TFs with sulindac (a cyclooxygenase inhibitor) showed some synergetic effect. Oral intubation of theaflavin mixture also inhibited TPA-induced ear edema (local inflammation). These observations suggested that theaflavin or its metabolites may be absorbed through the intestine and is transported to mouse ears. Feeding black tea extract enriched with theaflavin mixture (about 28%) in water, the sole source of drinking fluid, to female C57BL/6J Min (Apc+/-) for 10 weeks inhibited the formation of the numbers of small and large intestinal tumors per mouse. However, 0.4% black tea extract had less inhibitory effects on both numbers of small and large intestinal tumors per mouse. More studies are needed to determine the bioavailability and anti-carcinogenic effects of black tea and theaflavin rich black tea.

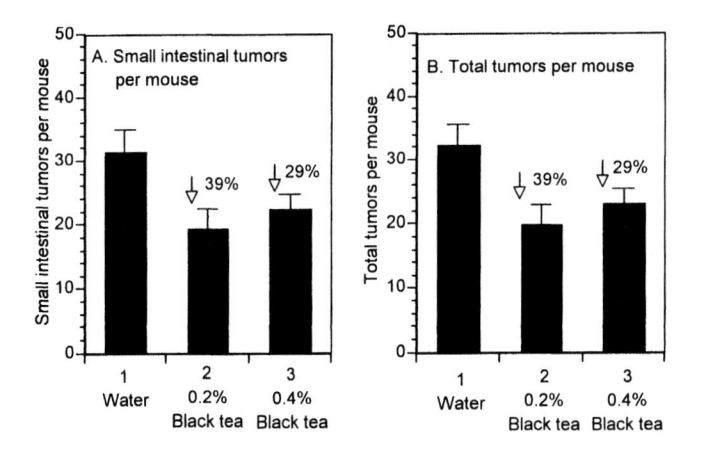

Figure 4. Effect of oral administration of black tea extract (enriched with theaflavins) on formation of small and large intestinal tumors in C57BL/6J Min mouse (Apc+/-). Female Min mouse (Apc+/-; 5-6 week old; 13 mice per group) were orally administered with water, 0.2% black tea and 0.4% black tea for 10 weeks. The mice were killed and corectal tumors were examined and counted. Data are expressed as the mean±SE as determined by Student's t test.

References

1. Graham, H.N. *Prev. Med.* **1992**, *21*, 334-345.
2. Balentine DA. In *Phenolic Compounds in Food and Their Effects on Health I: Analysis, Occurrences and Chemistry.* Ho, C.-T.; Lee, C.Y.; Huang, M.T. (Eds.), ACS Symp. Ser., 506, American Chemical Society, Washington, D.C., 1992, pp. 102-117.
3. Suzuki, N.; Hatate, H.; Itami, T.; Takahashi, Y.; Oguni, I.; Kanamori, N.; Nomoto, T.; Yoda, B. In *Food Factors for Cancer Prevention.* Ohigashi, H.; Osawa, T.; Terao, J.; Watanabe, S.; Yoshikawa, T. (Eds.), Sringer-Verlag: Tokyo, 1997, pp. 152-155.
4. Shiraki, M.; Hara, Y.; Osawa, T.; Kumon, H.; Nakayama, T.; Kawakish, S. *Mutat. Res.* **1994**, *323*, 29-34.
5. Liang, Y.C.; Tsai, T.C.; Lin-Shiau, S.Y.; Chen, C.F.; Ho, C.-T.; Lin, J.K. *Nutr. Cancer* **2002**, *42*, 217-223.
6. Liu, Y.; Rosen, R.T.; Ho, C.-T.; Ghai, G.; Huang, M.T. *Proc. Am. Asso. Cancer Res.* **2003**, *44*, 1101-1102.
7. Luceri, C.; Cadermi, G.; Sanna, A.; Dolara, P. *J. Nutr.* **2002**, *132*, 1376-1379.
8. Pan, M.H.; Lin-Shiau, S.Y.; Ho, C.-T.; Lin, J.H.; Lin, J.K. *Biochem. Pharmacol.* **2000**, *59*, 357–367.
9. Lin, Y.L.; Tsai, S.H.; Lin-Shiau, S.Y.; Ho, C.-T.; Lin, J.K. *Eur. J. Pharmacol.* **1999**, *19*, 379-388.
10. Lu, J.; Ho, C.-T.; Ghai, G.; Chen, K.Y. *Cancer Res.* **2000**, *60*, 6465-6471.
11. Sang, S.; Lambert, J.D.; Tian, S.; Hong, J.; Hou, Z.; Rye, J-H.; Stark, R.E.; Rosen, R.T.; Huang, M-T.; Yang, C.S.; Ho, C.-T. *Bioorg. Med. Chem.* **2004**, *12*, 459-467.

Chapter 25

Shea Butter: Chemistry, Quality, and New Market Potentials

Hisham Moharram, Julie Ray, Sibel Ozbas, Hector Juliani, and James Simon

New Use Agriculture and Natural Plant Products Program, Department of Plant Biology and Pathology, Rutgers, The State University of New Jersey, New Brunswick, NJ 08901

Shea butter from the Shea tree (*Vitellaria paradoxa* C.F. Gaertn.) is an important plant fat rapidly gaining popularity in cosmetics, personal care products and foods. Shea butter is becoming more popular because of its unsaturated fatty acids composition as well as the potential utility of its unsaponifiables fraction now being used in cosmeceutical, pharmaceutical and nutraceutical applications. The chemical constituents, physical and chemical properties, quality impacting factors, quality control issues, as well as the current and potential new uses of Shea butter by different industries are reviewed.

Introduction

The Shea tree (*Vitellaria paradoxa* C.F. Gaertn.) is an important multi-purpose tree of sub-Sahara and sub-tropical Africa. Shea has long been known for many different traditional uses in foods, medicines and personal care products by indigenous African peoples and, more recently, has received increasing interest by Western industries and consumers, alike. Despite increasing demand, Shea butter, the principle commercial product from nut of the Shea tree, is still facing difficulties becoming a major botanical commodity in global trade and major industries. The objective of this review is to present the chemistry of shea butter and discuss issues of quality facing the variety of industrial applications as foods, edible oils, nutraceticals, cosmetic and in pharmaceuticals. Pharmacological studies and the commercialization processes in Shea butter are also presented.

Taxonomy and Genetics

The current accepted taxonomic classification of the Shea tree is *Vitellaria paradoxa* C.F. Gaertn. Earlier references to Shea include *Butyrospermum parkii* (G.Don) Kotschy and *B. paradoxum* (Gaertn.) Hepper, (*1*). There are two recognized subspecies, *V. paradoxa* subspecies *paradoxa* and *V. paradoxa* subspecies *nilotica* which differ mostly in the properties of the butter they produce (*1*). The Shea tree is long-lived (up to 300 years) and indigenous to savannah and forest regions of sub-Saharan Africa from Senegal and Mali on the West coast to Ethiopia in the East Africa, and from Sudan to Uganda. Besides the subspecies differences, distinct genetic variations exist and the genetic and environment interaction results in different biochemical properties of the butter. Recent investigations (*2-3*) have shown that this can manifest itself in compounds that have important commercial implications and potential.

Shea Butter Extraction Procedures

The fruit of the Shea tree contains a nut which protects the live kernel inside. Traditional processing to remove the nut and procure the butter involves 1) removal of the fruit pulp (usually by fermentation and manual peeling), 2) boiling the nuts followed by drying in the sun, 3) shelling (to remove the kernels from the nut), 4) roasting the kernels to dry them, 5) pounding the kernels to a paste, 6) emulsification of the fat (by kneading and hand-beating of paste) and 7) boiling the paste to separate the fat from the Shea nut cake. During these steps, many factors can cause degradation and deterioration of the Shea butter quality. Introduction of mechanization helps reduce the time required for processing and extraction, but the process is still not fully mechanized as to allow greater industrial acceptance. This results in variations in butter quality from region to region and from batch to batch which present very real problems to receiving industries.

Shea Butter Constituents

Fatty Acids

Recent fatty acid profile which we conducted for shea butter in our laboratories yielded the following results: *palmitic acid* (C16:0) ~3.4%; *stearic acid* (C18:0) ~44%; *oleic Acid* (C18:2) & *linolenic acid* (C18:3) ~46.4%; *linoleic acid* (C18:1) ~6.2%; and, in traditionally processed Ghanaian Shea butter, a very small amount of *arachidic acid* (C20:0) (unpublished data). For industries, one of the most attractive parameters of Shea butter is the high ratio of unsaturated to saturated fatty acids. Maranz and Wiesman (4) analyzed Shea butter from the across the full range of distribution (including the *nilotica* subspecies) and found wide ranges for these same fatty acids. Palmitic ranged from 3.8-4.7%, stearic from 31.2-45.1%, oleic from 43.3-56.5%, linoleic from 6.2-8.5% and arachidic from 0.2-1.4%.

Triglycerides

A study by Maranz *et al.* (5) indicated that the principle triglycerides in Shea butter was *stearic-oleic-stearic*, or SOS, and *stearic-oleic-oleic*, or SOO. SOS ranged from 13% of total triglycerides in Ugandan Shea butter, to 45% in Burkina Faso, while SOO was highest (28-30%) in Ugandan Shea butter and some Malian Shea butter. The SOS to SOO ratio is an important determinant of melting point of plant butters (6). Maranz et al. (5) found that the ratio of SOS to SOO ranged from 0.5 in Ugandan Shea butter to 2.3 in the Burkina Faso Shea butter, over a 12°C difference in melting point.

Phenolic compounds (antioxidant activity)

Maranz *et al.* (2) reported on the phenolic constituents of shea kernels and found eight catechin compounds; gallic acid, catechin, epicatechin, epicatechin gallate, gallocatechin, epigallocatechin, gallocatechin gallate, and epigallocatechin gallate. They also found quercetin and *trans*-cinnamic acid. They report a mean kernel content of the eight catechin compounds at 4000 ppm with a range of 2100-9500 ppm. However, the relative proportions of these compounds varied from region to region in the Maranz *et al.* (2) study. Gallic acid, the major phenolic compound, comprised 27% of the measured total phenols, but exceeded 70% in some populations.

Tocopherol

Maranz and Wiesman (3) analyzed the *tocopherol* content of shea butter using an HPLC procedure and found high variability between provenances tested and a significant effect of climate on the α–*tocopherol* levels was detected. Total tocopherol content (α, β, γ, δ) in the samples they tested, from eleven different countries, ranged from 29-805 µg/g shea butter. Maranz and Wiesman (3) reported that α–*tocopherol* averaged 64% of the total tocopherol content. Their findings suggest that shea butter from hot, dry areas, such as N'Djamena in Chad, had the highest levels of α–*tocopherol* (a mean of 414 µg/g), while the lowest came from samples from cool highland areas such as northern Uganda (a mean of 29 µg/g).

Unsaponifiable compounds

The European Union study (INCO) found a total triterpenes range of between 7 and 12.5% across the Shea tree distribution zone. The two major triterpenes were acetyl triterpenes (1.6-5.4%) and cinnamyl triterpenes (2-7%). Other compounds found in the unsaponifiables fraction include α and β amyrine, butyrospermol, and lupeol. The sterolic fraction includes α-spinasterol and Δ7-stigmasterol (7), campesterol, and β-sitosterol (8). Findings by Ogunwande *et al.* (9) indicate that unsaponifiable compounds of Shea butter may impart antimicrobial activity. Their study found that methanol extracts of Shea butter inhibited *Pseudomonus aeruginosa, Salmonella typhi, Staphylococcus aureus, Bacillus cereus, Fusarium oxysporium,* and *Candida albicans.*

Physico-Chemical Properties of Shea Butter

Properly processed extracted and stored Shea butter appears to be very stable, low in free fatty acid (FFA) levels and highly creamy in consistency. Improper storage conditions can cause very high peroxide levels. Peroxides result from autoxidation catalyzed by heat, heavy metals, and ultraviolet radiation and may be precursors of volatile and malodorous compounds such as aldehydes or ketones. Table I. shows the chemical parameters of Shea butter, and Table II. shows the physical and organoleptic parameters.

Traditionally prepared Shea butter may be highly variable in color and percent unsaponifiables matter. The European Union (INCO) study of shea nut trees and shea butter covered the whole shea tree region from Senegal to Chad and included vastly different genetic material (combined the values of the two subspecies, *nilotica* and *paradoxa*) found in very different climatic (and edaphic) conditions. Thus, the values given by the European study for most of these parameters are, in fact, wide ranges reflecting the large genetic diversity found in natural stands of shea nut tree (See http://prokarite.org/vitellaria-dbase-en/index.html).

Table I. Chemical Parameters and Ranges Most Commonly Reported in Commercial Literature.

Parameter	*Range*
Fatty acid profile	Palmitic 3-8%; stearic 40-46%; oleic + linolenic 42-49%; linoleic 4-8%; arachidic 0-3%
Peroxide value	0-5 meq/kg, or <1 mmole/kg
Iodine value	45-71 centigram iodine/g Shea butter
Free fatty acids	<0.5%
% Un-saponifiables	4-9%
Saponification value	175-185 mg KOH/g Shea butter

Table II. Physical and Organoleptic Parameters and Ranges Most Commonly Reported in Commercial Literature.

Parameter	*Range*
Refractive index	1.4628 – 1.4680
Melting point	28-34 degree C
Color	Pale yellow to off white
Odor	Slight musty smell
Consistency or texture	Creamy to grainy

Shea Butter Quality Degradation Factors

There are several factors which may lead to the degradation of Shea butter such as microbiological contamination, chemical degradation, refining processes, and enzymatic degradation, and most occur post the collection stage during handling, processing and storage.

Microbiological Contamination

One of the major challenges facing Shea butter imports and trade (even the importation of raw or dried Shea nuts) into the US is microbiological contamination and quality degradation. Sources of contamination are many but can include the following:
- Fruit diseases and surface mold cause fruit decay. Microorganism contaminations reduce fruit quality and negatively impact shea butter color.
- Leaving the ground and crushed material for too long (even just over night) before hot water extraction. This allows lipase enzymes to breakdown fatty acids into free fatty acids and promotes oxidative degradation.
- Using contaminated water in the extraction process introduces contaminants to the extracted shea butter.
- Open containers in wet or damp areas allow contaminants to enter shea butter.
- Non-sterile packaging by wholesalers and retailers also introduces contaminants.
- Time before boiling, roasting and drying increases fungal infections. Improved efficiency is required at these steps.

Chemical Degradation (due to oxidation and hydrolysis)

- Agents or factors that result in reduction or destruction of cinnamic acid in the non-saponifiable fraction reduce the anti-histaminic and anti-inflammatory activity of shea butter.
- Exposure of boiled shea nuts to air for too long results in higher fungal attacks and peroxide levels.
- UV damage due to prolonged and inefficient sun drying promotes peroxide damage.
- Oxidation of unsaturated fatty acids causes the rancidity of shea butter and reduces quality.
- Exposure of shea butter to factors such as heat, light, and traces of metals particularly copper and iron promote rancidity.
- Microorganisms producing oxidizing enzymes also promote oxidative rancidity.

Refining

Refining is the normative process of Shea in Europe, but only because the European market has historically incorporated Shea butter as a cocoa butter substitute. Refining of Shea butter results in the loss of a significant portion, or even the entire unsaponifiable fraction. The unsaponifiable fractions are undesirable when the product is to be used as a cocoa butter substitute and when the product needs to be bleached and uniform in color and consistency. However, those same products which are removed in that refining process are also the groups of compounds considered to contribute to quality when the Shea butter is used for alternative markets, such as those which focus on the healing properties critical to pharmaceutical, cosmeceutical and personal care uses.

Enzymatic Degradation

When extraction is done by cold pressing, enzymes released from the tissue of the Shea nut can advance decay and degradation. Heating the Shea butter is necessary to denature these enzymes. Additionally, the longer the time before boiling, roasting or drying begins, the greater the likelihood of enzymatic degradation.

Butter Color

Shea butter color can be a determining quality factor for some industries. Final Shea color is related to quality of the kernels processed. The presence of fungal infections (black nuts) and 'charring' increases the yellowness of the butter and can be prevented or reduced by more efficient drying and roasting techniques. Generally, the best color range for un-refined Shea butter is light yellow to pale beige or off white. Apart from dyes that may traditionally be added, the sources of Shea butter yellowness are: fungal filaments or spores, tannins from kernels and kernel-coating, and contaminants formed during heating or roasting.

Quality Control Considerations

Contamination enters Shea butter at the processing and preparation stages. To avoid microbiological contamination of Shea butter, extraction should use clean well (or sterilized) water. Storage should be in air tight, cool and dry containers; and packaging should be in sterilized containers. Post processing, chemical contaminants come from contaminated water and/or unclean containers used during extraction or for storage. Rancidity is another quality reducing factor in Shea butter and results from oxidation of the unsaturated fatty acids. Polyunsaturated fatty acids are particularly susceptible to oxidative rancidity because they contain one or more double bonds which are, by their nature, highly reactive. The level of polyunsaturated fatty acids is at least one of the determinants of shelf life. Thus, factors which accelerate the reaction (such as heat, light, traces of 'pro-oxidant' metals such as copper and iron) contribute directly to degradation of quality and reduction of shelf life. Microorganisms produce oxidizing enzymes (insensitive to antioxidants), oxygen, high temperatures and oxidative products such as lipid peroxides, oxidative aldehydes and alkenals, and free fatty acids from breakdown of unsaturated fatty acids all promote the onset of rancidity of (unrefined) Shea butter. Recent analyses of Ghanaian Shea butter in our laboratory concluded that several interventions in the Shea butter primary processing are critical to lowering chances of degradation and attaining higher quality Shea butter. These interventions include:

- First, the practice of using heaped versus fresh fruit shows that there was no significant improvement due to heaping and that using fresh fruit will yield acceptable results. However, logistics and practical aspects of the extraction process may preclude this from becoming the accepted norm.
- Second, raw nuts yield much better quality butter than boiled nuts. This would also be a time and resource saving modification in the process. Boiling damages cell walls and makes nuts more prone to heat caused quality degradation. In addition, temperatures in the nuts may not reach high enough to denature all lipases and this may allow these degrading enzymes to become active, yet the temperatures may sufficiently damage membranes as to cause greater lipase release than would normally occur.
- Third, drying the boiled nuts on an open courtyard gave the worst quality shea butter within boiled nuts. This is probably due to the boiled nuts being open microbiological degradation. Nuts dried in closed containers gave better quality nuts probably due to minimal exposure to micro-organisms. However, the best quality she abutter at this stage came from raw nuts dried on the open courtyard. This can be easily explained by the protection the flesh of the fruit gives the nuts, plus the preservation of nut integrity due to not boiling it (which allows microorganisms to enter). Intact fresh, raw nuts dry well and keep the kernel safe from micro-biological degradation. Fried nuts gave lower quality shea within raw nuts than from non fried nuts. This can also be explained by the degrading effects of the heat (speeds up oxidation reactions), and that many nuts will have cracked open due to the heat and become accessible to micro-organism degradation.

- Fourth, frying the nuts within the boiling treatment gives better quality shea butter (probably due to the heat preventing any microbial degradation of the boiled nuts). However, if the nuts are used raw and not boiled, not frying the nuts gives the lower degradation and better shea butter quality.

The overall best quality shea butter comes from fresh fruit, raw nuts, courtyard dried kernels and no frying. Low quality water and containers with metals that catalyze chemical degradation reactions could be additional factors in lowering quality shea of butter. Though our analysis shows higher quality for certain treatments, realizing successful adoption by the processors (typically women) may be difficult due to the lower yield of butter that is achievable. Higher quality comes at a cost of greater effort and lesser return in form of shea butter extracted.

Our investigation of organoleptic properties found that forty percent of individuals who were given unrefined shea butter to assess reported unfavorable reactions to the odor. This negative response to the odor is a major problem in US (and European) public acceptance of unrefined shea butter. Refining shea butter removes all phenolic compounds (degraded and undegraded) and removes the 'unpleasant' aroma. Yet, unrefined Shea butter also contains the unsaponifiable fraction that is associated with its medicinal and healing properties such as the antioxidant activity of several phenolic compounds.

Our results also showed that the percent unsaponifiables is inversely correlated to the peroxide value, while FFA is directly and positively correlated to peroxide value. This is expected since both peroxide value and FFA are indicators of degradation. The same degradation processes usually also cause reduction in percent unsaponifiables in shea butter. Figures 1 and 2 show these relationships quite clearly.

Commercial Uses of Shea butter

Food Industry

Though used for centuries in Africa, and more recently in Europe, as a food ingredient (e.g. edible cooking fat and ingredient in sweets) and recognized as GRAS (generally recognized as safe) by the US Food and Drug Administration, safety and toxicity studies on shea butter to-date are few. Kitamura *et al.* (*10*) investigated the subchronic toxicity effects of the flavonoids in shea nut color (often used as a food-coloring agent) on Wistar rats in 13-week study. They investigated the effects on hematology, serum biochemistry, organ weight and histopathology and concluded that there were no significant toxicological changes in any parameter measured and that the no adverse effects dose of shea

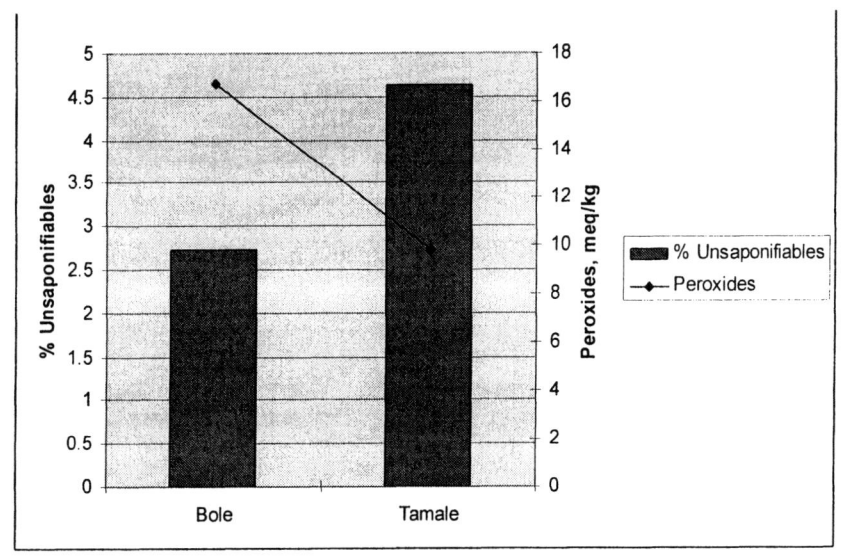

Figure 1. Relation between % Unsaponifiables and peroxide content.

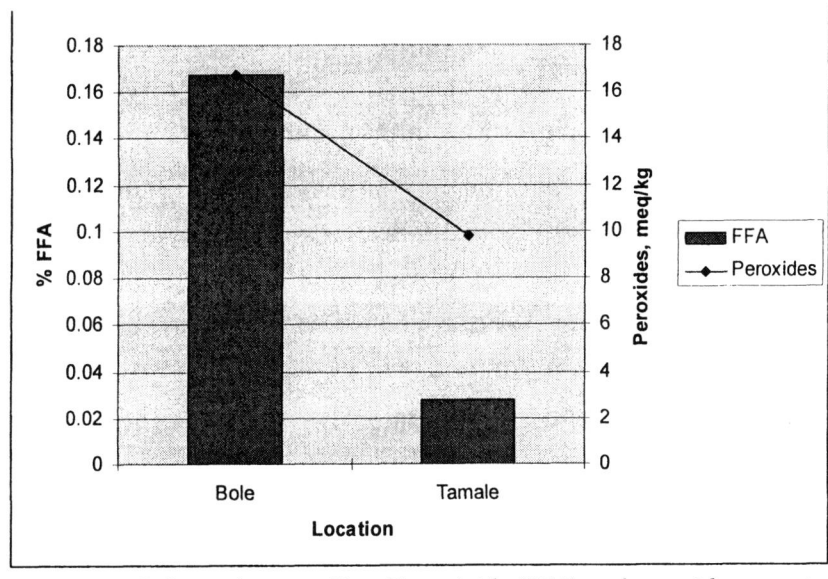

Figure 2. Relation between Free Fatty Acids (FFA) and peroxide content.

nut color was greater than 5% for both female and male rats (3775.5 mg/kg/day and 4387.7 mg/kg/day, respectively). Another 13-week study by Earl et al. (11) looked at the effects of feeding Colworth-Wistar rats shea oleine and hardened shea oleine. The study included palm and soya bean oils for comparison and found similar biological effects due to shea oleine as to palm and soya bean oils. Rats fed shea oleine exhibited a slightly reduced body weight gain, a raised alkaline phosphatase and an increase in food intake compared to palm or soya bean oils. The study found no adverse effects due to consumption of shea oleine in the diet. Earl et al. (12) also investigated absorption and excretion of shea oleine sterols in rat and man. Shea oleine and margarine contain high levels of 4,4-dimethysterols (4,4-DMS; up to 8%), mostly as cinnamic acid esters. There was no preferential absorption of any dimethysterols in either rat or human subjects, with rats releasing 27-52% of the 4,4-DMS in their feces and humans releasing 13-49%. Carthew et al. (13) investigated the carcinogenic potential of shea oleine in rats. They noted no adverse effects and, in comparison to other commercial edible oils, shea oleine showed no tumorigenic potential following dietary administration at 7.5 g/kg/day in rats.

The largest commercial use of shea butter in the food industry is as a cocoa butter substitute or cocoa butter equivalent (CBE), mainly in chocolates and confectionary (14). A cursory search for patents on Shea butter finds records such as patent #WO0030463 which presents a "fat substitute formulation and methods for utilizing same" (15). Maranz et al. (16) also found significant variation in the ratio of stearic to oleic acid and of SOS to SOO in Shea butter, particularly from West African sources. This wide variation opens the door to targeted use in different segments of the Food industry.

Cosmetics and Personal Care Products Industries

Perhaps the industry second to food in Shea butter utilization, but probably exceeds it in growth rate, is the cosmetic industry. Cosmetic uses of Shea butter are exploding in the US. Numerous patents are being filed for Shea butter uses and preparations. The US Patent and Trademark Office lists patent 20030095936 for a Shea butter based lip gloss formulation (17). European patent office Shea butter patent filings include; FR2811224 for "Cosmetic composition for care of sensitive skin includes oleanic acid or vegetable extract rich in oleanic acid, and at least one other vegetable extract chosen from Shea-butter flower and *Solanum lycocarpum*" (18), DE19958612 for "Lipid composition, used in cosmetic or pharmaceutical formulation or textile, paper or leather production, contains n-octadecanoic acid (12-(-1-oxooctadecyl)oxy) ester, neopentanoic ester and/or di-n-octyl ether as solvent" (19), WO9963963 for "composition for cosmetic or dermopharmaceutical use containing a mixture of green coffee and Shea butter extracts" (*20*), WO9805294 for "an oil-in-water emulsion for use on human skin for cleansing, preserving or improving the condition of the skin" (*21*). Shea butter products are commonly found in specialty salons to traditional supermarkets with an increasingly wider array of shea-based products from skin creams to shampoo, from lip gloss to soaps, all including Shea butter or its derivatives or components in their ingredients list. The use of Shea as an ingredient in personal health care products is often associated by consumers as a new natural product with health promoting attributes. Thioune *et al.* (*22*) published on the development of a Shea butter ointment and found satisfactory results when comparing Shea butter to lanoline and petroleum jelly. Loden and Andersson (*23*) found sterol rich canola oil gave surfactant irritated skin better protection than all the other oils including Shea butter. However, in the light of the great diversity unveiled by the comprehensive European study's published findings, it is very likely that the sterol rich Shea butter coming from specific provenances may have a good potential in this respect as well.

Pharmaceutical Industry

Despite empirical determination of the presence of several bioactive compounds in the unsaponifiables fraction of Shea butter (*2,3,9*), significant utilization of the medicinal properties of Shea butter is yet to be seen in the pharmaceutical industry. In fact, the pharmaceutical products industry is probably the least developed industry in terms of Shea butter use. There is a lot of hesitation by the pharmaceutical industry to make large purchases of Shea butter due to the varying and inconsistent quality of Shea butter coming out of African countries that produce it. Historically, Europeans have overcome this by purchasing only the whole nuts and conducting the extraction, purification, refining and fractionation of the Shea butter in their own industrial scale production facilities in Europe. American industries, thus far, prefer to address the problem at the source and imposing strict quality grades and standards on the sellers at the time of purchase. However, due to a number of logistical, practical and knowledge-based problems, this is still proving very difficult. To-date, no grades and standards exist for Shea butter, not even in the U.S. This will be even more difficult to achieve now that the results of the four year European INCO study have been published and show such a wide range in attributes and ranges of the same. Nonetheless, publications on medicinal and therapeutic properties of Shea butter constituents can be found. As early as 1979, Tella (*24*) published on the nasal decongestion activity of Shea butter. Thioune *et al.* (*25*) published on the use of Shea butter in preparation of an anti-inflammatory ointment. Patents based on Shea butter for pharmaceutical uses and preparations are also being filed in these areas. European patent office filings include; WO02055087 which is for "dihydro-triterpenes in the treatment of viral infections, cardiovascular disease, inflammation, hypersensitivity, or pain" (*26*), WO2004002504 for saponin rich "plant extracts" (*27*), WO0206205 for "method for preparing a fatty substance ester and use thereof in pharmaceutics, cosmetics or food industry" (*28*), and WO0103712 for "composition containing extracts of *Butyrospermum parkii* and the use as medicament or dietary supplement" (*29*). It would appear that grades and standards for Shea butter can be developed but need to be end product specific.

Potential New Uses

The recently completed European Union (INCO) funded study resulted in publication of data that shows a significant potential for novel utilizations of Shea butter from different provenances. The study found distinct regions across the distribution zone. Some were characterized by high unsaponifiables, gallic acid, and tocopherol content. Other regions were distinctly high in vitamin E but of soft triglyceride nature, or extremely high in gallic acid and Tocopherol but of hard triglycerides nature. This suggests that industries can target Shea butter from specific regions to purchase Shea butter of the right quality parameters to suit the products they make. Thus, cosmetic, cosmeceuticals and pharmaceutical industries could purchase from the regions producing Shea butter of high percent of unsaponifiables, high in anti-oxidant compounds, predominantly soft triglycerides and high Tocopherol content. Food and neutraceticals industries would purchase from regions producing Shea butter of both hard and soft triglycerides, high in vitamin E and low in % unsaponifiables. We envision that in addition to the multitude of new health care applications of Shea, future work in the incorporation of Shea butter into food products such as margarines and other specialty edible oils should be considered as well as further work using the Shea butter as a carrier for other bioactive and nutraceutical products. A study comparing the effects of a test fat high in myristic acid versus one high in stearic acid (Shea butter) on cholesterol concentration in blood plasma (*30*) found that blood plasma cholesterol concentration (high density lipoproteins, HDL) was higher after myristic acid than stearic acid (Shea butter) and myristic acid resulted in higher increase in postprandial HDL triacylglycerol than stearic acid (Shea butter). Earlier, Tholstrup *et al.* (*31*) found that intake of Shea butter high in stearic acid favorably affects blood lipids and factor VII coagulant activity in young men, compared with fats high in saturated fatty acids. Maranz *et al.* (*2*), also found significant variation inn the ratio of stearic to oleic acid and of SOS to SOO in Shea butter, particularly from West African sources. Findings such as

Tholstrup et al. (*30*) and Maranz et al. (*2*) are likely to have major impact on increased Shea butter use by the Food and Neutraceutical industries.

Acknowledgements

This work was conducted as part of our Agri-Business in Sustainable African Natural Plant Products Program (ASNAPP) with funding from the USAID (Contract Award No. HFM-O-00-01-00116), our Partnership in Food and Industry Development for Natural Products (PFID/NP) Program funded also by the USAID (Contract Award No. AEG-A-00-04-00012-00), TechnoServe-Ghana, the New Use Agriculture and Natural Plant Products Program and the New Jersey Agricultural Experiment Station, Rutgers University. Authors wish to express their appreciation to Julie Asante-Dartey and Daniel Acquaye of ASNAPP; to Kodzo Gbewonyo and to Dr. Peter Lovett for their important feedback.

References

1. Hall, J.B.; Aebischer, D.P.; Tomlinson, H.F.; Osei-Amaning, E.; Hindle, J.R. *Vitellria paradoxa*: a monograph. School of Agricultural and Forest Sciences, University of Wales, Bangor, UK, 1996; 105 pp.
2. Maranz, S.; Wiesman, Z.; Gart, N. *J. Agric. Food Chem.* **2003**, *51*, 6268-6273.
3. Maranz, S.; Wiesman, Z. *J. Agric. Food Chem.* **2004**, *52*, 2934-2937.
4. Maranz, S.; Wiesman, Z. *J. Biogeography* **2003**, *30*, 1505-1516.
5. Maranz, S.; Wiesman, Z.; Bisgaard, J.; Bianchi. G. *Agroforestry Systems* **2004**, *60*, 71-76.
6. Beckett, S. The science of chocolate. The Royal Society of Chemistry: Cambridge, 2000; 175 pp.
7. Dencausse, L.; Ntsourankoua, H.; Artaud, J.; Clamou, J-L. *Oleagineux Corps Gras Lipides* **1995**, *2*, 143-147.
8. Badifu, G.I.O. *J. Food Comp. Anal.* **1989**, *2*, 238-244.
9. Ogunwande, I.A.; Bello, M.O.; Olawore, O.N.; Muili, K.A.. *Fitoterapia* **2001**, *72*, 54-56.
10. Kitamura, Y.; Nishikawa, A.; Furukawa, F.; Nakamura, H.; Okazaki, K.; Umemura, T.; Imazawa, T.; Hirose, M. *Food Chem. Toxicol.* **2003**, *41*, 1537-1542.
11. Earl, L.K.; Baldrick, P.; Hepburn, P.A. *Internat. J. Toxicol.* **2002**. *21*, 13-22.

12. Earl, L.K.; Baldrick, P.; Hepburn, P.A. *Internat. J. Toxicol.* **2002**. *21*, 353-359.
13. Carthew, P.; Baldrick, P.; Hepburn, P.A. *Food Chem. Toxicol.* **2001**, *39*, 807-815.
14. Spangenber, J.E.; Dionisi, F. *J. Agric. Food Chem.* **2001**, *49*, 4271-4277
15. Kepplinger, J.; Brian, G. European Patent Office. **2000**, Patent Number WO0030463.
16. Maranz, S.; Kpikpi, W.; Wiesman, Z.; De Saint Sauveur, A.; Chapagain, B. *Economic Botany* **2004**, *58*, 588-600.
17. Light, O. **2003**, US PTO 20030095936.
18. Courtin, O. **2002**, European Patent Office. Patent Number FR2811224.
19. Huebner, G. **2001**, European Patent Office. Patent Number DE19958612.
20. Lintner, K. **1999**, European Patent Office. Patent Number WO9963963
21. Hyldgaard, J.; Jimmi, L.; Severin, J.A. **1998**, European Patent Office. Patent Number WO9805294.
22. Thioune, O.; Khouma, B.; Diarra, M.; Diop A.B.; Lo, I. *J. Pharma. Belgique* **2003**, *58*, 81-84.
23. Loden, M.; Andersson, A.C. *Brit. J. Dermatology* **1996**, *134*, 215-220.
24. Tella, A. *Brit. J. Clin. Pharma.* **1979**, *7*, 495-497.
25. Thioune, O.; Ahodikpe, D.; Dieng, M.; Diop, A.B.; Ngom, S.; Lo, I. *Dakar Medica* **2000**, *45*, 113-116.
26. Weidner, M.S. **2002**, European Patent Office. Patent Number WO02055087.
27. Hansen, O.K. **2004**. European Patent Office. Patent Number WO2004002504
28. Barrault, J.; Mickael, B.; Yannick, P.; Antoine, P. **2002**, European Patent Office. Patent Number WO0206205.
29. Weidner, M.S. **2001**, European Patent Office. Patent Number WO0103712.
30. Tholstrup, T.; Vessby B.; Sandstrom, B. *Eur. J. Clin. Nutr.* **2003**, *57*, 735-742.
31. Tholstrup, T.; Marckmann, P.; Jespersen, J.; Sandstrom, B. *Amer. J. Clin. Nutr.* **1994**, *59*, 371-377.

Indexes

Author Index

Acquaye, Dan, 126
An, Tianying, 143
Ang, Catharina Y. W., 55
Asante-Dartey, Julie, 126
Bai, Naisheng, 195
Bernart, Matthew W., 157
Betz, Joseph M., 2
Chandra, Amitabh, 110
Chang, Leing-Chung, 224
Chao, Pei-Dawn Lee, 212
Chen, Hang, 195
Cui, Chengbin, 14
Cui, Yanyan, 55
DiNovi, Michael, 55
Fu, Hui-Yin, 299
Ghai, Geetha, 314
Guo, Y.-W., 73
He, Xiangjiu, 14
Heinze, Thomas M., 55
Hirooka, Saori, 290
Ho, Chi-Tang, 39, 117, 185, 195, 240, 266, 299, 314
Hong, Di, 143
Hou, Yu-Chi, 212
Hsiu, Su-Lan, 212
Hu, Ke, 14
Hu, Lihong, 55, 143, 170
Huang, Mou-Tuan, 314
Huang, Tzou-Chi, 299
Huang, X.-C., 73
Huang, Yi-Tsau, 281
Hwang, Lucy Sun, vii
Inoue, Toshio, 290
Jhoo, Jin-Woo, 55
Juliani, Hector, 126, 326
Kakuda, Yukio, 254
Koroch, Adolfina R., 126
Kuo, George, 224
Lai, Po-Yong, 224
Lang, Qingyong, 55

Lee, Ting-Fang, 281
Lee, Tung-Ching, 224
Li, Jia, 143
Li, Shiming, 195
Li, Shiming, 240
Liang, Chia-Pei, 39
Lin, Yun-Lian, 281
Liu, Jianghua, 14
Liu, Yue, 314
Liu, Zhihua, 117
Masuda, Hideki, 290
Matchanickal, Resmi Ann, 92
Mattia, Antonia, 55
Mihalov, Jeremy J., 55
Moharram, Hisham, 126, 326
Ozbas, Sibel, 326
Pan, Min-Hsiung, 299
Rafi, Mohamed M., 92, 266
Ramji, Divya, 314
Rana, Jatinder, 110
Ray, Julie, 326
Rosen, Robert T., 117, 314
Sang, Shengmin, 117, 185, 314
Seeram, Navindra P., 25
Shi, John, 254
Shiao, Young-Ji, 281
Simon, James E., 39, 126, 326
Sun, Y.-Q., 73
Tsou, Samson C. S., 224
Wai, Chien M., 55
Wang, Mingfu, 39, 126
Wang, Nali, 14
Wu, Qing-Li, 39
Yang, Chung S., 314
Yang, Ray-Yu, 224
Yao, Xinsheng, 14
Yin, Feng, 170
Yu, Haiqing, 266
Zhang, W., 73
Zhou, Zhu, 195

Subject Index

A

Abbot, protein tyrosine phosphatases 1B inhibitor discovery, 146f
Absorption, effect of sugar structure on glycoside, 219
Aconitum spp. (Wu Tou), bioactives acting on different targets, 19t
Activating protein-1 (AP-1) activity curcumin, 96
  herbs affecting formation and activation of AP-1 proteins, 267
Activation of phase I/II enzymes, curcumin, 100–101
Adhyperforin. *See* St. John's wort
Adulteration, PC SPES herbal mixture, 124
Affective sensory evaluation, *Lippia multiflora* tea, 136, 137t
Aglycones, determination in biological fluids, 216–217
Albert Einstein College of Medicine, protein tyrosine phosphatases 1B inhibitor discovery, 147f
Alcohol intake, association between intake of catechins and, 259, 260t
Alizarin-1-methyl ether, known compound from noni roots, 191f
Alkaloid content standards, United States Pharmacopeia (USP), 4
Alkamides
  chemical structures of isobutylamides in *Echinacea* spp., 112f
  constituents in roots of various *Echinacea* spp., 113
  identification in *Echinacea* species, 115, 116f
  See also Echinacea

Allantoin, Jiaogulan, 171
Aloe-emodin, *Rheum palmatum*, 270, 271f
Aloysia citriodora, verbascoside, 135
Amphidinium species. *See* Marine dinoflagellates
Analytical challenges
  botanicals, 5–6
  developing, validating and disseminating methods, 11
  recommending reference standards, 52–53
Analytical methods validation, Office of Dietary Supplements, 8
Analytical Methods Workshop, Office of Dietary Supplements, 7
Analytical testing protocol, certification, 9–10
Androgen-independent prostate cancer (AIPC), PC SPES herbal mixture, 122–124
Andrographis paniculata
  andrographolide and neoandrographolide, 273–274
  anti-inflammatory activity, 273–274
  Kan Jang, 272–273
  King of Bitters, 272
  See also Asian herbs targeting inflammation
Andrographolide, anti-inflammatory activity, 273–274
Anthocyanins, bilberry extract, 43–46
Anthraquinones, *Rheum palmatum*, 270, 271f
Antibacterial activity, verbascoside, 135–136
Anticancer medicine, traditional Chinese medicine ZSG, 22f, 23f, 24

345

Anti-inflammatory activity
  black tea theaflavins, 315–316, 324
  herbs, 266–267
  See also Asian herbs targeting inflammation; Black tea theaflavins
Antimicrobial activity, cinnamaldehyde, 303
Antioxidant activity
  curcumin, 101–102
  digestion effect for antioxidants of Moringa species, 234, 235*f*
  flavonoids from *Ixeris denticulata*, 204, 205*t*
  four Moringa species, 227–229
  importance of structural elements of flavonoids, 210*t*
  Lippia multiflora, 129, 131, 132*t*, 135
  Moringa leaves, stems, and Seeds, 231, 233*f*
  Moringa leaves before and after acid hydrolysis, 231, 232*f*
  shea butter, 328
  tea catechins, 257
  See also Moringa species
Antiviral activity, verbascoside, 135–136
AOAC International review panel
  analytical methods for St. John's wort, 4
  Dietary Supplements Task Group (DSTG), 6
Apigenin
  antioxidant activity, 204, 205*t*, 210*t*
  fluorescence decay curve, 209*f*
  free radical scavenging activity, 208*f*
  isolation and identification from *Ixeris denticulata*, 200, 202*f*
  structure and DPPH and ORAC values, 206*f*
Arachidonic acid metabolism, effect of topical application of theaflavin mixture (TFs) on TPA-induced, in mouse ears, 319, 321*f*
Ardisia japonica, *p*-benzoquinonoid inhibitors from, 145
Aristolochic acid method, validation, 7
Arnebia euchroma (Ruan Zi Cao), bioactives acting on different targets, 19*t*
Artemisi a capillaris (Ying Chen Hao), bioactives acting on different targets, 19*t*
Artichoke
  leaf extracts, 42
  phenolic compounds, 41–42
  quality of commercial extracts, 42–43
  structure of 1,5-dicaffeoylquinic acid, 42*f*
  See also Instrumental analysis of botanicals
Asian herbs targeting inflammation
  activation of transcription factors, 266–267, 276–277
  Andrographis paniculata, 272–274
  Polygonum Hypoleucum Ohwi, 271–272
  possible targets of nutraceuticals, 276, 278*f*
  Rheum palmatum, 270, 271*f*
  Scutellaria baicalensis Georgi (Huangqin), 267–270
  signaling events, 276, 278*f*
  Tripterygium wilfordii Hook f., 274–276

B

Babylonia japonica (Bei), bioactives acting on different targets, 19*t*
Baicalein
  determination in biological fluids, 216
  metabolism by gut/liver, 217, 218

targeting inflammation, 268, 269f
Baicalin
    bioavailability, 213
    structure, 214f
    targeting inflammation, 269f, 270
Bax protein, curcumin, 97, 98f
Bcl-2 family, curcumin, 97, 98f
Bei (*Babylonia japonica*), bioactives acting on different targets, 19t
Benzoquinonoids, inhibitors from *Ardisia japonica* and *Hypericum erectum*, 145
Benzyl isothiocyanate (BITC)
    analysis methods, 159–160
    biological importance, 158–159
    chromatographic efficiency, 162–163
    enzyme reaction using glucotropaeolin, 158
    experimental, 160–162
    high performance liquid chromatography (HPLC) method, 160
    HPLC chromatogram of BITC standard in acetonitrile, 162f
    method development, 164–166
    method optimization, 166
    stability indication, 163–164
    standard and sample preparation, 162
    standard curve, 163f
    See also Maca
Beverages. See Functional beverages
Bilberry
    anthocyanins, 43–44
    colorimetric method, 45, 46f
    high pressure liquid chromatography (HPLC) analysis of anthocyanins, 45, 46f, 47f
    total anthocyanin contents of commercial samples, 45, 46f
    use of extracts as dietary supplement, 45

See also Instrumental analysis of botanicals
Bilobalide. *See Ginkgo biloba* extract (GBE)
Bioactivity-oriented isolation, screening bioactive components of traditional Chinese medicines, 16
Bioassay-guided extraction and isolation, PC SPES herbal mixture, 119–120
Bioavailability
    curcumin, 104–105
    glycosides, 213, 215
    iron, in cooking with Moringa species, 236–237
Biological activity, protein tyrosine phosphatases, 154
Biotransformation, curcumin, 103–104
Black pepper, piperine, 104
Black tea
    consumption worldwide, 315
    theaflavin constituents, 315
Black tea extracts
    anti-caries effect, 291, 296
    See also Dental caries
Black tea theaflavins
    animals and chemicals, 316–317
    anti-inflammatory activity, 315–316, 324
    effect of topical application of TFs (theaflavin mixture) on 12-O-tetradecanoylphorbol-13-acetate (TPA)-induced arachidonic acid metabolism in ears of CD-1 mice, 319, 321f
    effect of topical application of TFs on TPA-induced ear inflammation mouse model, 319, 320f
    ELISA assay procedure, 318
    inhibitory effect of oral administration of, on colorectal carcinogenesis in mice model, 323

inhibitory effect of oral administration of TFs on TPA-induced edema of mouse ears, 319, 322f
materials and methods, 316–318
preparation of ear homogenates, 318
synthesis of theaflavin mixtures, 317–318
Blocking agents, chemoprevention, 94–95
Boston Globe, herbs, 4
Botanical products
analytical challenges, 5–6
chemical fingerprinting, 41
needing reference materials and standards, 52–53
quality issues, 40–41
United States, 40
See also Instrumental analysis of botanicals
Brevicoryne brassicae, thioglucosidase isolation, 166
Broussonetia papyrifera, flavanoid inhibitors from, 148, 149f
Bruguiera gymnorrhiza
macrocyclic polydisulfide from, 74–77
See also Gymnorrhizol
Bulgarian tribulus
extract evaluation by LC/MS, 51, 52f
quality, 48
saponins, 50, 51f
See also Tribulus
Butter. See Shea butter

C

Cafetaric acid
constituents of Echinacea, 111–112
See also Echinacea
Caffeic acid
identification in Echinacea species, 113–114
phenolic acid, 28, 29f
Caffeoylquinic derivatives, quality control of artichoke leaf extracts, 42
Camellia sinensis, tea consumption, 255
Cancer
advances in diagnosis and treatment, 93
chemoprevention and diet, 93–95
Chinese herb *Ligusticum chuanxiong* (LC), 282, 287
Jiaogulan, 170–171
oral administration of black tea extract on colorectal carcinogenesis, 323
preventive activity of tea, 255, 258–259
resveratrol, 287
verbascoside, 135
See also Curcumin; Human leukemia cell lines
β-Carotene, association between intake of catechins and, 259, 260t
Catechin
antioxidant activity, 204, 205t, 210t, 257
epigallocatechin gallate (EGCG) degradation in mild alkaline solution, 243, 244f
fluorescence decay curve, 209f
free radical scavenging activity, 208f
major components of tea, 241
structure, 242f
structure and DPPH and ORAC values, 207f
See also Tea catechins
Center for Food Safety and Applied Nutrition (CFSAN), ephedrine-type alkaloids, 6
Chemical degradation, shea butter, 331

Chemoprevention
  cancer, and diet, 93–95
  cancer and edible phytochemicals, 94–95
China
  herbal medicine use, 40
  Jiaogulan, 170–171
  use of marine medicinal agents, 73–74
  See also Ixeris denticulata
Chinese medicines
  combination of seven Chinese herbs for prostate cancer, 117–118
  See also PC SPES herbal mixture; Traditional Chinese medicines (TCM)
Chinese tribulus
  extract, 50
  See also Tribulus
Chlorogenic acid
  artichoke, 42
  constituent of Echinacea, 111–112
  different pH buffer solutions, 246
  experimental conditions and procedures, 246, 247f
  flow chart of experiments, 247f
  health benefits, 241
  high performance liquid chromatography (HPLC) traces of, and isomers, 248f
  kinetic study of, transformation, 250–251
  NMR (nuclear magnetic resonance) spectra, 249f
  NMR spectra of cryptochlorogenic acid, 250
  separation and structure elucidation of isomers, 248–250
  structure, 242f
  transformation at pH 11.2, 252f
  transformation at pH 8.5, 251f
  transformation study, 246–251
  See also Echinacea

Cholesta-5,22-dien-3-ol, known compound from noni roots, 192f
Chronic diseases, tea catechins protecting against, 256–257
Chrysophanol, *Rheum palmatum*, 270, 271f
Cichoric acid
  constituent of Echinacea, 111–112
  See also Echinacea
Cinnamaldehyde
  antimicrobial and insecticidal activity, 303
  chemical structure, 305f
  DNA fragmentation of leukemic cell lines, 306–311
  effect on morphological changes of K562 human leukemia cell, 305
  inhibitory concentration on human leukemia cell lines, 304t
  mechanism by induced apoptosis in K562 cells, 311f
  See also Cinnamon
Cinnamomum osmophloeum clones. See Cinnamon
Cinnamon
  antimicrobial and insecticidal activity of cinnamaldehyde, 303
  chemical structures of cinnamaldehyde, eugenol, and linalool, 305f
  cinnamaldehyde-induced DNA fragmentation of leukemic cell lines, 306–311
  concentration-dependent inhibition on NADH oxidation, 309
  cytotoxic properties of leaf essential oil, 303–304
  degree of DNA fragmentation in cinnamaldehyde treated K562 cells, 306–307
  effect of cinnamaldehyde on intracellular superoxide radicals, 307, 308f, 309f

effect of cinnamaldehyde on morphological changes of K562 cell, 305
effect of drying on essential oil recovery from leaves, 302f
flavor and fragrance industries, 300
inhibitory concentration (50%) on human leukemia cell lines, 304t
levels of total essential oil in leaf samples by distillation, 301–302
molecular mechanisms of apoptosis, 310–311
oils from leaves of three types, 300–301
principal component analysis (PCA), 302, 303f
structural changes of cells under oxidative stress, 310
volatile composition of, essential oil, 300–302
Clinical effects, tea catechins, 257–259
Color, shea butter, 332
Colorectal carcinogenesis, oral administration of black tea extract, 323
Colorimetric method, anthocyanins in bilberry, 45, 46f
Cooking, iron bioavailability in Moringa species, 236–237
Corosolic acid, *Lagerstroemia speciosa* leaf, 150–151
Cosmetics industry, uses of shea butter, 336–337
Curcumin
  activating protein-1 (AP-1) activity, 96
  activation of phase I/II enzymes, 100–101
  antioxidant/prooxidant activity, 101–102
  Bax protein, 97
  Bcl protein family, 97
  bioavailability, 104–105
  cancer chemoprevention and diet, 93–95
  chemoprevention, 94–95
  Curcuma longa, 94
  degradation products, 102–103
  degradation study, 243, 245f, 246
  inhibition of apoptosis, 97
  light sensitivity, 102
  mechanisms of action *in vitro*, 95–102
  metabolism, 103–104
  metabolism, pharmacokinetics, tissue distribution and excretion, 102–105
  molecular targets of, 98f
  non-steroidal anti-inflammatory drugs (NSAID) vs., 100
  p21 gene suppression, 97
  p53 tumor suppressor gene, 97
  phytochemicals, 95
  structure, 242f
  structures of curcuminoids, 94f
  suppression of cyclooxygenase (COX-2), 99–100
  suppression of eukaryotic transcription factors NK-κB and AP-1, 95–96
  suppression of mitogen activated protein kinase (MAPK), 98–99
  suppression of tumor cells, 97
  turmeric as source, 93–94
Cyanidin, anthocyanidin, 26, 27f
Cyclic disulfides. *See* Gymnorrhizol
Cyclooxygenase (COX-2)
  herbs targeting inflammation, 267
  suppression by curcumin, 99–100

D

Da Huang (*Rheum coreanum* and *R. palmatum*), bioactives acting on different targets, 19t
Daidzein

determination in biological fluids, 216
metabolism by gut/liver, 218
Daidzin
bioavailability, 213, 215
structure, 214f
Damnacanthal, known compound from noni roots, 191f
Degradation
curcumin, 102–103
shea butter, 330–332
Dendranthema morifolium Tzvel, PC SPES herbal mixture, 117–118
Dental caries
activity of green, oolong, and black tea, 291
animals and treatment, 291
caries score of rats fed diet containing 5-methylthiopentyl isothiocyanate and/or black tea extract, 293f
comparing 5-methylthiopentyl isothiocyanate and benzyl isothiocyanate, 293, 295f, 296
data and statistical analysis, 291–292
experimental, 291–292
inhibitory concentration isobologram for sugar dependent adherence, 293, 295f
inhibitory effect on sucrose dependence adherence by growing cells of *Streptococcus sobrinus* 6715, 292
isothiocyanates, 290
materials, 291
pathway of formation, 290–291
photograph of rat molars with diet containing 5-methylthiopentyl isothiocyanate and/or black tea extract, 294f
synergistic effects, 296
Deoxyribonucleic acid (DNA) fragmentation

cinnamaldehyde-induced, of leukemic cell lines, 306–311
degree of, in cinnamaldehyde treated K562 cells, 306–307
induction by cinnamaldehyde, 307f
intracellular superoxide radical levels, 307, 308f, 309f
mechanism by cinnamaldehyde induced apoptosis in K562 cells, 311f
Diabetes
lifestyle management, 143–144
protein tyrosine phosphatases 1B as research target, 144
type 2, 143
See also Protein tyrosine phosphatases 1B (PTP1B) inhibitors
1,5-Dicaffeoylquinic acid, artichoke, 42
Diet, cancer chemoprevention and, 93–95
Dietary constituents, curcumin, 104
Dietary Supplement Health and Education Act (DSHEA)
defining dietary supplement, 3
labeling regulations, 5
quality, 5
Dietary Supplement Methods and Reference Materials Program
methods challenge, 11
Office of Dietary Supplements, 7
Dietary supplements
definitions, 3
framework for Food and Drug Administration (FDA) regulation, 3
growth of marketplace, 3–4
product quality, 4–5
study of ingredients, 10–11
Dietary Supplements Task Group (DSTG), AOAC International, 6
Digestion, effect on antioxidants of *Moringa* species, 234, 235f

1,3-Dimethoxy-2-
  methoxymethylanthraquinone,
  known compound from noni roots,
  191*f*
Dinoflagellates
  chemistry, 78
  lingshuiols, 74, 78, 79*f*
  promising source of bioactive
    substance, 74
  See also Marine dinoflagellates;
    Marine medicinal plants
1,1-Diphenyl-2-picrylhydrazyl radical
  (DPPH), DPPH assay, 198
Drug companies, protein tyrosine
  phosphatases 1B inhibitor
  discovery, 146*f*, 147*f*
Drug Registration and Administration
  Law, revision, 15
Dysidesterols
  absolute stereochemistry, 90
  chemical structure, 87*f*
  computer generated drawing, 89*f*
  crystal data, 88–90
  spectroscopic properties, 87–88
  uncommon steroids from sponge
    Dysidea sp., 86
  See also Marine medicinal plants

E

Echinacea
  alkamide constitution in roots, 113
  alkamides, 115, 116*f*
  cafetaric acid, 112*f*
  chemical structures of major
    isobutylamides, 112*f*
  chemical structures of polyphenols,
    112*f*
  chlorogenic acid, 112*f*
  cichoric acid, 112*f*
  Echinacea angustifolia root, 111–
    112
  echinacoside, 112*f*
  E. pallida herb, 112–113

E. purpurea herb, 111
genus, 111
high performance liquid
  chromatography (HPLC) profile
  of polyphenols in *E.*
  angustifolia, 114*f*
HPLC profile of alkamides in *E.*
  angustifolia, 115*f*
HPLC profile of alkamides in *E.*
  purpurea, 116*f*
HPLC profile of polyphenols in *E.*
  purpurea, 114*f*
market, 111
phenolic and caffeic acids, 113–
  114
phytochemical constituents of
  species, 111
species identification by HPLC,
  113–115
term, 111
uses, 111
Echinacoside
  constituents of Echinacea, 111–112
  See also Echinacea
Elderberry, high pressure liquid
  chromatography (HPLC), 46, 47*f*
Ellagitannins (ETs)
  biological activities, 34–35
  categories, 33
  dietary sources, 33–34
  formation of ellagic acid, 34*f*
  See also Tannins
Emodin
  anthraquinone in *Polygonum*
    Hypoleucum Ohwi, 271, 272*f*
  Rheum palmatum, 270, 271*f*
Emodin 1-β-O-glucoside, *Polygonum*
  Hypoleucum Ohwi, 271, 272*f*
Enzymatic degradation, shea butter,
  332
Enzymes, phase I/II, activation by
  curcumin, 100–101
Ephedrine-type alkaloids, funding for
  validation of analytical methods, 6–
  7

Epicatechin (EC)
  antioxidant activity, 204, 205t, 210t
  fluorescence decay curve, 209f
  free radical scavenging activity, 208f
  structure and DPPH and ORAC values, 207f
  synergistic effects, 259
  tea catechin, 256f
Epicatechin gallate (ECG)
  synergistic effects, 259
  tea catechin, 256f
Epigallocatechin (EGC)
  synergistic effects, 259
  tea catechin, 256f
Epigallocatechin gallate (EGCG)
  abundance in tea, 241
  degradation in mild alkaline solution, 243, 244f
  flavanol, 26, 27f
  synergistic effects of EGCG with Sulindac and Tamoxifen, 259, 260, 261f
  tea catechin, 256f
Eriobotrya japonica, triterpenoid inhibitors from, 148, 150
Essential oils
  Lippia multiflora, 128, 134
  volatile composition of cinnamon, 300–302
  See also Cinnamon
Ethinyl estradiol (EE)
  PC SPES herbal mixture, 121–122
  quantitative determination of EE by gas chromatography/mass spectroscopy (GC/MS), 120
Eugenol
  chemical structure, 305f
  inhibitory concentration on human leukemia cell lines, 304t
  See also Cinnamon
Eukaryotic transcription factors, suppression by curcumin, 95–96
Excretion, flavonoid conjugates into bile and urine, 218
Expert Review Panels (ERP)
  AOAC International, 9
  establishment and maintenance, 8
Extraction
  bioassay-guided, of PC SPES herbal mixture, 119–120
  PC SPES herbal mixture, 121
  shea butter, 327

F

Fatty acids
  association between intake of catechins and, 259, 260t
  shea butter, 328
Fiber, association between intake of catechins and, 259, 260t
Flavanoid inhibitors, *Broussonetia papyrifera* and *Ginkgo biloba*, 148, 149f
Flavonoids
  excretion of, conjugates into bile and urine, 218
  fluorescence decay curves from ORAC assay, 209f
  free radical scavenging activity in DPPH assay, 208f
  structure-activity relationship of, from *Ixeris denticulata*, 205, 210
  structures of, from *I. denticulata*, 206f, 207f
  subclass of polyphenols, 26, 27f
  See also Ixeris denticulata; Polyphenols
Food and Drug Administration (FDA), regulation of dietary supplements, 3
Food industry, uses of shea butter, 334, 336
Foods, interactions of catechins with, 259–260
2-Formylanthraquinone, known compound from noni roots, 190f

French Paradox, proanthocyanidins, 32
Functional beverages
analysis of ginkgo constituents in teas and fruit-flavored beverages, 64
analysis of ginkgo terpene trilactones in fruit-flavored drinks and hot water infusion of tea bags containing *Ginkgo biloba*, 65t
analysis of kava constituents in teas and fruit-flavored beverages, 67–70
analysis of St. John's wort (SJW) constituents in fruit-flavored beverages, 61
analysis of SJW constituents in tea, 57–58, 60
biological activities and phytochemical constituents of SJW, 57
biological activities of gingko leaf extracts, 61, 63
biological and toxicological activities of kava, 66
changes of SJW constituents in fruit-flavored drink during storage, 62f
gas chromatography–flame ionization detector (GC–FID) chromatograms of fruit-flavored kava beverages, 70f
ginkgo in, 61, 63–65
high performance liquid chromatography (HPLC) of hot water infusion of tea-bag products, 69f
kava in, 66–70
kavalactone analysis of fruit-flavored drinks using GC–FID, 68, 70
kavalactone analysis of tea-bag products using HPLC, 67–68
kavalactone content in tea-bag and fruit-flavored beverages, 69t
kavalactones in kava root, 67f
LC (liquid chromatography) of hot water infusion of tea bag, 59f
LC-ELSD analysis of ginkgo terpene lactones in fruit-flavored drinks, 65f
phytochemical constituents and analysis of ginkgo terpene trilactones, 63–64
phytochemical constituents and analysis of kava, 66–67
SJW constituents in tea bags extracted with methanol or hot-water infusion, 60t
SJW in, 57–61
structures of ginkgo terpene trilactones, 63f
structures of major active constituents of SJW, 58f

G

Gallic acid, phenolic acid, 28, 29f
Ganoderma lucidium Karst, PC SPES herbal mixture, 117–118
Genetics, Shea tree, 327
Genistein
isoflavone, 26, 27f
structure, 242f
Ghana. See *Lippia multiflora*
Ginkgo biloba extract (GBE)
flavanoid inhibitors from, 148, 149f
analysis of ginkgo constituents in teas and fruit-flavored beverages, 64
analysis of ginkgo terpene trilactones in fruit-flavored drinks and hot water infusion of tea bags containing *Ginkgo biloba*, 65t
biological activities of gingko leaf extracts, 61, 63

flavanoid inhibitors from, 148, 149f
ginkgo in functional beverages, 61,
  63–65
ginkgolides (A, B, C, and J) and
  bilobalide, 63f
LC–ELSD analysis of ginkgo
  terpene lactones in fruit-flavored
  drinks, 65f
phytochemical constituents and
  analysis of ginkgo terpene
  trilactones, 63–64
structures of ginkgo terpene
  trilactones, 63f
See also Functional beverages
Glucosinolates
  analysis methods, 159–160
  major, of maca, 159f
  See also Maca
Glucotropaeolin
  quantification in maca, 166–167
  structure, 159f
  See also Maca
Glycosides
  bioavailability of, 213, 215
  chemical structures, 214f
  determination of aglycones and
    conjugated metabolites in
    biological fluids, 216–217
  effect of sugar structure on
    glycoside absorption, 219
  excretion of flavonoid conjugates
    into bile and urine, 218
  metabolic pharmacokinetics of,
    219–220
  metabolism of phenolic aglycones
    by gut/liver, 217–218
  natural products in Chinese herbs,
    213
  presystemic metabolism in
    gastrointestinal tract, 215–216
Glycyrrhiza uraensis Fisch, PC SPES
  herbal mixture, 117–118
Glycyrrhizin
  metabolism by gut/liver, 218
  structure, 214f

Gylongiposide I, Jiaogulan, 171
Gymnorrhizol
  chemical structure, 75f
  isolation from Bruguiera
    gymnorrhiza, 74
  plausible retrosynthetic pathway,
    77
  properties, 75
  structural elucidation, 75–76
Gynostemma pentaphyllum. See
  Jiaogulan
Gypenosides, Jiaogulan, 171

H

Heart disease, protective effects of
  catechins, 258
Hepatic fibrosis, Chinese herb
  Ligusticum chuanxiong (LC), 282,
  287
Hepatic stellate cells (HSC). See
  Ligusticum chuanxiong (LC)
Herbal mixture for prostate cancer.
  See PC SPES herbal mixture
Herbal products, marketing, 40
Herbal teas
  health promoting properties, 127
  See also Lippia multiflora
Herbs
  anti-inflammatory properties, 266–
    267
  Asian countries, 266
  mainstream media, 4
  See also Asian herbs targeting
    inflammation
Hesperetin
  determination in biological fluids,
    216
  metabolism by gut/liver, 217–218
Hesperidin
  bioavailability, 213
  structure, 214f
High performance liquid
  chromatography (HPLC)

alkamides, 115, 116f
identification of Echinacea species, 113–115
kavalactone analysis of tea-bag products using, 67–68
Moringa species phenolics vs. antioxidant activity, 231, 232f
phenolic and caffeic acids, 113–114
High pressure liquid chromatography (HPLC)
anthocyanins in bilberry extracts, 45, 46f, 47f
elderberry extract, 46, 47f
quality control of natural products, 52
quality of commercial artichoke extracts, 42–43, 44f
soy isoflavone extract, 48, 49f
tribulus extracts, 51, 52f
High-throughput screening assays, libraries for traditional Chinese medicines, 20, 21f
Human Genome Project, differences amongst individuals, 94
Human leukemia cell lines
cinnamaldehyde-induced DNA fragmentation of, 306–311
cytotoxic properties of cinnamon leaf essential oil, 303–304
effects of cinnamaldehyde on morphological changes of K562 cells, 305
inhibitory concentration of essential oils, 304t
schematic of mechanism of cinnamaldehyde induced apoptosis in, 311f
See also Cinnamon
Hydrolysis, shea butter, 331
1-Hydroxy-2-formylanthraquinone, known compound from noni roots, 190f

1-Hydroxy-2-methylanthraquinone, known compound from noni roots, 190f
Hyperforin. See St. John's wort
Hypericin. See St. John's wort
Hypericum erectum, p-benzoquinonoid inhibitors from, 145
Hypericum perforatum L. See St. John's wort

I

Ibericin, known compound from noni roots, 192f
India, herbal medicine use, 40
Indian tribulus
extract, 50
See also Tribulus
Inducible nitric oxide synthase (iNOS), inhibition by phytochemicals, 267
Inflammation. See Asian herbs targeting inflammation
Ingredient Ranking Subcommittee, soliciting methods, 9
Ingredients, study of dietary supplements, 10–11
Inhibition of apoptosis
cinnamon, 310–311
curcumin, 97, 98f
Ligusticum chuanxiong, 284, 285f
traditional Chinese medicines, 16–17
Insecticidal activity, cinnamaldehyde, 303
Instrumental analysis of botanicals
artichoke, 41–43
bilberry, 43–46
botanical products, 41
materials and instruments, 41
needing reference materials and standards, 52–53
soy isoflavones, 47–48, 49f

tribulus, 48–51, 52f
Iron bioavailability, cooking with Moringa species, 236–237
Isatis indigotica Fort, PC SPES herbal mixture, 117–118
Isobutylamides, chemical structures of, in *Echinacea* spp., 112f
Isoflavones, soy, 47–48, 49f
Isolation of components
  bioassay-guided, of PC SPES herbal mixture, 119–120
  PC SPES herbal mixture, 121
  procedures for traditional Chinese medicines with multiple effects, 21f
  screening of traditional Chinese medicines, 16
Isothiocyanates
  anticaries activity, 291
  See also Dental caries
Ixeris denticulata
  antioxidant activity of flavonoids from, 204, 205t
  apigenin, 200, 202f
  chemical structures of isolated compounds, 202f, 203f
  column chromatography, 197
  1,1-diphenyl-2-picrylhydrazyl radical (DPPH) assay, 198
  extraction method, 197
  fluorescence decay curves from ORAC assay, 209f
  free radical scavenging activity of flavonoids in DPPH assay, 208f
  identification of compounds isolated from, 200–201, 204
  importance of structural elements of flavonoids to antioxidant activity, 210t
  isolation of compounds from, 199
  luteolin, 200, 202f
  mass spectrometry (MS), 197
  materials and methods, 196–199
  nuclear magnetic resonance (NMR), 198
  oxygen radical absorbance capacity (ORAC) assay, 198–199
  reagents, 196
  rehmaglutin D, 201, 203f, 204
  β-sitosterol, 200, 202f
  species, 196
  stigmasterol, 201, 202f
  structure-activity relationship, 205, 210
  structures of flavonoids and their DPPH and ORAC values, 206f, 207f
  use as popular Chinese folk medicine, 196

J

Japan
  herbal medicine use, 40
  Jiaogulan, 170
Japan Tobacco, protein tyrosine phosphatases 1B inhibitor discovery, 146f
Jiaogulan
  aglycons, 181
  ^{13}C nuclear magnetic resonance (NMR) data of new compounds, 177–181
  Gymnostemma pentaphyllum, 170
  gypenosides, gylongiposide, allantoin, and vitexin, 171
  health-promoting properties, 170–171
  plausible biogenetical pathway for saponins from *G. pentaphyllum*, 182
  products in Chinese markets, 171
  structures of new compounds, 171–176

K

Kaempferol
  antioxidant activity, 204, 205t, 210t

fluorescence decay curve, 209f
free radical scavenging activity, 208f
structure and antioxidant activity assay values, 206f
Kava
  analysis of kava constituents in teas and fruit-flavored beverages, 67–70
  biological and toxicological activities of kava, 66
  gas chromatography–flame ionization detector (GC–FID) chromatograms of fruit-flavored kava beverages, 70f
  high performance liquid chromatography (HPLC) of hot water infusion of tea-bag products, 69f
  kava in functional beverages, 66–70
  kavalactone analysis of fruit-flavored drinks using GC–FID, 68, 70
  kavalactone analysis of tea-bag products using HPLC, 67–68
  kavalactone content in tea-bag and fruit-flavored beverages, 69t
  kavalactones in kava root, 67f
  phytochemical constituents and analysis of kava, 66–67
  See also Functional beverages
Kavalactones. See Kava
Kinetics, chlorogenic acid transformations, 250–251, 252f
Korea, Jiaogulan, 170
Korea Research Institute of Chemical Technology, protein tyrosine phosphatases 1B inhibitor discovery, 147f

L

Labeling regulations, dietary supplements, 5

Lagerstroemia speciosa leaf, corosolic acid, 150–151
Lantana camara, verbascoside, 135
Lepidium meyenii. See Maca
Leukemia. See Human leukemia cell lines
Libraries, extracts, fractions and compounds from traditional Chinese medicines, 20, 21f
Light sensitivity, curcumin, 102
Lignans, subclass of polyphenols, 28, 29f
Ligusticum chuanxiong (LC)
  activity of tetramethylpyrazine, 282
  altering expression of cell cycle related proteins in hepatic stellate cells (HSC–T6), 285f
  development of hepatic fibrosis, 282, 287
  effect on HSC–T6 (hepatic stellate cells) cell growth, 283–284
  effects on activation of caspase-3 in HSC–T6 cells, 286
  hepatic fibrosis, 282
  HSC–T6 and dosage response of LC, 283f
  induction of HSC–T6 apoptosis and expression of p21 and p27 by, 284, 285f
  lactate dehydrogenase release in cultured LC-treated HSC–T6, 284t
  LC inducing HSC–T6 apoptosis, 285f
  popular Chinese herb, 282, 287
  resveratrol, 287
Linalool
  chemical structure, 305f
  inhibitory concentration on human leukemia cell lines, 304t
  See also Cinnamon
Lingshuiols
  chemical structures, 79f
  isolation from *Amphidinium* sp., 78
  partial structures, 78, 80, 81f

spectroscopic analysis, 80, 82–83
See also Marine dinoflagellates
Lippia multiflora
  affective sensory evaluation of different varieties, 136, 137*t*
  aromatic plant family, 127
  chemical and antioxidant activity analysis, 129
  chemical structure of verbascoside, 132*f*
  contents of major phenols, verbascoside, and nuomiside, 131–132
  distribution, 127
  essential oils, 134
  folk medicine of West Africa, 127–128
  high performance liquid chromatography (HPLC) profile of methanolic extract, 133*f*
  materials and methods, 128–130
  pharmacological activities, 134–136
  plant material, 128–129
  qualitative HPLC–MS analyses, 129–130
  quality control analysis, 129
  quality control parameters for different varieties, 131*t*
  quality standards, 130–131
  quality standards for dried leaves, 138*t*
  quantitative HPLC analysis, 130
  tea tasting, 136, 138
  tea tasting method, 129
  total phenols and antioxidant activities, 131
  wild-gathered by local communities, 128
Los Angeles Times, herbs, 4
Luteolin
  antioxidant activity, 204, 205*t*, 210*t*
  flavone, 26, 27*f*
  fluorescence decay curve, 209*f*
  free radical scavenging activity, 208*f*
  isolation and identification from *Ixeris denticulata*, 200, 202*f*
  structure and DPPH and ORAC values, 206*f*

M

Maca
  benzyl isothiocyanate (BITC) from glucotropaeolin, 158
  biological importance of BITC, 158–159
  BITC standard curve, 163*f*
  chromatographic conditions, 161
  chromatographic efficiency, 162–163
  development of method, 164–166
  experimental, 160–162
  high performance liquid chromatography (HPLC) of BITC standard in acetonitrile, 162*f*
  instrumentation and equipment, 161
  Lepidium meyenii Walpers, 157
  linear gradient elution of BITC, 161*t*
  major glucosinolates of, 159*f*
  materials, 160–161
  method optimization, 166
  methods of BITC and glucosinolate analysis, 159–160
  popularity as nutritional supplement, 157–158
  quantification of glucotropaeolin in, 166–167
  safety, 160
  stability indication, 163–164
  standard and sample preparation, 162
  See also Benzyl isothiocyanate (BITC)

Mangrove *Bruguiera gymnorrhiza*
  macrocyclic polydisulfide from, 74–77
  See also Gymnorrhizol
Marine dinoflagellates
  amphidinol, 82–83
  isolation of lingshuiols, 78
  polyhydroxypolyene compounds from *Amphidinium* sp., 78–83
  spectroscopic properties of lingshuiols, 83
  stereochemistry of tetrahydropyran rings of lingshuiol, 82*f*
  structure of lingshuiols, 78, 79*f*, 80–83
  See also Marine medicinal plants
Marine invertebrates
  steroids from gorgonian *Subergorgia reticulata*, 84–85, 86*f*
  steroids from sponge *Dysidea* sp., 86–90
  See also Dysidesterols; Suberrestisteroids
Marine medicinal plants
  Bruguiera gymnorrhiza, 74–77
  dysidesterols, 86–90
  gymnorrhizol, 75–77
  lingshuiols, 74, 78–83
  macrocyclic polydisulfide from mangrove, 74–77
  marine dinoflagellates, 74
  polyhydroxypolyene compounds from *Amphidinium* sp., 78–83
  suberretisteroids, 84–85, 86*f*
  uncommon steroids from sponge *Dysidea* sp., 86–90
  use in China, 73–74
Medical components, association between intake of catechins and, 260, 261*f*
Medicinal plants. *See* Marine medicinal plants

Merk–Frosst, protein tyrosine phosphatases 1B inhibitor discovery, 147*f*
Metabolism
  curcumin, 103–104
  phenolic aglycones by gut/liver, 217–218
  tea catechins, 257–259
Methods development, interagency agreement, 8
1-Methoxy-3-hydroxyanthraquinone, known compound from noni roots, 191*f*
1-Methyl-3-hydroxyanthraquinone, known compound from noni roots, 191*f*
5-Methylthiopentyl isothiocyanate
  anti-caries effect, 291, 296
  See also Dental caries
Microalgae, marine biological system, 78
Microbiological contamination, shea butter, 331
Mitogen activated protein kinase (MAPK), suppression by curcumin, 98–99
Molecular targets, curcumin, 97, 98*f*
Morin, metabolism by gut/liver, 217–218
Morinda citrifolia. See Noni
Morindone-5-methylether, known compound from noni roots, 191*f*
Moringa species
  antioxidant activities (AOA) methods, 226
  antioxidant (AO) contents, 226
  antioxidant contents of Moringa, 228–229
  AOA methods and water/methanol (W/M) extractions, 227
  AOA of leaves, stems and *Seed*s, 231, 233*f*
  compositions of W/M extracts, 227–228

contribution of AO to AOA, 229, 230f
cooking enhancing effect, 236
digestion effect on AOA, 234, 235f
four species in study, 226t
HPLC profiles of phenolics vs. AOA, 231, 232f
in vitro iron bioavailability, 236–237
materials and methods, 225–227
nutritional quality, 226
nutritional quality of four species, 235, 236t
objectives of study, 225
plant materials and preparation, 225–226
ranking by averaged AOA, 227
simulated digestion, 226–227
temperature effect on AOA, 231, 233
vegetable combinations, 237
Mouse model. *See* Black tea theaflavins
Multiple targets
components of traditional Chinese medicines working on, 17–19
screening traditional Chinese medicines, 20

N

Naringenin
determination in biological fluids, 216
flavanone, 26, 27f
metabolism by gut/liver, 217–218
Naringin
bioavailability, 213
structure, 214f
National Research Council (Canada) of Certified

Reference Materials projects within Office of Dietary Supplements, 10
Neoandrographolide, anti-inflammatory activity, 273–274
Neophellamuretin, determination in biological fluids, 216
Neutraceutical industry, potential uses of shea butter, 338–339
New drug development, traditional Chinese medicine as resource, 15–20
Noni
chemicals, 186
description, 185–186
extraction and isolation procedure for roots, 187, 189
general spectroscopic procedures, 187
materials and methods, 186–187
Morinda citrifolia, 185
schematic of extraction procedure for roots, 188f
structures of known compounds from roots, 190f, 191f, 192f
structures of new compounds from roots, 189f, 190f
Non-steroidal anti-inflammatory drugs (NSAID), curcumin versus, 100
Nor-damnacanthal, known compound from noni roots, 191f
Nuclear factor-kappa B (NF-κB) transcription factor
herbs targeting inflammation, 267
suppression by curcumin, 95–96
Nuomiside
content in *Lippia multiflora*, 131–132
HPLC profile of methanolic extract of *L. multiflora*, 133f
See also Lippia multiflora
Nutritional quality
Moringa leaves, 235, 236t
testing method, 226

O

Office of Dietary Supplements (ODS)
  analytical methods validation, 8
  Analytical Methods Workshop, 7
  certification of analytical testing protocol, 9–10
  challenge in developing, validating and disseminating methods, 11
  collaborative study process, 8–9
  Dietary Supplement Methods and Reference Materials Program, 7
  establishment, 3
  Expert Review Panels (ERP), 8–9
  Ingredient Ranking Subcommittee, 9
  methods development, 8
  role, 6–11
  Single Laboratory Validation studies, 8, 10
  Standard Reference Materials (SRM™), 9–10
  studying ingredients, 10–11
  validating analytical methods, 9–10
Oleanolic acid
  activity of lowering blood glucose level, 150–151
  inhibitory activity of potent derivatives, 153–154
  masking carboxylic acid of, by methyl group, 151–152
  structures and inhibitory activity of, C-28 long-chain acid derivatives against protein tyrosine phosphatases 1B (PTP1B) enzyme, 153t
  structures and inhibitory activity of, C-28 long-chain peptide derivatives against PTP1B enzyme, 152t
Ontogen, protein tyrosine phosphatases 1B inhibitor discovery, 146f
Orobanche hederae, verbascoside, 135
Oroxylin A, targeting inflammation, 269f, 270
Oxidation, shea butter, 331
Oxygen radical absorbance capacity (ORAC), ORAC assay, 198–199

P

p21 gene, curcumin, 97, 98f
p53 tumor suppressor gene, curcumin, 97, 98f
Paeoniflorin
  determination in biological fluids, 216
  metabolism by gut/liver, 218
  structure, 214f
Panax pseudo-ginseng Wall, PC SPES herbal mixture, 117–118
PC SPES herbal mixture
  adulteration and product recall, 124
  androgen-independent prostate cancer (AIPC), 122–124
  bioassay-guided extraction and isolation, 119–120
  combination of herbs, 117–118
  ethinyl estrodiol (EE) detection, 122
  extraction and isolation procedure, 121f
  gas chromatography (GC) of estradiol-diTMS and ethinylestradiol-monoTMS, 123f
  general experimental procedures, 118–119
  identification of EE, neobaicalein and tenaxin I, 121
  materials and methods, 118–120
  prostate cancer patients, 117–118
  quantitative determination of EE by GC–mass spectrometry (GC–MS) analysis, 120
  structure of EE, neobaicalein and tenaxin I, 122f

People's Republic of China. *See*
　Traditional Chinese medicines
　(TCM)
Personal care products industry, uses
　of shea butter, 336–337
pH. *See* Chlorogenic acid
Pharmaceutical industry, uses of shea
　butter, 337–338
Pharmacia, protein tyrosine
　phosphatases 1B inhibitor
　discovery, 147*f*
Pharmacokinetics
　curcumin, 102, 104
　metabolic, of glycosides, 219–220
Pharmacological activities, *Lippia
　multiflora*, 134–136
Phase I/II enzymes, activation by
　curcumin, 100–101
Phellamurin
　bioavailability, 213, 215
　structure, 214*f*
Phenolic acids
　identification in Echinacea species,
　　113–114
　subclass of polyphenols, 28, 29*f*
Phenolic aglycones, metabolism by
　gut/liver, 217–218
Phenolic compounds
　artichoke, 41–42
　biological activities, 241
　chlorogenic acid, 241
　degradation study of curcumin,
　　243, 245*f*, 246
　degradation study of tea catechins,
　　241, 243
　health-benefiting properties, 241
　shea butter, 328
　structures of polyphenolic
　　compounds, 242*f*
　tea catechins, 241, 242*f*
　transformation study of chlorogenic
　　acid, 246–251
Phenols, total, and antioxidant
　activities in *Lippia multiflora*, 131,
　132*t*

Physcion
　anthraquinone in *Polygonum
　　Hypoleucum Ohwi*, 271, 272*f*
　Rheum palmatum, 270, 271*f*
Physcion 1-β-O-glucoside, *Polygonum
　Hypoleucum Ohwi*, 271, 272*f*
Physico-chemical properties, shea
　butter, 329–330
Phytochemicals
　definition, 95
　polyphenols, 26
Piper methysticum. *See* Kava
Piperine, black pepper, 104
Polydisulfides. *See* Gymnorrhizol
Polygonum Hypoleucum Ohwi
　anthraquinones, 271–272
　anti-inflammatory activity, 272
　anti-proliferatory activity, 271
　See also Asian herbs targeting
　　inflammation
Polyhydroxypolyene compounds. *See*
　Marine dinoflagellates
Polyphenols
　chemical structures of Echinacea,
　　112*f*
　flavonoids, 26, 27*f*
　identification in Echinacea species,
　　114*f*
　lignans, 28, 29*f*
　phenolic acids, 28, 29*f*
　stilbenes, 28, 29*f*
　types of dietary, 26, 28
　See also Echinacea; Tannins
Polyunsaturated fatty acids,
　association between intake of
　catechins and, 259, 260*t*
Proanthocyanidins (PAs)
　antioxidant properties, 32–33
　basic structure, 31*f*
　condensed tannins, 31–33
　estimated consumption, 32
　French Paradox, 32
　See also Tannins
Product quality
　dietary supplements, 4–5

See also Quality; Quality control
Product recall, PC SPES herbal mixture, 124
Prooxidant activity, curcumin, 101–102
Prostate cancer. *See* PC SPES herbal mixture
Protein tyrosine phosphatases 1B (PTP1B) inhibitors
  p-benzoquinonoid inhibitors from *Ardisia japonica* and *Hypericum erectum*, 145
  biological activity evaluation, 154
  carboxylic acid of oleanolic acid, 151–152
  companies pursuing development, 144, 146*f*, 147*f*
  corosolic acid from *Lagerstroemia speciosa* leaf, 150–151
  discovery from traditional Chinese medicine, 144–150
  flavanoid inhibitors from *Broussonetia papyrifera* and *Ginkgo biloba*, 148, 149*f*
  inhibitory activity of natural compounds against PTP1B enzyme, 150*t*
  inhibitory activity of potent oleanolic acid derivatives, 153–154
  series of oleanolic acid C-28 long-chain aliphatic acid derivatives, 152–153
  structural modification of oleanolic acid, 150–154
  structures and inhibitory activity of oleanolic acid C-28 long-chain acid derivatives against PTP1B enzyme, 153*t*
  structures and inhibitory activity of oleanolic acid C-28 long-chain peptide derivatives against PTP1B enzyme, 152*t*
  triterpenoid inhibitors from *Eriobotrya japonica*, 148, 150
  type 2 diabetes drug research, 144
Protodioscin, saponin in tribulus, 50, 51*f*
Prototribestin, saponin in tribulus, 50, 51*f*
Pseudocinnamomum, *Cinnamomum osmophloeum* Kaneh, 300
Pseudohypericin. *See* St. John's wort
PTP1B inhibitors. *See* Protein tyrosine phosphatases 1B (PTP1B) inhibitors

Q

Quality
  bilberry extracts, 45, 46*f*
  dietary supplements, 4–5
Quality control
  artichoke leaf extracts, 42
  botanical products, 41
  commercial artichoke extracts, 42–43
  Lippia multiflora, 129, 130–131
  shea butter, 332–334
Quality degradation factors, shea butter, 330–332
Quercetin
  antioxidant activity, 204, 205*t*, 210*t*
  determination in biological fluids, 216
  effect of sugar structure on absorption, 219
  flavonol, 26, 27*f*
  fluorescence decay curve, 209*f*
  free radical scavenging activity, 208*f*
  metabolism by gut/liver, 217–218
  structure, 242*f*
  structure and DPPH and ORAC values, 206*f*

R

Rabdosia rubescens Hara, PC SPES herbal mixture, 117–118
Rats. *See* Dental caries
Reactive oxygen species (ROS)
  aging and free radicals, 196
  curcumin, 98f, 101
  role in inflammatory process, 268
Refining, shea butter, 332
Rehmaglutin D, isolation and identification from *Ixeris denticulata*, 201, 203f, 204
Research and development, proposal for traditional Chinese medicines, 21, 24
Resveratrol
  activity with cancer cell lines, 287
  cancer prevention, 241
  stilbene, 28, 29f
  structure, 242f
Rheum coreanum and *R. palmatum* (Da Huang), bioactives acting on different targets, 19t
Rheum palmatum
  anti-inflammatory activity, 270, 271f
  See also Asian herbs targeting inflammation
Rhubarb, anti-inflammatory activity, 270
Ruan Zi Cao (*Arnebia euchroma*), bioactives acting on different targets, 19t
Rubiadin, known compound from noni roots, 191f
Rutin, structure, 214f

S

St. John's wort
  analysis of St. John's wort (SJW) constituents in fruit-flavored beverages, 61
  analysis of SJW constituents in tea, 57–58, 60
  analytical methods for constituents, 4
  biological activities and phytochemical constituents of SJW, 57
  changes of SJW constituents in fruit-flavored drink during storage, 62f
  depression, 3
  LC (liquid chromatography) of hot water infusion of tea bag, 59f
  market, 3–4
  SJW constituents in tea bags extracted with methanol or hot-water infusion, 60t
  SJW in functional beverages, 57–61
  structures of major active constituents of SJW, 58f
  See also Functional beverages
Saponins
  isolation and characterization of new, from Jiaogulan, 171–176, 181–182
  soysaponins, 48
  tribulus extracts, 49–51
Saturated fatty acids, association between intake of catechins and, 259, 260t
Saw palmetto, PC SPES herbal mixture, 117–118
Screening procedures, bioactive components of traditional Chinese medicines, 16–17
Scullcapflavone I and II, targeting inflammation, 268, 269f
Scutellaria baicalensis Georgi
  major bioactive constituents, 269f
  PC SPES herbal mixture, 117–118
  targeting inflammation, 267–270
  See also Asian herbs targeting inflammation
Secoisolariciresinol, lignan, 28, 29f

Sennosides A and B, *Rheum palmatum*, 270, 271*f*
Serenoa repens, PC SPES herbal mixture, 117–118
Shea butter
  antioxidant activity, 328
  butter color, 332
  chemical degradation, 331
  commercial uses, 334, 336–338
  commonly reported chemical parameters and ranges, 330*t*
  constituents, 328–329
  cosmetics and personal care products industries, 336–337
  enzymatic degradation, 332
  extraction procedures, 327
  fatty acids, 328
  food industry, 334, 336
  microbiological contamination, 331
  oxidation and hydrolysis, 331
  pharmaceutical industry, 337–338
  phenolic compounds, 328
  physical and organoleptic parameters and ranges, 330*t*
  physico-chemical properties, 329–330
  potential new uses, 338–339
  quality control considerations, 332–334
  quality degradation factors, 330–332
  refining, 332
  relation between free fatty acids and peroxide content, 335*f*
  relation between unsaponifiables and peroxide content, 335*f*
  Shea tree, 326–327
  taxonomy and genetics of Shea tree, 327
  tocopherol, 329
  triglycerides, 328
  unsaponifiable compounds, 329
Sideritis lycia, verbascoside, 136
Single Laboratory Validation AOAC short course, 8
  funding, 10
β-Sitosterol, isolation and identification from *Ixeris denticulata*, 200, 202*f*
Southeast Asia, Jiaogulan, 170
Soy isoflavones
  dietary supplements, 47
  groups, 47
  HPLC method, 48
  HPLC profile of typical and questionable extract samples, 49*f*
  soy saponins, 48
  See also Instrumental analysis of botanicals
Spices
  healing tools, 93
  tumeric, 93–94
  See also Curcumin
Stachytarpheta jamaicensis, verbascoside, 135
Staphylococcus aureus, activity of verbascoside against, 135–136
Sterols. *See* Dysidesterols; Suberrestisteroids
Stigmasterol, isolation and identification from *Ixeris denticulata*, 201, 202*f*
Stilbenes, subclass of polyphenols, 28, 29*f*
Storage, St. John's wort constituents in fruit-flavored drink, 62*f*
Streptococcus sobrinus. *See* Dental caries
Structure-activity relationship
  flavonoids from *Ixeris denticulata*, 205, 210
  structures of flavonoids from *I. denticulata*, 206*f*, 207*f*
Suberretisteroids
  computer generated drawing, 86*f*
  molecular formulae, 84–85
  spectroscopic properties, 85
  uncommon steroids from *Subergorgia reticulata*, 84

See also Marine medicinal plants
Sugen, protein tyrosine phosphatases 1B inhibitor discovery, 146*f*
Sulindac, synergistic effects of epigallocatechin gallate (EGCG) with, 260, 261*f*
Synergism. *See* Dental caries
Systematic isolation, screening bioactive components of traditional Chinese medicines, 16

T

Takeda, protein tyrosine phosphatases 1B inhibitor discovery, 146*f*
Tamoxifen, synergistic effects of epigallocatechin gallate (EGCG) with, 260, 261*f*
Tannins
　analytical techniques, 30
　challenge of ellagitannin (ET) research, 33–34
　chemical and biological properties, 30
　classes, 29
　condensed tannins: proanthocyanidins (PAs), 31–33
　dietary sources, 28
　estimating PA consumption level, 32
　formation of ellagic acid from hydrolyzable tannins, 34*f*
　French Paradox, 32
　gallotannins (GTs), 33
　hydrolyzable tannins: ellagitannins (ETs), 33–35
　polymers of polyphenols, 28
　potent antioxidant properties of PAs, 32–33
　potent biological activities of ETs and ellagic acid (EA), 34–35
　roles in plants, 29–30
　structure of PAs, 31*f*
　See also Polyphenols

Taxonomy, Shea tree, 327
Tea catechins
　affinity for proline rich proteins, 259
　antioxidant activity, 257
　cancer prevention, 258–259
　correlation coefficients between total catechin intake and dietary factors, 260*t*
　degradation products of epigallocatechin gallate (EGCG), 244*f*
　degradation study, 241, 243
　EGCG in mild alkaline solutions, 243, 244*f*
　health benefits, 255
　interaction with foods and medical components, 259–260
　major, in tea, 256*f*
　metabolism and clinical effects, 257–259
　physiological and synergistic effects, 256–259
　protective effects against heart disease, 258
　strong affinity for Fe^{2+} and Fe^{3+}, 259
　synergistic effects, 259
　synergistic effects of EGCG with Sulindac and Tamoxifen, 260, 261*f*
　See also Black tea theaflavins; Catechin
Teas
　cancer prevention, 255
　See also Black tea theaflavins; Herbal teas; *Lippia multiflora*
Tectoquinone, known compound from noni roots, 190*f*
Temperature, effect on antioxidants of *Moringa* species, 231, 233
12-O-Tetradecanoylphorbol-13-acetate (TPA)
　inducing inflammation, 315–316

topical application of theaflavin mixture (TFs) on TPA-induced arachidonic acid metabolism mouse ears, 319, 321f
topical application of theaflavin mixture (TFs) on TPA-induced ear inflammation mouse model, 319, 320f
See also Black tea theaflavins
Tetramethylpyrazine, *Ligusticum chuanxiong*, 282
Theaflavins. *See* Black tea theaflavins
Thioglucosidase
  enzyme reaction using glucotropaeolin substrate, 158
  high performance liquid chromatography (HPLC) chromatogram from maca extract incubation with, 164f
  isolation from *Brevicoryne brassicae*, 166
  quantification of maca glucosinolates, 160
  Sinapis alba, 164
  See also Benzyl isothiocyanate (BITC); Maca
Tissue distribution, curcumin, 102–104
Tocopherol, shea butter, 329
Traditional Chinese medicines (TCM)
  Actonium and different targets, 18, 19t
  Arnebia euchroma, 18–19
  Bei (*Babylonia japonica*), 19t
  bioactivity-oriented isolation, 16
  Da Huang (*Rheum coreanum* and *R. palmatum*), 19t
  diverse components working on multiple targets, 17–19
  diversity of chemical components, 16
  induction of apoptosis and inhibition of cell cycle, 16–17
  investigation on TCM formulas, 19

isolation and bioactive compounds of anticancer ZSG, 22f, 23f, 24
libraries for extracts, fractions and compounds from, 20, 21f
major difficulties for research on bioactive components, 17
multiple-target and high-throughput screening assays, 20
procedures for screening bioactive components from, 16–17
proposal of TCM research and development, 21, 24
relationship between diverse components and multiple targets, 18
resource for new drug development, 15–20
revised "Drug Registration and Administration Law", 15
Ruan Zi Cao (*Arnebia euchroma*), 19t
systematic isolation, 16
Wu Tou (*Aconitum* spp.), 19t
Ying Chen Hao (*Artemisi a capillaris*), 19t
See also Protein tyrosine phosphatases 1B (PTP1B) inhibitors
Transcription factors NF-κB and AP-1, suppression by curcumin, 95–96
Transformation study. *See* Chlorogenic acid
Tribulus
  Bulgarian, 50
  Chinese and Indian, 50
  HPLC/MS method for saponin identification, 51, 52f
  methods to quantitate saponins in, 50
  protodioscin and prototribestin, 50, 51f
  quality of, from Bulgaria, 48
  saponins and bioactivity, 49–50
  use world-wide, 48

See also Instrumental analysis of botanicals
Triglycerides, shea butter, 328
Tripterygium wilfordii Hook f.
  anti-inflammatory effect, 274
  terpenoids, 275
  triptolides A and B, 275–276
  See also Asian herbs targeting inflammation
Triptolides A and B, anti-inflammatory effect, 275–276
Triterpenoid inhibitors, *Eriobotrya japonica*, 148, 150
Tumor cells, curcumin, 97, 98*f*
Tumor suppressor agents, chemoprevention, 94–95
Turmeric
  potential of curcumin component, 93–94
  See also Curcumin

U

United States, botanical products, 40–41
United States Pharmacopeia (USP), alkaloid content standards, 4
Unsaponifiable compounds, shea butter, 329

V

Validation, analytical methods, 9–10
Vegetable combinations, cooking with Moringa leaves, 237
Verbanaceae
  family of aromatic plants, 127
  See also Lippia multiflora
Verbascoside
  chemical structure, 132*f*
  content in *Lippia multiflora*, 131–132
  HPLC profile of methanolic extract of *L. multiflora*, 133*f*
  pharmacological activities, 134–136
  See also Lippia multiflora
Verbena officinalis, verbascoside, 136
Vitamin C, association between intake of catechins and, 259, 260*t*
Vitamin E, association between intake of catechins and, 259, 260*t*
Vitellaria paradoxa
  Shea tree, 326–327
  See also Shea butter
Vitexin, Jiaogulan, 171

W

West Africa folk medicine
  herbal tea of *Lippia multiflora*, 127–128
  See also Lippia multiflora
Wogonin, targeting inflammation, 268, 269*f*
World Health Organization (WHO), herbal medicine use, 40
Wu Tou (*Aconitum spp.*), bioactives acting on different targets, 19*t*
Wyeth, protein tyrosine phosphatases 1B inhibitor discovery, 146*f*

Y

Ying Chen Hao (*Artemisi a capillaris*), bioactives acting on different targets, 19*t*

Z

ZSG
  bioactive compounds from extract, 23*f*
  isolation procedure, 22*f*
  traditional Chinese medicine, 24
  See also Traditional Chinese medicines (TCM)